Advanced Technologies and Societal Change

Series editor

VDE Verband der Elektrotechnik
Elektronik Informationstechnik e.V., Frankfurt, Germany

For further volumes:
http://www.springer.com/series/10038

Reiner Wichert · Helmut Klausing
Editors

Ambient Assisted Living

6. AAL-Kongress 2013 Berlin, Germany,
January 22–23, 2013

Editors
Reiner Wichert
Computer Graphics Research
Fraunhofer Institute for Computer
  Graphics Research
Darmstadt
Germany

Helmut Klausing
VDE-Verbandsgeschäftsstelle
Verband der Elektrotechnik Elektronik
  Informationstechnik e.V.
Frankfurt
Germany

ISSN 2191-6853 ISSN 2191-6861 (electronic)
ISBN 978-3-642-37987-1 ISBN 978-3-642-37988-8 (eBook)
DOI 10.1007/978-3-642-37988-8
Springer Heidelberg New York Dordrecht London

Library of Congress Control Number: 2013954025

© Springer-Verlag Berlin Heidelberg 2014

This work is subject to copyright. All rights are reserved by the Publisher, whether the whole or part of the material is concerned, specifically the rights of translation, reprinting, reuse of illustrations, recitation, broadcasting, reproduction on microfilms or in any other physical way, and transmission or information storage and retrieval, electronic adaptation, computer software, or by similar or dissimilar methodology now known or hereafter developed. Exempted from this legal reservation are brief excerpts in connection with reviews or scholarly analysis or material supplied specifically for the purpose of being entered and executed on a computer system, for exclusive use by the purchaser of the work. Duplication of this publication or parts thereof is permitted only under the provisions of the Copyright Law of the Publisher's location, in its current version, and permission for use must always be obtained from Springer. Permissions for use may be obtained through RightsLink at the Copyright Clearance Center. Violations are liable to prosecution under the respective Copyright Law.
The use of general descriptive names, registered names, trademarks, service marks, etc. in this publication does not imply, even in the absence of a specific statement, that such names are exempt from the relevant protective laws and regulations and therefore free for general use.
While the advice and information in this book are believed to be true and accurate at the date of publication, neither the authors nor the editors nor the publisher can accept any legal responsibility for any errors or omissions that may be made. The publisher makes no warranty, express or implied, with respect to the material contained herein.

Printed on acid-free paper

Springer is part of Springer Science+Business Media (www.springer.com)

Preface

To promote a healthier lifestyle and to support carers, families and care organizations the research field of Ambient Assisted Living (AAL) for ICT-enabled independent living by using products and services for better health and functional capability of older adults has been established. This research field has a huge impact on effecting requirements for ICT projects and solutions dependent on their lifestyle, physical or mental health. A wide variety of research projects has just delivered results or is working on it. The outcome of these projects contribute to achieving the overarching goal of the European Innovation Partnership on Active and Healthy Ageing (EIP AHA), which by 2020 aims to increase the number of healthy life years in Europe by two years.

On the other side there is still a lack of viable business models to be considered almost unanimously as the greatest market obstacle to a broad implementation of innovative AAL systems. Thus one of the main objectives is to strengthen the industrial base in Europe and enhance competitiveness for ICT products and care services that will assist the users to achieve the autonomy, independence and dignity appropriate to their needs and conditions. Due to these effects it is now more than ever essential to bring together all the necessary stakeholders and enable the very important networking between policy makers, developers, producers, service providers, end user organisations, designers, health professionals, sociologists, carers and older adults and other end user groups.

For this reason a conference series has been established as an annual showcase event for the people involved in this community: the AAL-Kongress (Congress for Ambient Assisted Living) with its purpose is to exhibit and demonstrate ICT solutions, promote networking within the community, provoke debate on various topics and highlight new or emerging developments in the area to inform the AAL community and discuss the problems and challenges we have to face in the common years. The first AAL Kongress 2008 had the focus on applications of intelligent assistive systems within the areas of "health & homecare", "safety & privacy", "maintenance & housework" und "social environment". At the second AAL-Kongress more than 520 participants attended. It focused on use cases to support the manufacturing of products adjusted to the needs of the user. In 2010 the third AAL-Kongress had been organized with close to 600 participants also with the focus on use cases. In 2011 it advanced to the leading congress for AAL with 870 participants. In 2012 the focus laid on technologies in a self-determined life and the number of participants passed over 1000, still addressing economic challenges and trendsetting applications on innovative technology.

In 2013 the sixth AAL-Kongress is focussing on "quality of life in times of changing demography and technology". From the large number of contributions from the call for papers a selection has been made with topics such as: smart homes, activities of daily living, telemonitoring, AAL platforms, interaction and robotics. To underline the research priority the research papers have been evaluated more restrictive. 129 papers from 537 authors within 7 countries have been submitted to the sixth AAL-Kongress. After a solid review process 22 papers were accepted to be included in these scientific proceedings of the conference. Three independent reviewers were matched by their expertise area to the topic of each paper.

In closing I would like to thank the reviewers of the Reviewing Committee, the organizers of this event and all of the paper presenters and conference participants who helped to make the AAL-Kongress 2013 a success.

Reiner Wichert

Program Committee AAL-Kongress 2013

Alexander Viehweger	Verband Sächsischer Wohnungsgenossenschaften e.V., Dresden (Program Chair)
Uwe Fachinger	Universität Vechta (Chair—Economical Challenges)
Udo Gaden	Sozialwerk St. Georg e.V., Gelsenkirchen (Chair—Care)
Sibylle Meyer	SIBIS Institut für Sozialforschung und Projektberatung GmbH, Berlin (Chair—Social Aspects)
Reiner Wichert	Fraunhofer IGD, Darmstadt (Chair—Research)
Jan Alexandersson	DFKI, Saarbrücken
Martin Braecklein	Robert Bosch Healthcare GmbH, Stuttgart
Bernd Dechert	ZVEH, Frankfurt
Wolfgang Deiters	Fraunhofer ISST, Dortmund
Petra Friedrich	Hochschule Kempten
Sabine Häring	Miele & Cie. KG, Gütersloh
Armin Hartmann	Hartmann Real Estate, Bochum
Andreas Hein	Universität Oldenburg
Stefan Heusinger	DKE, Frankfurt
Benno Kotterba	md-pro GmbH, Karlsruhe
Harald Künemund	Universität Vechta
Joachim Latt	Bosch Sicherheitssysteme GmbH, Kassel
Heidrun Mollenkopf	BAGSO e.V., Expertenrat/Demenz Support, Stuttgart
Asarnusch Rashid	FZI Karlsruhe
Cord Schlötelburg	DGBMT, Frankfurt
Gudrun Stockmanns	Hochschule Niederrhein
Christine Weiß	VDI/VDE Innovation + Technik GmbH, Berlin
Ralph Welge	Leuphana Universität Lüneburg
Volker Wittpahl	Ingenieurs- und Innovationsbüro, Oldenburg
Anton Zahneisen	SOPHIA Consulting & Concept GmbH, Bamberg

Reviewing Committee (Technical Research Papers)

Jan Alexandersson	DFKI GmbH, Saarbrücken
Jürgen Brehm	Universität Hannover
Wolfgang Deiters	Fraunhofer ISST, Dortmund
Alois Ferscha	Johannes Kepler Universität, Linz, Austria
Rainer Günzler	HSG-IMIT, Villingen-Schwenningen
Andreas Hein	Universität Oldenburg
Sabine Häring	Miele & Cie. KG, Gütersloh
David Hradetzky	University of Applied Sciences Northwestern Switzerland
Lothar Mühlbach	Heinrich-Hertz-Institut, Berlin
Asarnusch Rashid	FZI Karlsruhe
Gudrun Stockmanns	Hochschule Niederrhein, Duisburg
Ralph Welge	Universität Lüneburg, Lüneburg
Reiner Wichert	Fraunhofer IGD, Darmstadt
Wolfgang Zagler	Technische Universität Wien, Austria

Contents

Part I Smart Homes

Functional Assessment in Elderlies' Homes: Early Results from a Field Trial 3
Enno-E. Steen, Thomas Frenken, Melina Frenken and Andreas Hein

LsW: Networked Home Automation in Living Environments 19
Frerk Müller, Peter Hoffmann, Melina Frenken, Andreas Hein and Otthein Herzog

Guiding Light for the Mobility Support of Seniors 35
Guido Kempter, Walter Ritter and Andreas Künz

Vibroacoustic Monitoring: Techniques for Human Gait Analysis in Smart Homes 47
Klaus Dobbler, Moritz Fišer, Maria Fellner and Bernhard Rettenbacher

Part II Activities of Daily Living

Unobtrusive Respiratory Rate Detection Within Homecare Scenarios 61
Bjoern-Helge Busch and Ralph Welge

Context-Enriched Personal Health Monitoring 79
Barbara Franz, Mario Buchmayr, Andreas Schuler and Werner Kurschl

Mneme: Telemonitoring for Medical Treatment-Support in Dementia 93
Torben Wallbaum, Melina Frenken, Jochen Meyer, Andreas Hein and Carsten Giehoff

PASSAge: *P*ersonalized *M*obility, *A*ssistance and *S*ervice *S*ystems in an *A*geing Society 109
Matthias Bähr, Sarah Klein, Stefan Diewald, Claus Haag,
Gebhard Hofstetter, Maher Khoury, Daniel Kurz, Andreas Winkler,
Andrea König, Nadine Holzer, Monika Siegrist, Axel Pressler,
Luis Roalter, Thomas Linner, Matthias Heuberger, Kerstin Wessig,
Matthias Kranz and Thomas Bock

Part III Telemonitoring

Facial Movement Dysfunctions: Conceptual Design of a Therapy-Accompanying Training System 123
Cornelia Dittmar, Joachim Denzler and Horst-Michael Gross

Detecting Activities of Daily Living with Smart Meters............ 143
Jana Clement, Joern Ploennigs and Klaus Kabitzsch

A Personalized and Context-Aware Mobile Assistance System for Cardiovascular Prevention and Rehabilitation 161
Alexandra Theobalt, Boris Feodoroff, Dirk Werth and Peter Loos

GlobalSensing: A Supervised Outdoor-Training in Cardiological Secondary Prevention......................... 175
Tim Janus, Torben Kohlmeier, Viktor Marinov, Janina Marks,
Christian Mikosch, Michael Nimbs, Thorsten Panke,
Jörn Störling, Oliver Dohndorf, Heiko Krumm,
Jan-Dirk Hoffmann, Anke Workowski and Detlev Willemsen

Part IV AAL Platforms

Representation of Integration Profiles Using an Ontology.......... 195
Ralph Welge, Bjoern-Helge Busch, Klaus Kabitzsch,
Janina Laurila-Epe, Stefan Heusinger, Myriam Lipprandt,
Marco Eichelberg, Elke Eichenberg, Heike Engelien,
Murat Goek, Guido Moritz and Andreas Hein

Methods and Tools for Ontology-Based Configuration Processes of AAL Environments...................................... 213
Tom Zentek, Alexander Marinc and Asarnusch Rashid

Contents xiii

**The Robot ALIAS as a Database for Health Monitoring
for Elderly People**... 225
Tobias Rehrl, Jürgen Geiger, Maja Golcar, Stefan Gentsch,
Jan Knobloch, Gerhard Rigoll, Katharina Scheibl,
Wolfram Schneider, Susanne Ihsen and Frank Wallhoff

Part V Interaction

**Visual and Haptic Perception of Surface Materials
for Direct Skin Contact in Human–Machine Interaction**............ 249
C. Brandl, A. Mertens, J. Sannemann, A. Kant,
M. Ph. Mayer and C. M. Schlick

**Human-Robot Interaction: Testing Distances that Humans
will Accept Between Themselves and a Robot Approaching
at Different Speeds**... 269
Alexander Mertens, Christopher Brandl, Iris Blotenberg,
Mathias Lüdtke, Theo Jacobs, Christina Bröhl,
Marcel Ph. Mayer and Christopher M. Schlick

Display of Emotions with the Robotic Platform ALIAS............ 287
Jürgen Geiger, Ibrahim Yenin, Frank Wallhoff and Gerhard Rigoll

Part VI Robotics

**Housing Enabling: Detection of Imminent Risk Areas
in Domestic Environments Using Mobile Service Robots**........... 301
Nils Volkening, Andreas Hein, Melvin Isken, Thomas Frenken
and Melina Brell

**Mobile Video Phone Communication Carried
by a NAO Robot**.. 315
Paul Panek, Georg Edelmayer, Peter Mayer, Christian Beck
and Wolfgang L. Zagler

The Robot ALIAS as a Gaming Platform for Elderly Persons...... 327
Jürgen Geiger, Thomas Leykauf, Tobias Rehrl, Frank Wallhoff
and Gerhard Rigoll

Part VII Community Conclusions

**Can the Market Breakthrough in AAL be Achieved
by a Large Scale Pilot?** 343
Mohammad Reza Tazari and Reiner Wichert

Part I
Smart Homes

Functional Assessment in Elderlies' Homes: Early Results from a Field Trial

Enno-E. Steen, Thomas Frenken, Melina Frenken and Andreas Hein

Abstract Early results from a field trial regarding the assessment of functional status relevant to self-care ability in domestic environments are presented. A previously developed technical system for unobtrusively recording location information using home automation data was installed in the homes of five participants aged 64–84 years over a period of partially more than nine month. The recordings are manually evaluated to check whether items of geriatric assessment tests relevant to self-care ability can be assessed using the sensor recordings. The evaluation is a preliminary step to develop an automatic assessment algorithm and to develop a model for mapping domestic assessment results to result scales of clinical assessment tests. The mapping is required since most clinical assessment tests are not suitable for execution in domestic environments and thus new approaches are required which do also account for the difference between performance and capacity in functional abilities as proposed within WHO's ICF.

Keywords Functional assessment · Geriatric assessment · Home automation · Abstracted room model · Mobility · Field trial

1 Introduction

In order to comply with the wish of many elderly people to live safely and independently in their own homes as long as possible, early prevention of diseases, long-term rehabilitation, and the detection of acute incidences are essential especially when regarding the demographic change. However, since elderly people

---

E.-E. Steen (✉) · T. Frenken · M. Frenken
OFFIS—Institute for Information Technology, Escherweg 2, 26121 Oldenburg, Germany
e-mail: enno-edzard.steen@offis.de

A. Hein
School of Medicine and Health Sciences, University of Oldenburg, Ammerländer Heerstraße 114–118, 26129 Oldenburg, Germany

are a diverse target group provision of those services requires in-depth knowledge of their individual characteristics i.e. their capabilities, health status, and habits. In professional care facilities this information and deviations from typical conditions are retrieved within the so called geriatric assessment. Various assessment tests are used to assess especially elderly people in relevant domains. However, those assessment tests are standardized to and designed for clinical environments. When aiming at performing assessments in domestic environments of people, the main problem is that clinical assessment tests are designed to be explicit test-situations which are supervised and evaluated by health care professionals. Despite that, supervised assessments of each patient at home will not be possible due to cost reasons, even when using tele-care systems, and many people will not want to perform several tests at home over a longer period of time. Therefore, a more suitable approach for domestic assessment tests is to perform those continuously throughout the day without having the inhabitants perform an explicit test.

Performing assessments continuously in domestic environments requires the use of unobtrusive information technology to be cost-effective especially when regarding the potential number of elderly people in need of care due to the demographic change. In our previous research we have already presented a technical approach to measure required sensor data for implementing domestic assessment tests [1]. The system was recently used within a field trial conducted in Oldenburg, Germany over a period of partially more than nine month in five households of elderly people aged 64–84 years. One of the field trial's objectives was to investigate whether home automation recordings can be used to transfer items of clinical assessment tests from various domains to domestic environments without having to re-implement the test-situation. On the long-term our objective is to infer a general model of domestic assessment tests which describes a mapping of their results to the scales of clinical assessment tests and thus allows health care professionals to assess people's functional status directly in their homes. The model may allow for new possibilities in early prevention and long-term rehabilitation. This paper focuses on early results from the field trial and on investigating which assessment test items could be implemented at home by using home automation technology.

2 Medical Motivation

Health care professionals require knowledge about the health status, individual capabilities, and habits of patients in order to provide prevention of diseases or effective rehabilitation or assistance for activities of daily living. Within this paper, such information is summarized under the term characteristics. In professional care facilities the individual characteristics of especially elderly patients are retrieved within the so called geriatric assessment which is a "multidimensional process designed to assess an elderly person's functional ability, physical health, cognitive and mental health, and socio-environmental situation" [2]. During the geriatric assessment various standardized assessment tests are used to assess the individual characteristics of patients in various domains relevant to pursuing an independent lifestyle. Frequently used assessment tests are the Barthel Index [3] or the Instrumented Activities of Daily Living test [4] for assessing self-care ability and personal hygiene, the Social Situation Test according to Nikolaus in the field of social activity, or the Mini-Mental State Examination [5] for measuring cognitive capabilities. One example from the mobility domain is the frequently used Timed Up&Go (TUG) test [6]. Another important geriatric assessment instrument is the Resident Assessment Instrument—Home Care (RAI-HC 2.0) [7] which focuses on care-relevant aspects.

Although those assessment tests are frequently used and have been proven to be able to assess functional ability of patients and to detect risk of losing self-care ability or being at risk of falling they are only validated and standardized in professional environments which are free of obstacles or other environmental influences. Additionally, several problems regarding clinical assessments are known [8]. Among those are subjective execution limiting comparability of results, time-intensiveness due to manual documentation, infrequent execution only after acute incidents leading to timely constricted results, and perception as a test-situation making patients perform at their best. Additionally, clinical assessments do measure the capacity of patients in their domains only since all environmental influences are ignored. However, according to WHO's ICF measuring a patient's performance is required as well in order to distinguish between environmental and personal factors limiting his or her functioning or health. Performance can only be measured in the every-day environments of patients.

Therefore, new approaches to performing assessments in domestic environments are required which make clinical assessment tests suitable for continuous execution and enhance their acceptance in daily living. Such approaches will have to utilize information technology in order to be cost-effective. A continuous unobtrusive execution allows to assess the performance of patients without creating a test-situation. In order to compare the results from domestic and clinical environments models for mapping the results and for describing the context factors under which the results have been obtained are required. Helmer et al. [9] recently presented a suitable, not yet sufficiently detailed model called 3DLC.

3 State of the Art

Various technical approaches to monitoring people in their domestic environments have been presented previously. Most approaches focus on either tracking the location, the movements, or the interaction of people with objects. From a technical point of view the approaches can be distinguished into those utilizing ambient sensors and those using body-worn sensors [10]. Since our target group are elderly and also demented people, we focus on ambient sensor technologies since we assume those being more suitable for unobtrusive monitoring and since they do not require explicit usage by the person.

Generally spoken, two kinds of ambient sensor technologies have been mainly used for domestic monitoring: special purpose sensors and home automation sensors [11]. Regarding special purpose sensors, Pallejà et al. [12] have presented an approach to basic gait analysis on a straight walking path utilizing a laser range scanner (LRS). Within our own work [13] we have demonstrated the use of a LRS for precise assessment of self-selected gait velocity in domestic environments without restrictions to the walking paths. Stone and Skubic [14] have developed an approach using two Kinect sensors for domestic movement analysis but found limitations regarding clothing not reflecting light in the infrared spectrum and for measurements in which persons move too close to walls or furniture. An approach presented by Poland et al. [15] utilizes a camera attached to the ceiling for locating the inhabitant. The approach requires dividing and marking the floor within the coverage of the camera evenly into rectangles called virtual sensors. An ultrasonic ring consisting of 15 ultrasonic sensors was used by Steen et al. [16] to locate users when placed on the ceiling of a room. Steinhage et al. [17] introduced a system based on a smart underlay with capacitive proximity sensors consisting of conductive textiles. By tracking the usage of electrical devices Hein et al. [18] recognize a person's position within an environment and infer activities executed which are associated with the usage of electrical devices.

Among those approaches using home automation sensors, Pavel et al. [19] developed a system based on passive motion sensors covering various rooms of a flat. Gait velocity could be computed by dividing known distances between coverages by measured transition times. Placing three passive motion sensors in a sufficient long corridor makes those computations more reliable. Within our own work [1] we have recently presented a new approach based on the definition of motion patterns by usage of available presence events generated from any kind of sensor. By providing an abstracted definition of the environment, physically possible walking paths can be computed and monitored automatically. Floeck et al. [20] utilize motion detectors, door contacts, and light switches to track the global location of people in their smart homes and compute location probabilities to generate emergency alarms.

Although some of the mentioned approaches try to infer functional status of inhabitants from their measurings, only few have tried to map their results to established scales like those of clinical assessment tests. Within our own research we have recently presented a new approach to perform the TUG test unsupervised by use of ambient sensor technologies and without the need to create an explicit test-situation [21].

4 Approach

Within our own approach, previously presented [1], we utilize mainly ambient sensor technologies i.e. home automation sensors in order to localize inhabitants within environments and to compute traversal times between sensors as well as approximate walking speed using the known distance between deployed sensors. The main advantage of the system about previous approaches is its adaptability to changing environments which is why the system could be installed in five flats of participants during a field trial with ten sensors per site in under eight working hours. Since the system has already been described in a previous publication [1], we only briefly summarize its main concepts and working steps:

- **Definition of an Abstracted Room Model:** First an abstracted room model is generated by analyzing the floor plan (manually drawn by the user or loaded from a CAD-file) of the domestic environment and the definitions of the available sensors. This model contains, among other things, the information about the rooms, obstacles for a person's motion and the coverage areas of the sensors.
- **Computation of a Sensor-Graph:** The abstracted room model enables the generation of a sensor-graph which consists of the sensors and the adjacency relations between these sensors respectively between their coverage areas. Two sensors are adjacent if there is a path between the corresponding coverage areas that does not include an area of another sensor. To automatically find these adjacent relations the system uses a path-planning algorithm, namely a discrete version of the potential field method.
- **Feature Generation:** After the determination of the sensor-graph the system generates two features for each adjacency relation, one for each direction. A feature corresponds to a traversal of a person from the measurement range of one to an adjacent sensor. Furthermore, additional complex features can be defined manually.
- **Measurement of Features:** The defined features are used to find features in the sequence of sensor-events caused by the motion of a person inside the domestic environment. For this purpose each feature definition is transformed into a finite state machine. Every generated sensor-event is first validated by means of the sensor-graph. After that, a valid sensor-event is sent to each state machine. If a state machine detects the belonging feature analyzing the current event and previously received events it creates a feature instance and stores it for further analysis.
- **Computation of Location and Walking Speeds:** Using the detected features and the abstracted room model it is possible to compute the most likely location of an inhabitant within his or her environment for each point in time. A location at a certain time is assumed to be near the location of the end sensor of the lastly recognized feature within the abstracted room model. Additionally, using the length of detected features and the traversal time between the two sensors comprising the feature an approximate walking speed of the inhabitant can be computed.

5 Experiment

An experiment conducted during a field trial had two main aims: (1) to proof the general feasibility of the approach to monitor elderly people over a longer period of time unobtrusively and (2) to gather enough sensor data for checking which items of clinical assessment tests can be implemented by using presence information in domestic environments. However, there is no automatic checking of assessment items implemented yet, so all evaluations regarding this point have been done manually. For this paper we focus on the second objective for assessment tests from the domains of mobility, personal hygiene, social activity and sleeping.

The figures presented in the following sections are generated from evaluation of two of the five participants. Participant A is male, aged 70 years and lives alone in a flat with four rooms. He leaves the house only a few times per week. His living habits are rather untypical since he is up till early morning and sleeps till early afternoon. He has huge overweight (more than 130 kg). He does not cook warm meals for himself but eats in a social kitchen most days in the early afternoon. In the evening he prepares cold meals. Participant B is female and aged 76 years. She lives alone in a flat with six rooms and is very active. She leaves the house several times a day and is socially very active. She has a very mannered and self-dependent lifestyle and goes to bed at 23 o'clock while standing up latest at 7 o'clock.

5.1 Methods

The field trial started 2011/10/10. Five community-dwelling elderly people aged 64–84 years (2 male, 3 female) living alone and mostly independent participated. The participants were divided into a short-term and a long-term group with three and two members. The short-term group was only monitored for a single month while the long-term group is still monitored until at least 12 months are reached. Home automation (HA) sensors, five light barriers and five reed contacts, were installed in all flats. Figure 1 shows an abstracted room model of a flat including sensor placements (grey boxes, LB = light barrier, RC = reed contact) and computed walking paths (lines). 170 different walking paths were defined in all flats together.

During the field trial overall 176,183 activations of HA sensors were recorded from which 53,443 traversed walking paths could be detected between rooms. Additional walking paths can be detected within rooms by making use of additional sensors which were placed inside rooms. However, for this evaluation we do only use features which represent walking paths between different rooms. Sensor recordings from LB and RC were processed as described within the approach section.

Fig. 1 Visualization of an abstracted room model

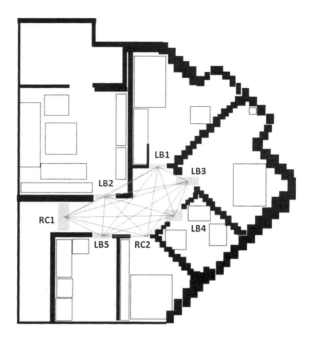

5.2 Results on Mobility

Mobility is known to be one of the most important functional requirements for pursuing an independent lifestyle. Therefore, assessment tests in this domain are mainly about whether a patient is still able to move i.e. to change body positions and to move around and about how far he or she is able to move.

Within our approach the number of movements an inhabitant performs is directly correlated to the number of features detected by the system. Since the length of each feature is known from the path planning step we are also able to compute the distance an inhabitant moves. Figure 2 shows the average number of recognized features per hour over a period of seven months for participant A. In average the inhabitant reached approx. 112 features per day which corresponds to walking approx. 118 m. The number of features recognized varies with the hour of the day. Obviously, this is due to living habits the person has. This inhabitant normally slept between 5 and 13 o'clock and thus did only generate features when walking to the bathroom. Figure 2 shows peaks of features at approx. 4–5 o'clock when the inhabitant went to bed, at 13–15 o'clock when he stood up and again at 19 o'clock when the inhabitant prepared meals and walked to the living room to watch TV until going to bed. Therefore, the number of features recognized per hour does also reflect the person's circadian rhythm and is very typical across all months shown.

Similar results were found for all other patients. The only difference was the specific number of features recognized and the hours of the day peaks were found.

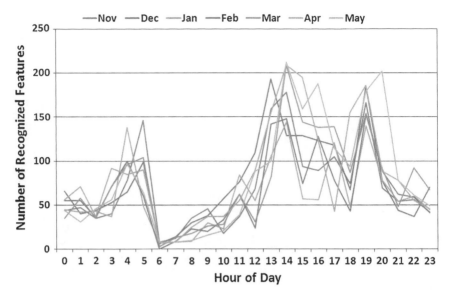

Fig. 2 Participant A: average number of features recognized per hour of day in 7 months

However, for all participants 3–4 peaks per day could be found which strongly reflect the people circadian rhythm and were very stable across all weeks and months. This makes us conclude that the number of features per day can not only be used to assess the people's mobility (in walking) but may also be used to distinguish between typical days and those a person moves untypically often or sparse.

5.3 Results on Personal Hygiene

Personal hygiene is an essential part of an independent lifestyle. Therefore, assessment tests in this domain mainly include whether a person is able to wash himself or herself and to go to toilet without requiring help by another person.

Neither washing nor toileting can be detected directly by using the deployed home automation sensors. The strongest hint to executing these actions is given in our approach by the number and the duration of presences within the bathroom. Figure 3 shows the number and the duration of presences in the bathroom per day for participant A. Duration categories have been chosen to reflect toileting with a duration under 10 min, taking a shower with a duration of 10–20 min, and taking a bath or cleaning the bathroom with more than 20 but under 60 min. All other presences are categorized as more than 60 min and are assumed to be measurement errors. Most bathroom visits have a duration of less than 10 min. The participant shown in Fig. 3 has an average number of these visits of approx. eight per day. We assume these visits to be toileting. Presences with a duration of more than 10 but

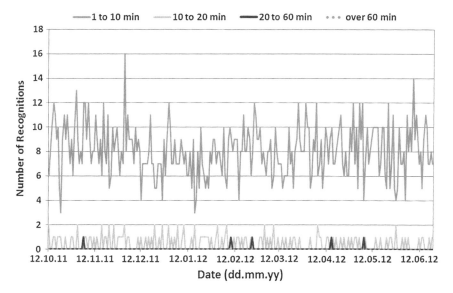

Fig. 3 Participant A: average number of bathroom visits and their duration per day

less than 20 min represent taking a shower or washing himself with water from the basin. In average, he has only 1–2 visits with a duration between 10 and 20 min every second day. Additionally, there is approximately one presence per week which has a duration of more than 20 min. We assume that these entries represent taking a bath.

Results from other participants were very similar. Nearly all participants had an average number of eight visits per day with a duration of under 10 min. The number of visits between 10 and 20 min and between 20 and 60 min varied among participants depending on whether they had a shower or a bath tub and whether they used to take baths or preferred to take showers every day or only wash themselves in the basin. However, the results bring us to the conclusion that especially toileting can be detected reliably using presence times in the bathroom. Cleaning in general can be detected as well but deciding how exactly the participant cleaned himself or herself is difficult. Nevertheless, it has to be regarded that a direct detection of all hygiene actions is not possible.

5.4 Results on Social Activity

Social activity is not only important for self-care ability but also for cognitive fitness. Social assessment include whether a person leaves the house for a sufficient number of times and meets other people. Leaving the house may also point out being able to buy food.

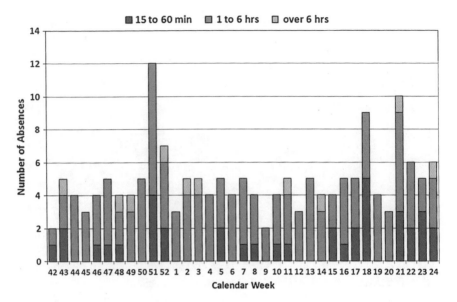

Fig. 4 Participant A: number of absences from the flat and their duration per week

Using the recorded sensor data, especially information from a reed contact at the front door, and a special outdoor feature it can easily be decided whether a participant is at home or not. This feature consists of closing the front door as start-event and opening the door as end-event i.e. no events are generated by the other sensors between closing and opening the front door. Each detected outdoor feature whose duration is more than 15 min is regarded as absence from the flat. Figure 4 shows the number of absences and their duration in three timeslots per week for participant A. In average he leaves the house four times a week. Most absences have a duration of more than one hour. From interviews we know that most of these absences represent eating out in a social kitchen. Absences with a duration of less than an hour but more than 15 min represent buying food in a grocery store or bringing clothes to the laundry. In irregular intervals he visits his brother who lives more than one hour driving time away and often stays overnight. These visits are recognizable by duration of more than 6 h.

The number of absences and their duration varies strongly across the participants. This is also due to the fact that people with very different social connections participated. While participant A was socially rather inactive, participant B had many social activities and had many days in which she was more hours away from home than at home. For another participant regular absences with a duration of 2–3 h were detected. An interview revealed that these absences were visits at the doctor. In general, using the proposed system it seems to be possible to detect social activity by simply counting the number of absences which lasted longer than one hour. All absences below often represented buying food or making general errands.

5.5 Results on Sleeping

Sleeping is related to physical and mental health. Sleeping irregularly and too short often gives hints to possible problems. However, only few assessment tests deal with this issue since it takes too much time and effort in clinical environments to monitor sleep and since many people do not feel well in hospitals and thus do not sleep very well.

Within our approach sleeping is best detected by computing the presence times within the bedroom. This does exclude napping or sleeping in other rooms for now. The Figs. 5 and 6 show the average presence in minutes per room and outdoors per hour of the day for participants A and B. As mentioned earlier, participant A has a rather untypical daily rhythm since he sleeps from approx. 5 to 13 o'clock every day. This is shown in Fig. 5 since his presence per hour in the mentioned timeslot is close to 60 min. Lower values indicate that he sometimes does not sleep or at least not sleep in the bedroom. However, the usage of the bedroom corresponds to the sleeping times as according to interviews. Participant B has typical sleeping habits and goes to bed between 23 and 24 o'clock while standing up at approx. 7 o'clock. Additionally, she used to nap between 13 and 15 o'clock for about an hour which is clearly shown in Fig. 6.

The printed figures also give hints to other habits of the participants such as meal preparation. This may be concluded from e.g. participant B often using the kitchen between 8 and 9, 11 and 13, and 18 and 20 o'clock. Additionally, most social activities happen either between breakfast and lunch or between lunch and

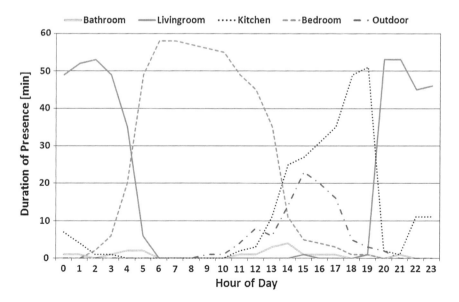

Fig. 5 Participant A: average duration of minutes for different rooms per hour of day

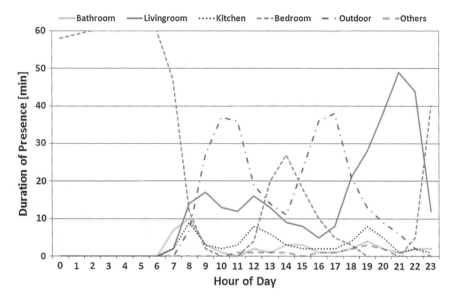

Fig. 6 Participant B: average duration of minutes for different rooms per hour of day

dinner. This may be taken from the fact that the average minutes outside in the timeslots between 9 and 12 o'clock and between 15 and 19 o'clock are rather high. In summary, sleeping habits could be detected very reliably for all participants. Since many activities in daily life are executed in certain rooms we also conclude that average presence times per room may also give valuable information about activities executed and may be used to model a typical behavior of people in their domestic environments.

6 Discussion

Within this paper only early results from our evaluation could be presented. However, we were able to show that assessment tests or at least some items from these may be assessable by using only recordings from light barriers and reed contacts placed between rooms. Especially an inhabitant's mobility, social activity, sleeping habits and general activity should be assessable with reasonable reliability. The manual evaluation was a preliminary step to start detecting items of assessment tests in sensor data automatically.

Although sensor recordings from light barriers seem to be suitable to measure self-selected gait velocity [21] and thus to assess items from the mobility domain, using only light barriers has serious limitations when it comes to assessing other domains: Using single light barriers it is not possible to reliably detect the direction in which a person walked. This does also mean that a person may enter a light barrier and turn around and there is no possibility to detect whether a person really entered or left a room. However, this problem may be compensated by combining light barriers with additional sensors covering areas within rooms such as motion sensors. By using those sensors moving direction may be indirectly detected. Reed contacts were used within the field trial but only partially evaluated for this paper.

Despite all technical problems, it remains future work to clarify the relationship between results obtained from clinical assessment tests and those performed in domestic environments. Currently, there is only very little research available on how any assessment results at home are related to results from clinical environments or how performance and capacity in various domains may be compared. Therefore, our objective is to infer a model on how results from domestic assessment tests could be mapped to the result scales of clinical assessment tests. A new field trial in three cities in Lower Saxony, Germany is currently under preparation. At each site 10–15 flats will be equipped with sensors to further investigate unsupervised assessments while all participants will have to complete a set of clinical assessment tests regularly.

However, even if the automatic assessment of people's characteristics in their own homes will be available soon, health care professionals will not be able to review the assessment results continuously. In order to really save costs respectively to enable caregivers to take care of more patients in the same time, technical systems will have to provide hints to problems or changes in the functional or health status. Therefore, part of our future work will also be to infer a model to learn individually normal characteristics of persons at home and to alert caregivers in case of untypical deviations.

7 Conclusion

Many elderly people want to live at home even in high age and despite of diseases they may have. In order to comply with this wish early prevention and long-term rehabilitation are essential. In order to provide these services effectively health care professionals require knowledge about the functional and health status of their patients. In clinical environments these characteristics are obtained by using various assessment tests. However, such clinical tests are not suitable for domestic assessments which should run continuously and without creating a test-situation. This saves costs and increases compliance by inhabitants. Implementation of those domestic environments is only possible by use of information technology.

Several approaches to technical systems monitoring people's actions and movements at home exist using either body-worn or ambient sensors. However, only very few of those systems deal with mapping their results to clinical assessment tests' result scales which makes them relevant to clinical decisions. We have developed such a technical system as well and utilized it during a field trial in order to investigate which assessment test items are assessable by use of home automation sensors. Using the data from five elderly participants aged 64–84 years obtained over partially more than nine month we have shown that several items from assessment tests from the domains of mobility, personal hygiene, social activity, and sleeping are assessable. Although data analysis was performed manually for this paper we think that this will work automatically soon.

Future work is to infer two models. The first model is about the mapping of domestic assessment results to clinical result scales in order to make those measurements relevant to clinical decision making. The second model will be used to describe individually normal functional status of persons and to detect changes in these relevant to endangering persons' self-care ability. This will alter health care professionals by only having to analyze available assessment results in case of anomalies. In order to gather data for those two models a new field trial will be conducted in three cities in Lower Saxony, Germany.

Acknowledgments This work was in part funded by the German Ministry of Education and Research within the research project PAGE (grant 01FCO8044) and in part by the Ministry for Science and Culture of Lower Saxony within the Research Network "Design of Environments for Ageing" through the "Niedersächsisches Vorab" grant programme (grant ZN 2701).

References

1. Frenken, T., Steen, E.E., Brell, F., Nebel, W., Hein, A.: Motion pattern generation and recognition for mobility assessments in domestic environments. In: Proceedings of the 1st International Living Usability Lab Workshop on AAL Latest Solutions, Trends and Applications. In conjunction with BIOSTEC 2011, pp. 3–12. SciTePress, Rome, 28–29 Jan 2011, ISBN 978-989-8425-39-3
2. Beers, M.H., Berkow, R. (eds.): The Merck Manual of Geriatrics. Merck & Co., Inc., NJ (2006), http://www.merck.com/mkgr/mmg/home.jsp
3. Mahoney, F.I., Barthel, D.W.: Functional evaluation: the barthel index. Md State Med. J. **14**, 61–65 (1965)
4. Lawton, M.P., Brody, E.M.: Assessment of older people: self-maintaining and instrumental activities of daily living. Gerontologist **9**(3), 179–186 (1969)
5. Folstein, M.F., Folstein, S.E., McHugh, P.R.: "Mini-mental state". A practical method for grading the cognitive state of patients for the clinician. J. Psychiatr. Res. **12**(3), 189–198 (1975)
6. Podsiadlo, D., Richardson, S.: The timed "up & go": a test of basic functional mobility for frail elderly persons. J. Am. Geriatr. Soc. **39**(2), 142–148 (1991)
7. Garms-Homolová, V.: Assessment für die häusliche Versorgung und Pflege: Resident Assessment Instrument- Home Care(RAI HC 2.0). Huber (2002)
8. Frenken, T., Vester, B., Brell, M., Hein, A.: aTUG: Fully-automated timed up and go assessment using ambient sensor technologies. In: Proceedings of 5th Int Pervasive Computing Technologies for Healthcare (PervasiveHealth) Conference, pp. 55–62 (2011), ISBN 978-1-61284-767-2
9. Helmer, A., Lipprandt, M., Frenken, T., Eichelberg, M., Hein, A.: 3DLC: A Comprehensive Model for Personal Health Records Supporting New Types of Medical Applications. Journal of Healthcare Engineering 2(3), 321–336 (September 2011), ISSN 2040-2295. doi: 10.1260/2040-2295.2.3.321

10. Scanaill, C.N., Carew, S., Barralon, P., Noury, N., Lyons, D., Lyons, G.M.: A review of approaches to mobility telemonitoring of the elderly in their living environment. Ann. Biomed. Eng. **34**(4), 547–563 (2006)
11. Bauer, P., Rodner, T., Litz, L.: AAL-Eignung von HomeAutomation-Sensorik—Anforderungen und Realität. In: Wichert, R., Eberhardt, B. (eds.) Technik für ein selbstbestimmtes Leben (AAL 2012). VDE Verlag (2012), ISBN 978-3-8007-3400-9
12. Pallejà, T., Teixidó, M., Tresanchez, M., Palacín, J.: Measuring gait using a ground laser range sensor. Sensors **9**(11), 9133–9146 (2009), http://www.mdpi.com/1424-8220/9/11/9133/
13. Frenken, T., Baumgartner, H., Scheve, C., Meyer, J., Ulbts, J., Hein, A.: Selbstbestimmt im letzten Lebensjahr: AAL-Technologien im Kontext der End-of-Life-Care. In: Proceedings 3. Deutscher AAL-Kongress: Technologien—Anwendungen—Management, Ambient Assisted Living. VDE Verlag (2010), ISBN 978-3-8007-3209-8
14. Stone, E.E., Skubic, M.: Passive, in-home gait measurement using an inexpensive depth camera: initial results. In: Proceedings of 6th International Pervasive Computing Technologies for Healthcare (PervasiveHealth) Conference (2012), BEST PAPER AWARD
15. Poland, M.P., Gueldenring, D., Nugent, C.D., Wang, H., Chen, L.: Spatiotemporal data acquisition modalities for smart home inhabitant movement behavioural analysis. In: ICOST '09: Proceedings of the 7th International Conference on Smart Homes and Health Telematics, vol. 5597, pp. 294–298. Springer, London (2009)
16. Steen, E.E., Eichelberg, M., Nebel, W., Hein, A.: A novel indoor localization approach using dynamic changes in ultrasonic echoes. In: Wichert, R., Eberhardt, B. (eds.) Ambient Assisted Living. Advanced Technologies and Societal Change, pp. 61–76. Springer, Hedelberg (2012)
17. Steinhage, A., Lauterbach, C.: SensFloor: Ein AAL Sensorsystem für Sicherheit, Homecare und Komfort. In: Ambient Assisted Living, pp. 37–42. VDE-Verlag (2008)
18. Hein, A., Winkelbach, S., Martens, B., Wilken, O., Eichelberg, M., Spehr, J., Gietzelt, M., Wolf, K.H., Büsching, F., Hülsken-Giesler, M., Meis, M., Okken, P.: Monitoring systems for the support of home care. Inform. Health. Soc. Care **35**(3–4), 157–176 (2010), http://dx.doi.org/10.3109/17538157.2010.528243
19. Pavel, M., Hayes, T.L., Adami, A., Jimison, H., Kaye, J.: Unobtrusive assessment of mobility. In: Proceedings of 28th Annual International Conference of the IEEE Engineering in Medicine and Biology Society EMBS '06, pp. 6277–6280, Aug 2006
20. Floeck, M., Litz, L., Rodner, T.: An ambient approach to emergency detection based on location tracking. In: Abdulrazak, B., Giroux, S., Bouchard, B., Pigot, H., Mokhtari, M. (eds.) Toward Useful Services for Elderly and People with Disabilities. Lecture Notes in Computer Science, vol. 6719, pp. 296–302. Springer, Heidelberg (2011)
21. Frenken, T., Lipprandt, M., Brell, M., Wegel, S., Gövercin, M., Steinhagen-Thiessen, E., Hein, A.: Novel approach to unsupervised mobility assessment tests: field trial for aTUG. In: Proceedings 6th International Pervasive Computing Technologies for Healthcare (PervasiveHealth) Conference (2012), BEST PAPER AWARD

LsW: Networked Home Automation in Living Environments

Frerk Müller, Peter Hoffmann, Melina Frenken, Andreas Hein and Otthein Herzog

Abstract Due to demographic changes a lot of research in the field of Ambient Assisted Living (AAL) has been done within the last years. Many of the project systems never left the status of a research prototype nor reached real home environments. The "Länger selbstbestimmt Wohnen" (LsW) project is one step beyond the targets of typical research projects and is integrating preexisting (home automation) technologies in home environments to support the elderly residents. Requirements, selection of possible scenarios and evaluation were done in close cooperation with the residents. The integrated AAL system will remain within the flats of the residents beyond the end of the LsW project. Main points of the LsW project were the analysis of current mobile devices like tablets as human machine interfaces for AAL systems, the integration of existing home automation technologies in existing buildings, the connection between the tablet and the home automation components, and the interoperation between several flats.

F. Müller (✉) · M. Frenken
OFFIS Institute for Information Technology, Oldenburg, Germany
e-mail: frerk.mueller@offis.de

M. Frenken
e-mail: melina.frenken@offis.de

P. Hoffmann · O. Herzog
Center for Computing and Communication Technologies, University Bremen, Bremen, Germany
e-mail: phoff@tzi.de

O. Herzog
e-mail: oh@tzi.de

A. Hein
School of Medicine and Health Sciences, University Oldenburg, Oldenburg, Germany
e-mail: andreas.hein@uni-oldenburg.de

1 Motivation for LsW

Due to demographic changes a lot of research in the field of Ambient Assisted Living has been done within the last years. The objective of the "Länger selbstbestimmt Wohnen" (LsW) project is one step beyond the targets of typical research projects. Within this project it is expected that supporting systems will be developed which are integrated into real life environments of elderly people participating in the project and will be left within their flats after the project ends.

This leads to the fact that the project has to develop systems which are closer to real products than to research prototypes and need to be maintainable by external companies after project completion. One of the major key points of this project is not to focus on high-end research prototypes of the next generation, but on the needs of the residents and the support that can be offered by state-of-the-art technologies. Therefore the project tries to combine and extend available market solutions to create new supporting features close to available market products. Next to this these systems will be evaluated to figure out the real benefit of the daily usage. This should help the housing companies to decide on further installations for other apartments not related to a research project. So the motivation of the project is to develop AAL support for elderly, evaluate it and have it directly ready for the market at the end of the project.

2 Principles of AAL Approaches

There are two approaches for AAL technologies which are applied often to AAL projects. One is focusing on home automation with the objective of improving the living situation at home by adapting technical installations for the needs of elderly people. The second one is focusing on the use of mobile IT systems with the objective of offering a kind of assistive companion for the elderly.

2.1 Home Automation Technology

Existing home automation technologies like EIB, LON, FS20, Homematic, Zigbee, and a lot more offer several possibilities to bring extra services or non-invasive health status monitoring possibilities to homes to support elderly people [1]. Others list key points. Challenges of AAL technologies are:

- An extended healthy, active and dignified life for the elderly that is also widely accessible to the low-income strata of society,
- Implementation by using simple low-cost sensor technology, which also makes it affordable to the lower income people,
- Adaptively retrofitting existing home structures with minimal impact, modification and cost,

- The system should be customizable to the individual's needs, as well as to different cultural needs.

Home automation devices are an integral part of AAL technology strategies. To enable AAL research on smart homes, living labs have been established. Those living labs are laboratories that try to emulate real home environments to develop best suited technologies. In Germany, one of the most popular living labs is the Fraunhofer *inHaus* center in Duisburg. Technologies developed here focus on home automation and electronic assistance. The Living lab *ProPotsdam Komfortwohnung* is specialized on comfort, safety and living for elderly. The *DAI Labor* laboratory concentrates on connectivity and networking of home automation to provide access via a home service platform. Also the IDEAAL apartment at OFFIS in Oldenburg is such a living lab concentrating on developing and evaluating different AAL scenarios with real end users [2]. Within the European context, the Philips *HomeLab* in Eindhoven (NL), *LivingTomorrow* in Vilvoorde/Amsterdam (BE/NL), or the Social Informatics Lab—*SILab* in Newcastle (UK) can be mentioned as well. This is far from an exhausting list; websites like www.openlivinglabs.eu or www.aal-deutschland.de list a lot more labs all over Europe and the world. Those living labs provide examples of the use of home automation with slightly different focuses. An international overview of smart home technologies is given in [3] and [4]. Nevertheless, those are laboratories which can only try to imitate the real life homes. However, lots of new challenges arise in reality and therefore real life projects are essential to establish AAL systems well suited to the users.

2.2 Mobile Assistance

Up-to-date IT with small and networked devices increases the users' range of mobility. This includes mobility inside their home environment as well as outdoors. One goal is to monitor the individual status (e.g. medical status) to improve the quality of life and safety. Another important goal is to give the users the possibility to control their environment at any place and any time [5]. While the number of people using mobile devices like smartphones or tablets increases, also the number of APPs grows for controlling e.g. electric devices at home.

By combining strategies of adaption and awareness this kind of technology will eventually become usable as mobile assistants for handicapped users:

- As a passive assistant the system offers information about the devices at home while being away: For example users who have forgotten whether they switched off the oven at home or not are able to check it and control it from anywhere.
- As an active assistant the system can influence its environment and change states: For example the system could switch off the oven by itself if it detects that the user has left the flat.

Mobile assistance means the combination of (small) devices and context-aware applications. In working environments this kind of assistance is developed and used for supporting the worker with information needed in the current situation [6]. The central idea is to carefully select the information for the use in this situation [7].

The approach in AAL is similar: developing an ambient assistive environment based on existing IT-techniques like "Ambient Intelligence" [8] and "Intelligent Objects" [9]. The assistive environment monitors the individual status of the user and the state of the (technical) environment and derives resulting workflows for execution. The system predicts the safety status of the user and it informs him/her with a feedback about possible upcoming hazards, or influences the "environment" to protect the user before any hazardous situation can occur [10].

2.3 Ergonomics and Usability Design

As many technical and especially mobile systems and applications are developed for "younger" users without any handicaps, the design of an appropriate AAL system is still challenging. AAL systems are not simple lifestyle products but have to take into account the usability for handicapped users. Several aspects are essential for a successful development:

- Any device such as a piece of hardware must follow the standard rules of ergonomics.
- Any application or APP as a piece of software must follow the standard rules of software usability.
- Furthermore any application or APP as a piece of software must follow the ideas of accessibility as well as of individual customization.

As the target group for AAL systems consists mostly of elderly and/or handicapped users the approach of a "barrier free" design will be a good starting point for designing those systems. It is not sufficient to use buttons or displays for designing an AAL System. Standards in this field describe the various possible kinds of handicaps [11] and give ideas on how to solve these issues. Nevertheless it is not possible to consider all handicaps as the variety is too broad. The range is from physical handicaps like tremor over cognitive handicaps like blindness up to mental ones like dementia. So a detailed analysis of the target groups is essential.

Beside ergonomics and usability, esthetics are even more important for the acceptance of an AAL product. Especially for AAL systems this means that most users do not like systems which are obviously recognized as being designed for elderly or handicapped users. Examples are smartphones designed for elderly people with big buttons and special colors which usually are not a success in the market place [11].

3 LsW Environment Concept

The project "Länger selbstbestimmt Wohnen" (LsW) goes beyond a typical research project. Beside its short duration and its small budget some more aspects make the project special.

The origin of the project is not just technically driven but inspired by the residential market. A main partner of the project is a housing company (HC) which is aware of the demographic changes and the resulting needs of the residents. Therefore they plan to offer adapted flats to the elderly as a new target group. To acquire more knowledge about the special needs of this target group the potential users were involved into the development of this project. This was done by performing workshops during the project where the project partners and the prospective users collaborated to develop system requirements. Within a first workshop the users were asked to collect daily problems without discussing any possible solutions for that. Afterwards the research groups tried to identify technical solutions for some of these daily problems and presented these results in a second workshop to the prospective users. At the end there was a list of features which afterwards were implemented and integrated in the flats of the residents.

It is special to the project that its outcome is not a simple prototype or demonstrator but a real product which remains in the flat and stays with the users after the end of the project. That leads to the fact that the system must work stable for a long time. Furthermore it must be easily maintainable which means that the software has to be adaptable as well as the hardware components have to be replaceable by a technical service team and not only by the involved researchers.

3.1 The Living Situation

One more aspect makes the LsW project even more special. This is the living situation within the building LsW is integrated into. It is a house of 10 parties with an unusual high grade of social networking between these parties.

Social networking in this context means the real world interaction between the people living there. Everyone knows and helps each other if necessary. They have a shared flat for meetings and also guests. New residents are not selected by the HC but by the current residents to support this social spirit. This is quite interesting as the average age of the residents is higher than in other living environments. Most of them are already retired. This means special needs and therefore special scenarios to be developed in the LsW project, but also very specific opportunities for designing AAL services like in-house alarm systems to inform the neighbor in the case of emergencies or forgotten household devices left switched on.

3.2 Home Automation Concept

Originating from the workshop results several scenarios were extracted and decided to be realizable within the project. Most of the scenarios are based on home automation technologies which allow for the integration of multiple features with one set of home automation actors and sensors. As a result of the workshops the following scenarios were developed within the scope of home automation technologies.

3.2.1 Scenario: Everything Switched Off

The first scenario is called "Everything Switched Off". This means for the resident of the apartment just a new light switch next to the entrance turning off all lights, power outlets as well as the cooker. From the conceptual point of view this is more difficult. It has to be decided how to switch on the power outlets on arrival as well as to decide which devices should be switched off/on automatically and which devices need to be handled semi-automatically, as for example an electric iron. This device should be switched off on leaving the flat, but not switched on automatically when returning. Therefore a simple traffic light metaphor was created. All power outlets marked by a green spot are declared to be no safety issue. Therefore only devices like TVs or lights should be plugged in there. Power outlets without any mark are conventional and will be therefore not affected by the "Everything Switched Off" scenario. These power outlets may be used, e.g., for devices like a fridge, alarm clocks or phones. The last category of devices is connected to power outlets marked red. To this category belong all devices which must be switched off when leaving the apartment, but must not be switched on again automatically on arrival. The most important devices are the electric iron and the cooker. For example, if an electric iron was already forgotten on leaving the apartment it will stay most likely forgotten on arrival back in the apartment. This could lead to an fire in the apartment on arrival of the resident which would be as undesirable as a fire during the absence.

The electric iron will be often plugged into different power outlets and therefore it was decided to use a mobile adapter instead of the power outlet adapter which is permanently attached to the device and not to the power outlet in the wall. It also has a Switch-On-Button and therefore allows for switching it on again manually, even if the power was switched off automatically by the home automation system.

The cooker received its own Power-On-Button. This push button has two selectable push states. The first and probably normal state is "Switch on the cooker for 30 min". This will activate the power line of the cooker for 30 min and switches it off afterwards as normal cooking ends mostly before that timeout. If the cooker was switched off manually before, it will not be noticed by the resident at all. The second option for the push button is the "Permanently On"-option. This may be used for more complex meals which take more than 30 min. The Global-Switch-Off-Button next to the entrance will power off the cooker in both cases.

3.2.2 Scenario: Everything is Closed

Next to this Global-Switch-Off it was stated by the users that they often come back to their flats because of checking the windows once again. The first idea to address this issue was to close the windows automatically, but as a result of the workshop the residents wanted to do that on their own. This leads to the "Everything Is Closed"-scenario. If someone leaves the flat and presses the Global-Switch-Off-Button used for the "Everything Switched Off" scenario the "Everything Is Closed" scenario is started. It checks all the windows and presents a phrase in native language to the user stating in which rooms windows are left open. The sentence will be presented through a tablet mounted next to the entrance. Speech synthesis algorithms generate the spoken text.

3.2.3 Scenario: Everything is Save

The third scenario is called "Everything Is Save". It focuses on the handling of fire alerts which seem to occur pretty often because of forgotten cooking activities. Therefore fire alerts detected by smoke detectors are not only presented in the apartments of their occurrence, but are also forwarded to the other apartments. In these apartments the speech synthesis is used again. It gives repeating information about the location where the fire was detected until the user confirms the fire alert by pressing the Global-Switch-Off-Button in his/her apartment. Next to the audio notification within all apartments also all lights will be switched on to improve orientation on the way to the exit. This scenario is a good example of the benefit of the existing living situation in the house where everybody knows and trusts in each other.

Next to the features to be implemented for the elderly some simple features for maintenance have also taken into account. As all the home automation devices should be integrated in already existing living environments it is most likely that a least some of the devices will be powered by a battery. Therefore the maintenance crew needs to know about empty or damaged devices. Should one of the devices send a low battery state, it will cause an automatically generated mail which is sent to the maintenance center. This will make handling of the home automation much easier for the elderly and also for the housing company.

3.3 Mobile Assistance Concept

Beside the home automation support a mobile assistance system for the residents was also developed, i.e., a passive assistance inside their flats. As a central device a tablet was selected with an APP which gives the residents the control of the home automation as well as support on communication. The tablet was chosen as it can be carried around easily within the flat. Furthermore the touch interface of these devices appeared to be highly intuitive and understandable for the users.

The first scenario of mobile assistance to be considered for the concept was to inform about the state of the home automation: The resident is empowered to gather the information if a device in another room is switched on or off and to control it from anywhere. So he/she can avoid unnecessary movements. Mobility in this context means that the resident has the opportunity to control his/her environment from anywhere in the flat using the tablet and not being forced to go to a special place to do so.

In the workshops it was explained to be important that any control action has to be activated by the resident him/herself. The idea of letting "the system" control devices (e.g. lights, ovens) by itself was not appreciated by the users. It was not so much the fear of being patronized but the fear of getting more and more inactive and not longer self-activated. The residents know about the importance of self-activation for staying independent and wanted to have that included in the design of the system.

A second aspect that was seen as helpful for the residents inside their own flat is the support by mobile communication. This was achieved by two special sub-scenarios:

- The mobile assistant is connected to the "Everything is Save" scenario. Speech synthesis on this tablet is used to inform the resident if a fire alert occurs in one of the neighbors' flats. The goal was to give the resident a more secure feeling in the way that he/she can rescue themselves or prevent neighbors from being hurt and help them in any possible way.
- The mobile assistant supports the communication to the outside of the flat. On the workshops it was reported that it often takes a long time for the resident to react if a visitor rings the door bell. Due to physical handicaps some residents are slow in reaching the door and opening it. So visitors may leave thinking that no one is at home, while the resident is still on the way to the door. The idea to improve that situation was to connect the tablet to the door bell system including its video, voice and control subsystems: a camera at the front door transmits an image stream of the entrance to the tablet. So the resident sees the video of the visitor and can use the tablet to talk to him/her and to open the door if necessary. Again this is possible from any place inside the flat avoiding unnecessary walks for the resident.

A side effect in terms of communication is given from the basic functionality of tablets connected to the internet. It is the possibility to use standard tools like internet browsers or even Skype.

While the assistive functionality with its "logic" should stay in the background of the (software) system to not to irritate the user, the design of the user interface is an essential point in the project.

The first design was influenced by typical interaction design paradigms for tablets. A graphical user interface with elements to be activated by touch interaction was prototypically implemented. To control the flat the user first had to choose the room and in a second step to choose the switch or socket he/she liked to

LsW: Networked Home Automation in Living Environments

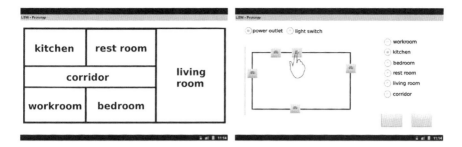

Fig. 1 Graphical concept of the user interface for the home automation subsystem of the LsW application

control (see Fig. 1). The idea was to give the user the total control on any single switch and socket in any single room. During the workshops it was figured out that this interface was much too complex for the residents. So it was replaced by a simpler and more textual design (see Fig. 2). Even if this leads to more text to read for the users, it made them feel more comfortable with the system.

One more challenge for the user interface of the LsW application was the complexity of the whole system. Not only the home automation but also communication in terms of the "Everything is Safe" scenario, the door bell and internet as an extra option had to be considered in the design. This lead to the idea of a central application which starts and shows those sub-applications which are currently used and hides all others (see Fig. 3).

Fig. 2 Home-24 App for tablet PCs [12]

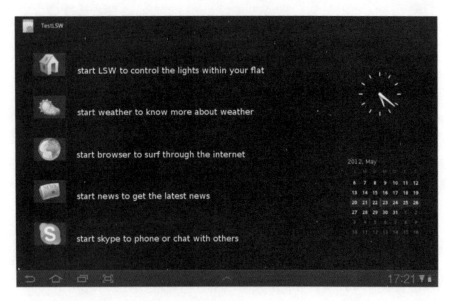

Fig. 3 Concept for the LsW main application

4 Realization of LsW

Due to the fact that the system has to be more stable than a prototype at the end of the project, it was decided to use mostly off-the-shelf technology already available in the market and extend it by the features needed in addition. Because the apartments are already existing and people are living there, radio based home automation devices were chosen. This makes the necessary modification of the apartments quicker and easier.

Figure 4 gives an overview of the overall architecture developed for the LsW scenario. The boxes marked as apartment A, B and C represent the three apartments to be used within LsW scenarios. Apartment A is shown in the figure in more detail than the other apartments whereas the functionality added is the same for all apartments. Due to safety and security reasons it was decided to use as much of the already available cable-dependent infrastructure as possible. All lines marked as wired (phone line and wired Ethernet) belong to the infrastructure which was available within the building before project start. Every apartment has its own phone line whereas the internet connection is shared between all parties of the building. This is in fact an untypical configuration for German buildings, but still a good base for the LsW scenarios. Every apartment got a wireless access point for distributing wireless LAN within a single apartment. This access point also distributes phone calls by using the SIP [13] standard or ordinary phone connectors. By setting up an AVM FritzBox 7390 [14] it is now possible to have an audio/visual connection from the door bell (lower middle) through the central router (center) to each wireless access point of each apartment finally ending at a

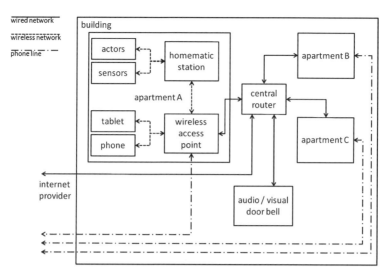

Fig. 4 Basic structure of components within the LsW scenarios

SIP-softphone installed on an Android tablet (e.g. in apartment A, upper left). Next to this the same software on the tablet can also be used to perform an ordinary phone call on the phone line. Crossing the wireless access point and the central router it is also possible to exchange messages between each Homematic [15] station placed within the building. Therefore, e.g., smoke detected by one of the smoke detectors placed in apartment A can notify a user in apartment B by switching the light on and playing spoken texts and alarming sounds within this apartment.

The following section describes the development of the scenarios in more detail.

4.1 Home Automation Realization

All home automation actors and sensors placed within the apartments are based on the Homematic system already introduced within the home automation technology section. Next to this an AVM FritzBox 7390 was used as wireless access point as this device supports also SIP phoning which is important for the communication scenario. As an Android tablet the Samsung Galaxy Tab 10.1 [16] was added to the scenario for speech synthesis and visual notifications within all home automation scenarios.

As already described power outlets to be switched off when leaving the apartment are marked green or red. By checking the devices within the flats it was noticed that all devices to be marked safety critical and therefore labeled red are

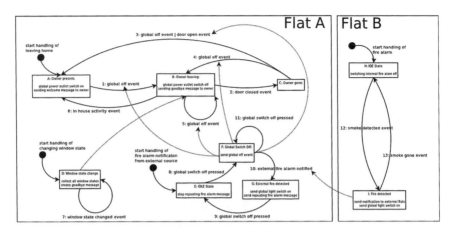

Fig. 5 Final state machine to describe the home automation scenarios developed in the LsW project

not fixed to a specific power outlet and as a result all critical devices got their own power adapter including a switch to reactivate the device after switching it off globally.

Figure 5 gives a brief overview of the final state machines which were implemented during the project. It consists of four different state machines for switching the light off, handling fire alerts in the users' flat and also in remote flats, and also handling window state changes. All these scenarios interact with the single Global-Switch-Off button at the entrance door. This will be described more in detail in the following section.

In the upper left corner the main state machine for controlling lights and power outlets is placed. It consists of three states: Owner-Present, Owner-Leaving, or Owner-Gone. In the state Owner-Present the power outlets are switched on and the lights are controllable. Once a Global-Off-Event was received the Owner-Leaving state is reached. This can only be the case if the Global-Switch-Off-Button was pressed and currently no alarm in the owners' or a remote flat is noticed by the system. Should the owner right after pressing the Switch-Off-Button close the entrance door it is expected that the resident left the apartment. Should this not be the case, because someone closes the door but continues to stay in the flat, every action within the flat noticed will bring the state machine back to the Owner-Present state and switch on the green power outlets. This could be switching on a light, opening a window or switching on the cooker. The only exception is another Global-Off-Event. This will bring the automaton back to Owner-Leaving state.

If a Global-Switch-Off-Event was recognized by the system not only the devices will be switched off, but also a text will be spoken informing about the window states within the whole apartment so to not forget any windows left open. For performance issues of the Homematic control center this sentence is prepared in advanced every time the state of a window changes. This allows for reducing time gaps between pressing the Global-Switch-Off-Button and the spoken reaction by the system. As this was a problem during the first development in the project the automaton in the lower left corner was introduced to the system.

As already described above, the Global-Switch-Off-Button is designed to be used for several scenarios. In the case of a remotely detected fire (in the figure caused in flat A by flat B) the automaton in the lower middle is triggered. This state machine will deactivate the Global-Switch-Off-Event first and replace it by the Remote-Fire-Alert-Noticed-Event and therefore will not automatically switch off the lights anymore on pressing during fire alert. Next to this all lights in the flat will be switched on by the event and an alarm will be presented generated by the tablet. This alarm is also spoken text next to an alarm signal which will be presented in loops until the Global-Switch-Off-Button is pressed. The spoken message informs the residents about the fire alert, especially about the location of the fire. This will allow for offering help and makes emergency calls even quicker. The communication between the different apartments is done via LAN. As all apartments use the same internet connection the data can be routed in between the building using the LAN. This is safer regarding the connection origination than Wireless LAN which would have been also a possible solution.

4.2 Mobile Assistance Realization

For the implementation of the mobile assistance a Samsung Galaxy Tab 10.1 was chosen as the hardware platform. One main reason was that the Android Operating System used on these tablets offers easy solutions for the implementation of mobile assistance due to well documented APIs and a complete set of sensor infrastructure.

Due to the chosen operating system the LsW "application" was implemented as a set of Android "APPs". As introduced in the concept this set has a central APP as its backbone which organizes the specialized APPs for home automation and communication. These "sub-APPs" are running all time in the background. If activated by the user the central APP gives the focus of the user interface to the activated APP. This allows for running all intended functionalities in parallel. An example should clarify the idea: The resident has his/her tablet with him in the living room. The LsW APP is running on it. A visitor rings the door bell and the LsW APP gives the focus to the APP for the door bell control. The resident can now see who wishes to visit him and may open the door using this APP. To make it easier for the visitor to find the way to the living room, the resident changes the focus to the home automation APP and switches the lights in the corridor on. He/she can do all that while sitting in the living room.

An advantage of this modular concept for the LsW system is that changes in functionalities can be implemented easily without influencing the other parts of the system. This advantage was used in the very first steps of implementation. APPs developed from third parties like the Home24 APP for controlling the home automation could be integrated and tested while other APPs were still in the design phase.

5 User Acceptance

The implementation of the project was done stepwise. While the concept of the home automation and the implementation of the mobile assistance could be done in parallel, the installation in the real environment in the flats of the involved residents was done in separate steps.

The experiences of the installation in the first flat including the first user (residents) experiences could be used to improve the installation in the second flat and after that in the third flat. This process improved the installation in terms of getting better and shorter by every flat. Unfortunately it had the disadvantage that the final user studies could start rather late in the project.

At the moment while this paper is written, the user study with the targeted number of residents was just started. So no report on the user acceptance can be given here and now, but will be presented in the talk at the AAL-Conference in January 2013.

6 Gained Experiences So Far

Even though the user studies are ongoing some experiences could already be gained. These can be divided into three aspects:

- Common individual user experiences,
- Technical experiences, and
- Design experiences.

Common individual user experiences

In general the first user experiences can mostly be seen as fighting some psychological fears. This was shown by the results of the first workshop where the idea and some technical demonstrations were presented to the residents. As many components have to be connected wirelessly one main fear was the effects of the so called electrosmog on the residents. There were also some other reported fears the technical system could somehow not work properly. This fear appeared to originate from bad experiences with other technical systems. Those common fears were accompanied by fears stemming from individual experiences having mostly their origin in the knowledge of their own age-dependant handicaps. So some residents felt threatened by the technique itself: they feared to be overburdened by it. Furthermore there was the fear of confusion by switches having different functionalities like a usual light switch, the "Everything OFF" switch and the timed switch for the oven.

Most of those fears could be dealt with or at least be reduced by intense follow-up workshops. But for future projects it is important to consider those and similar psychological aspects as early as possible in the projects.

Technical experiences

Two main points have to be reported so far. The first one is that existing and available components are sometimes quite too focused on some special situations and scenarios. This leads to problems that may occur when using those components in combination with other IT systems, which is one of the tasks to be performed to fulfill the objectives of the project. These issues might be solved in future when more systems are available in the marketplace and the needs for interoperability of those systems are getting stronger.

The second issue figured out so far is that all these home automation installations must be strongly individualized for every apartment and its residents. This might lead to the point that only a specialized company or in the first step a specialized employee will be able to install and maintain those systems such as an AAL integrator.

Design experiences

Closely related to the psychological experiences are the experiences in terms of design. This does not only mean the design of the system and its user interface but also the design (and the concept) of how to train the users on the system:

- Signs/Icons and/or texts of newly installed components must be easily understandable for the users. To design a system according to this requirement it must be taken into account that the targeted user group is elderly and maybe handicapped people.
- The design of the user interface must also respect that the target group might not really be used to devices of the newest technical generation. So the usage, e.g., of a tablet needs to be discussed before integrating it in similar projects.
- A challenge of the design in LsW was the synchronised combination of analogue and digital user interfaces. This was seen especially in the kitchen. The installed ovens often have analogue, mechanical butons to control the temperature of the oven. If an APP (from some mobile device) influences the oven, e.g., changes the temperature it is essential that not only the digital user interface (the APP) changes but also the analogue one. Both interfaces must be absolutely sychronised.
- A good introduction of the system in terms of an intensive training of the prospective users is also essential. This helps to make the user/resident accept the system.

Acknowledgments The "Länger selbstbestimmt Wohnen" project is funded and supported by the "Bremerhavener Gesellschaft für Investitionsförderung und Stadtentwicklung mbH"(bis). For the support and the partnership in the project we would like to thank especially the "STÄWOG Städtische Wohnungsgesellschaft Bremerhaven mbH" who offered the project a realistic environment and of course all residents of the involved social project. It was very inspiring to have all those workshops with them and to collaborate with them during the system implementation.

References

1. Marsh, A., Biniaris, C., Vergados, D., Eppler, A., Kavvadias, C., Bigalke, O., Robert, E., Jerabek, B., Caragiozidis, M.: An assited-living home architecture with integrated healthcare services for elderly people. Med Care Compunectics **5** (2008)
2. Kröger, T., Brell, M., Lipprandt, M., Müller, F., Helmer, A., Hein, A.: IDEAAL, der Mensch im Mittelpunkt, vol. 4. Deutscher AAL Kongress, Berlin (2011)
3. Cook, D.J., Das, S.K.: How smart are our environments? An updated look at the state of the art. Pervasive Mob Comput **3**(2) (2007)
4. Augusto, J., Nakashima, H., Aghajan, H.: Ambient intelligence and smart environments: A state of the art. In: Handbook of Ambient Intelligence and Smart Environments, pp. 3–31 (2010)
5. Schiele, B., Starner, T., Rhodes, B., Clarkson, B.P.A.: Situation aware computing with wearable computers. In: Fundamentals of Wearable Computers and Augmented Reality, Lawrence Erlbaum Press (2001)
6. Lawo, M., Herzog, O., Boronowsky, M., Knackfuß, P.: The open Warable computing group. Pervasive Comput, IEEE **10**(2), pp. 78–81 (2011)
7. Lukovic, P., Timm-Giel, A., Lawo, M., Herzog, O.: WearIT@work: Towward real-world industrial wearable computing. Pervasive Comput, IEEE (2007)
8. Edelkamp, S. (ed.): KI 2011: Advances in artificial intelligence. In: 24th Annual German Conference on AI. Lecture Notes in Artificial Intelligence (2011)
9. Herzog, O., Schildhauer, T.: Intelligente objekte. In: Technische Gestaltung—wirtschaftliche Verwertung—gesellschaftliche Wirkung, Berlin (2009)
10. Hoffmann, P., Lawo, M.: AAP—Ambient assitant protection. Von der klassischen Arbeitssicherheit zur intelligenten Arbeitssicherheitsassistenz. In: WCI, Wireless Communication and Information—Mobile Gesellschaft, Berlin, 25–26 Oct 2012
11. Hellbusch, J.E., Kerstin, P.: Barrierefreiheit verstehen und umsetzen. dpunkt Verlag GmbH, Heidelberg (2012)
12. Breternitz, M.: Home-24, [Online]. Available: http://www.home-24.net. Accessed 11 2012
13. Session Initiation Protocol Core Group: Session initiation protocol [Online]. Available: http://datatracker.ietf.org/wg/sipcore/charter. Accessed 11 2012
14. AVM GmbH: AVM Homepage, [Online]. Available: http://www.avm.de/en/. Accessed 11 2012
15. eQ-3 AG: Homematic Homepage, [Online]. Available: http://www.homematic.com/. Accessed 11 2012
16. Samsung Electronics Co. Ltd.: Samsung Electronics Co. Ltd. Homepage, [Online]. Available: http://www.samsung.com/. Accessed 11 2012

Guiding Light for the Mobility Support of Seniors

Guido Kempter, Walter Ritter and Andreas Künz

Abstract This paper shows how we develop and implement an intelligent light wayguidance system, which will attenuate age-related mobility impairments caused by reduced spatio-temporal orientation, worry about getting lost, and fear of falling. Guiding light consists of up to date lighting technologies, innovative intelligent control algorithms, smart mobility monitoring systems, and a distributed information system for mobility parameters of older persons. Together with end-users and all stakeholders these components can be combined with interpersonal care services. We will present work in progress of an ongoing project within AAL Joint Programme (AAL-2011-4-033).

1 Introduction

Light is used to meet visual needs of human, e.g. highlighting risks of falling [1], is applied for temporal orientation throughout the day, e.g. emphasizing day-night rhythm [2], for spatial navigation during activities of daily living, e.g. illumination of defined location areas [3] and is used as remembering as well as information signal, e.g. light spots and light signals [4]. Light, therefore, has great potential for attenuation of age-related mobility impairments caused by reduced spatio-temporal orientation, worry about getting lost, and fear of falling. This paper shows a new solution for Ambient Assisted Living, called Guiding Light, that pursues the following objectives.

Temporal orientation. Human time experience involves several aspects: simultaneity and successiveness, movement time, duration experience, and circadian rhythm. Orientation disorders of elderly often include difficulty in

---

G. Kempter (✉) · W. Ritter · A. Künz
University of Applied Sciences Vorarlberg, Dornbirn, Austria
e-mail: guido.kempter@fhv.at

temporal orientation. They may have problems with rough estimation of the time that had lapsed since last activity and with correct chronological classification of time of day, weekdays, and seasons. Especially, circadian rhythm, an endogenously driven roughly 24-hour cycle in humans, is affected by age [5]. Although circadian rhythms are endogenous they are adjusted to the environment by external cues, the primary one of which is light. Older people's light exposure might be insufficient for maintaining optimal circadian rhythm regulation [6]. With Guiding Light we help older people to enhance their temporal orientation by different light-based timing generators, e.g. periodically lighting fluctuations such as circadian-based light treatments [7].

Spatial orientation. Elderly people could have difficulty with spatial orientation and with questions like, where am I situated at the moment and how can I find a certain location [8]. Spatial orientation is a complex cognitive skill that enables wayfinding and is necessary for everyday functioning in the environment. Elderly people with orientation disorders may show spatial disorientation at familiar places or forget intended destinations, e.g. they can get lost in their own home and are unable to find the bathroom or bedroom [9]. With Guiding Light we facilitate spatial orientation and help elderly people to find their way, e.g. through different light quality coding of rooms and drawing the attention by salient illuminating subsequent location during locomotion.

Spatio-temporal orientation. Another aspect of orientation is to know what, where and when to perform different activities of daily living. With increasing age and together with some age-related diseases elderly people have problems to stay orientated to what's going on in their immediate environment. A common example is that elderly people with dementia may lose sense of time and locality, and either not remember to eat, or not remember that they just ate, and want another meal. With Guiding Light we help elderly people stay orientated to what's going on or to what they should do at a specific place and time of day, e.g. through directing individual attention in a timely manner (e.g. signaling the best time to go outdoor), by implementing orientation lights to reach the goals on time, and switching on and off other lightings at scheduled time automatically.

Individuality in orientation. Since humans show strong individuality in their daily routines, we need to facilitate orientation very carefully and prudently. Roters-Möller [10] shows that there is no uniform everyday structure among elderly people. Even in very limited age groups, a highly individualized organization of day exists (e.g. flexible sleeping time and extended breakfast). Therefore, standardisation of activities of daily living may lead to restricted self-determination of seniors and goes hand in hand with a loss of daily involvement, which can lead to apathy [11]. For this reason our assistance in spatial and temporal orientation (e.g. structuring activities of daily living) will operate individual and subliminal. With Guiding Light individual lighting assistance will be implemented by intelligent control loops in room automation (e.g. genetic algorithms).

Extent of personal mobility. Mobility aims to overcome the effect of spatial and temporal distances on human activities [12]. As the ability of individual temporal and spatial orientation of elderly people will be influenced by our new light wayguidance system, end-user (e.g. seniors, caregivers) might consider it important to know positive as well as negative variations of personal mobility. This knowledge might be very important as it will strengthen self-control and self-determination, setting up appropriate expectations, improving compliance, and supporting participation among seniors [13]. For caregivers this knowledge is decisive for adequate home care, nursing, and servicing. We expand Guiding Light system with an information system that gives information about senior's general motility, dynamics of body movement, and distances in indoor as well as outdoor locomotion.

2 System Description

Guiding Light system is conceptualized as an intelligent home automation system, on the one hand designed to monitor older person's behaviour and on the other hand to adaptively control all kind of lighting sources in their home, which guide them to more indoor and outdoor mobility. Electronic devices are commercial room sensors (e.g. sensors for movement, occupancy, position, light), different room actors (e.g. lighting devices, window blinds), and wireless body sensors (e.g. activity, position). Resident monitoring is also taking into account vital data from non-invasive measuring equipment that is commonly applied by elderly (e.g. pulse, blood pressure and blood sugar, body weight with percentage of body fat and body water) (Fig. 1).

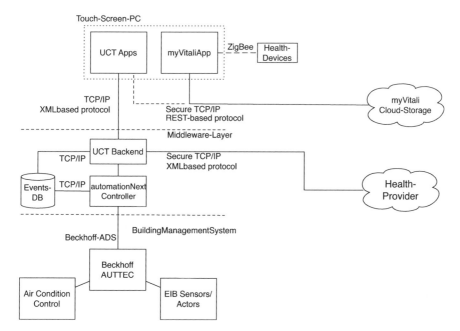

Fig. 1 Software architecture of guiding light system

I/O-devices are combined with Central Processing Unit (CPU) in different modes. With system-mode Guiding Light operates as client–server system with several peripheral computers and a Beckhoff device as CPU, which can be replaced by any industrial computer. Development of monitoring and control algorithms is done in this mode. On the other side the easy-mode does not require a peripheral computer. The mechanism is commonly defined on the basis of so-called functional blocks, channels and connection codes that describe the specific functionality of devices. Easy-mode devices are pre-programmed and loaded with a default set of parameters on CPU. This mode is primarily intended for practical use of Guiding Light system.

The architectural abstraction layers of Guiding Light is following Open Systems Interconnection (OSI) model which facilitates easy integration of additional hardware and software components into system as well as implementation of ready-proved solution into other buildings. Guiding Light uses freely selectable building management software (e.g. from Tridonic GmbH & Co KG or automation NEXT GmbH) on upper layers of home automation in controlling, monitoring, and finally optimizing the building facilities. Home automation also includes learning systems composed of intelligent algorithms.

Guiding Light allows individual and easy-to-use control from different user interfaces (e.g. touch screen panels, handheld or desktop computer, television) and allows selected remote access to a limited number of persons from the internet (e.g. family members, caregivers, physician). Senior's mobility measurements and interactive elements are visualized on these user interfaces following established usability and accessibility standards (e.g. ISO 9241, WCAG 2.0) and design-for-all approach. This allows concerned older persons as well as all other significant people to make judgements as to the mobility of residents. It is feasible to perform data analysing, storing, and visualisation within a protected shroud of privacy, where the sensed data about mobility are kept in the local control of their owners. For selected data exchange the cryptographic protocols such as encryption/decryption, digital signature, and hash code are used as protection mechanisms. In order to achieve a highly flexible, interoperable, and modular system Guiding Light:

- combines peer-to-peer network with server-based network to form a strong efficient portable and compatible network architecture,
- implements a multilayered, modular software architecture to ensure effective configuring, scaling, and servicing,
- takes care of hardware heterogeneity and allow different communication technologies like Ethernet, ZigBee, Bluetooth, WLAN,
- considers current bus systems like EIB/KNX, LM, DALI, EnOcean, BACnet, MODBUS, Beckhoff ADS and EtherCAT, and
- uses an open and documented XML-interface to give third-party developers the opportunity to create own applications (plugins) for our system.

3 Data Management

Lighting control

Within our approach lighting needs and preferences may change from hour to hour, from day to day, from room to room, and from one older person to another. This requires a highly flexible room lighting system able to provide variable lighting situations. Increasing control technologies, the steady diffusion of dimmable electronic ballasts for lamps, and evolution of interoperable building automation communication protocols brings us one step closer to the dynamic use of electric light. As a new lighting trend, dynamic lighting allows this kind of variations in those lighting parameters, which are most relevant for light quality to support spatio-temporal orientation and way finding.

Lighting quality is much more than just providing an appropriate quantity of light for specific tasks on chosen places within a room [3]. Other important factors that potentially support daily activity and last but not least mobility include time related change, luminance distribution, and light colour characteristics [7]. Taking into account these lighting parameters to control room lighting for a typical living day we will receive highly complex dynamic lighting profiles, such as shown in Fig. 2, as an example for timing of light intensity and light colour temperature for two different lighting devices in a living room.

An intelligent light control system, i.e. Guiding Light, continuously adapts light properties of living space (e.g. circadian lighting variations, orientation lights, light quality coding of rooms, salient illuminating intended goals) based on monitoring

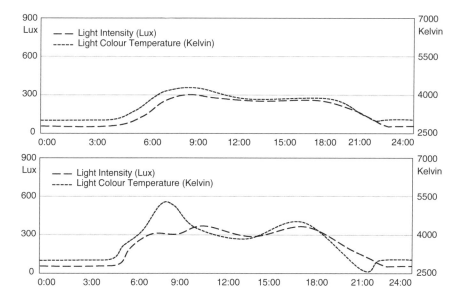

Fig. 2 Timing of light intensity and light colour temperature for two different lighting devices

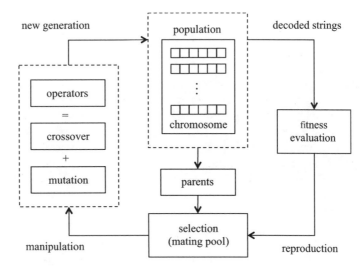

Fig. 3 Process flow diagram of automatic lighting control by means of genetic algorithm

results (see mobility monitoring in Chap. 4). Sophistically varying chronological states of the electronic systems in older person's home will help seniors in structuring their daily activities and improving mental orientation in time and space, resulting in improved mobility. We will use genetic algorithms (Fig. 3) for intelligent adaption's, which resemble evolution in nature and are often used to find good solutions within a huge search space.

A genetic algorithm is a stochastic search procedure in which a successor situation is generated by combining properties of two preceding situations. In our case a situation is defined by the state of lighting systems in older person's home. Each state is rated by the evaluation or fitness function, which is in our case the nature and scope of older person's mobility. According to this evaluation two states are selected for reproducing new states applying crossover and mutation procedures. Lighting characteristics of lamps will change automatically according to the personal daily routine of residents. At the same time mobility parameters of the residents are monitored (see next chapter) and the results of analysing these data are used to change the programming of light variations. The adjustment of light programming will be done automatically, nevertheless, residents can manually readjust their lights at any time.

Mobility monitoring

Mobility monitoring is achieved by means of embedded room sensors and miniaturized sensors carried on the body. The use of sensors in buildings should be possible with normal techniques and at relatively low cost. In our example each room consists of a motion detector with a detection angle of 360°. The 360° detection angle is divided into four sectors with independent passive infrared sensors for each sector. The sectors are 90° each and can be parameterised individually. For each movement sensor the range, timing, and the sensitivity can be set for each block via parameters.

Figure 4 shows motion patterns for 1 week in three different rooms of senior's flat (four sectors per room) and corresponding overall activity measures per day/night on the top. Activity measures for day and night result from aggregation of all sectors of motion sensors in living room, sleeping room, and bathroom/toilet room of older persons within relevant time period. Night time period has be defined starting at 10 o'clock p.m. and ending at 5 o'clock a.m., following by day time period up to 10 o'clock p.m. The first person (apartment no. 5) has been diagnosed with temporal disorientation (following Mini-Mental-State Examination). The second person (apartment no. 6) shows normal spatio-temporal orientation. According to this diagnosis the overall mobility profile of the first person shows less overall intensity and more night activity in relation to day activity than the second person. Second person shows well structured history in activity in sleeping room with highest motion intensity in the morning.

Both persons spend, however, nearly same time inside their apartment of same size. Additional mobility monitoring is performed in the form of trend analysis, pattern discovery, and association rules which is applied to data obtained from unobtrusive sensors to capture comprehensive information about what, where and

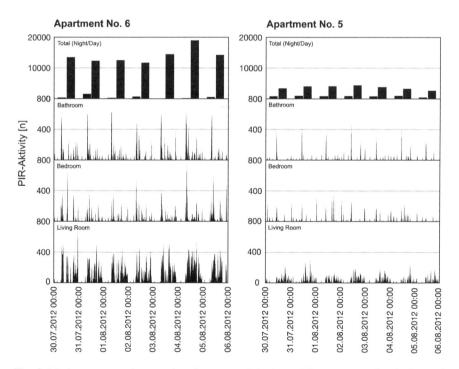

Fig. 4 Motion patterns of two seniors for one week in three different rooms of their flats and corresponding activity measures per day/night

when residents are performing different activities of daily living. Results from continuous monitoring are used for intelligent light control system and to derive certain measured values of individual mobility (e.g. general motility, dynamics of body movement, and distances in indoor as well as outdoor locomotion).

4 Pilot Application

Guiding Light is major part of an ongoing project within AAL Joint Programme (AAL-2011-4-033). The system will be validated and tested in real-life settings in different European countries. Evaluation is made in order to give evidence that our system is an essential support for the mobility of elderly persons. With the pilot application we will follow an approach of zonal lighting, which differs between zone of occupation and zone of locomotion depending on individual daily structure. The time related change of luminance distribution and light colour characteristics of room zones will be executed in a way, that residents are not restricted in carrying out any activities because of insufficient lighting. But they should have the feeling, that activities can more easily be done when following the subliminal stimulation by light way guidance system.

Before the actual implementation in selected households, Guiding Light will be tested iteratively with end-users to make sure that it really conforms to users' needs and requirements. For many of the components of the lighting way guidance system we can rely on components which are already developed and tested. For example, the interactive graphical user interface of my Vitali AG has already undergone extensive validation with elderly end-users. Similarly, the home automation solutions of Tridonic GmbH & Co KG and the lighting solutions of Bartenbach Lichtlabor GmbH has been tested with older adults in various test runs. Nevertheless, both still require further adaptation.

Our lighting way guidance system is a highly flexible, modular and scalable automation system that can integrate very heterogeneous hardware. This is necessary, since primary target group has very different mobility restrictions and a wide variety of technological preconditions in their living rooms. Thus configuration of pilot application will start with a focus group consisting of primary and secondary end-users. Within this focus group older person's individual characteristics of mobility, which might be vulnerable with short-term ageing process, and their technical facilities of accommodation are discovered.

An important task that has to be fulfilled in preparation for the tests is to define selection criteria. Apart from age (>70), mobility, and gender (50 % male, 50 % female, if possible), motivation and state of health are considered in selecting test persons. The two are closely related because a person who expects to benefit from the research, e.g. by improving their mobility, will also be highly motivated to participate. Besides, information about a person's habits such as the taking of certain medicines, consumption of alcohol, coffee and/or other stimulants, will also be obtained to make sure that the data gathered are not distorted.

Physicians will provide the medical and geronto-psychiatric expertise and their experience with the clearance of ethical issues to guarantee the smooth preparation and running of the implementation. The selection of appropriate test persons has to bear in mind the primary target groups for future marketing efforts. The elderly people should live at home or in assisted accommodation with our support facilities. They should spend most of their time in their apartment or house and regularly watch TV, read newspapers, or perform some visually demanding tasks. The daylight situation should be such that the person normally needs artificial light for a specific period of time to perform their daily routines.

Each pilot application will be configured according to the older person's individual characteristics of mobility (e.g. instability in walking, spatial disorientation) and technical facilities in their accommodation (e.g. conventional electrical installation or existing bus system). At the beginning, we have to make a decision, which of the older person's individual characteristics of mobility should be monitored and what changes of mobility should be stored and visualized on the information system. User monitoring is based on a relevant selection of room sensors (e.g. for movements, carbon dioxide), sensors continuously carried at the side of the body (e.g. for person's position, light exposure), and occasionally used measurement equipment for vital data (e.g. blood pressure, blood sugar). It is important to note that only those sensors will be used for pilot application which are well accepted by test persons. The central processing unit of Guiding Light registers all sensor data and analyses them with regard to all relevant quantitative parameters of mobility (e.g. general motility, dynamics of body movement, and distances in indoor as well as outdoor locomotion). Primary end-users will receive feedback about this ongoing analysis on their favoured device (e.g. touch screen panel, television, handheld, computer screen).

At any moment, the primary end-users have the possibility to allow for displaying this feedback information on devices of attached secondary end-users inside and outside the residential building (e.g. caregivers, family members, physician). The user interface of all components will, however, follow usability and accessibility standards. The information system does not provide recommendations for action as to the mobility of an elderly person, instead it provides the means for the person concerned or a caregiver to investigate and track a person's mobility. The system will rather facilitate face-to-face communication by updated information and promoting motivation of all involved people in order to support social inclusion and mobility of elderly persons.

The feedback system is not the primary instrument for improving indoor and outdoor mobility. We will achieve the main goal of Guiding Light—improving mobility—rather by ambient influencing temporal and spatial orientation of older persons through ambient stimuli. This is being done with supplying results from sensor data analysis to intelligent adaption method within a central processor unit (e.g. fieldbus processor, set-top box, computer). This method consists of data mining procedures (e.g. genetic algorithms), which automatically tries to find the optimal state of lighting systems and shading systems in an older person's home in a timely manner. This means, that lighting way guidance system will automatically redefine control parameters of electro-mechanical actuators within a home automation system. Genetic algorithms are defined by means of many configuration parameters, the effects of which on building automation we have investigated in previous projects [7]. Potential actuators are lighting solutions from a wide range (e.g. direct/indirect surface-mounted and pendant luminaires, light lines for emphasizing contours, miniaturized shelf lighting, recessed floor luminaires, table lamps and uplighters, emergency lighting as well as wall-mounted navigation sign luminaires) and other motor-controlled facilities (e.g. window blinds). The adaptations will usually happen in a way that is unobtrusive to users who can manually switch on or off lightings whenever they wish.

5 Discussion

Flexibility and adaptability of Guiding Light are of paramount importance to respond to differing social and organisational needs across Europe and to ensure user acceptance in different European markets. When it comes to technology, flexibility is guaranteed by the modular and open architecture of our new lighting wayguidance system. Customers can choose from a range of home automation options to suit their needs and wishes for supporting their mobility. Once calibrated, Guiding Light settings can either be retained or they can be altered depending on user needs, time of the year or changing personal circumstances—either manually by the user or automatically by the system, e.g. seasonal and time of day adaptation, which also saves energy. Later on, clients who purchase our lighting way guidance system, can decide if they want to keep the sensor equipment for further adaptations and visualisation of their mobility data or not. Also, clients will be able to choose from a variety of optional applications that can supplement the guiding light, among them:

- link to public and private transportation services,
- biofeedback for relaxation and brain jogging exercises for mental activation,
- health advice and recommendations in accordance with their vital data and state of health,
- a link-up to support to other health and care services.

Since light adaptation occurs mainly in line with the older person's behaviour, language only plays a role in certain applications such as the visualisation of mobility data on a screen, the online ratings facility etc. These will have to be adapted to local conditions and language requirements once it is has been assessed which applications are most suitable and in demand. This task will fall to the companies that want to market Guiding Light in different countries.

As far as the organisation of care is concerned, care settings and regulatory schemes vary widely across Europe and have an influence on people's living arrangements in old age. Most of the older adults prefer remaining in their own homes as they age. Residence in a nursing home or care home is not the norm and is increasingly regarded as a last resort. In the Netherlands, for instance, care of the elderly is increasingly moved away from nursing homes (intramural care) into the community (extramural care). From a public policy perspective, this is a cost-effective option provided that the home can be used as a platform to ensure people's general wellbeing, i.e. a home as a care environment. In Switzerland, however, many people are reluctant to give up their place in a nursing home for fear of losing their entitlements to paid care. In Austria, recent legislation which makes it possible to officially employ informal carers mainly from East European countries has strengthened the extramural trend. Guiding Light can be adapted to all care scenarios from private homes to senior residences or care in the community.

Acknowledgments The project Guiding Light (no AAL-2011-4-033) is funded under European Ambient Assisted Living Joint Programme (AAL JP). Consortium partner are Fachhochschule Vorarlberg (coordinator), Tridonic GmbH & Co KG, Bartenbach Lichtlabor GmbH, myVitali AG, Institut für Sozialforschung und Demoskopie OHG and YOUSE GmbH.

References

1. Härlein, J., Halfens, R.J.G., Dassen, T., Lahmann, N.A.: Falls in older hospital inpatients and the effect of cognitive impairment: A secondary analysis of prevalence studies. J. Clin. Nurs. **20**, 175–183 (2010)
2. Rea, M.S., Figueiro, M.G., Bierman, A., Bullough, J.D.: Circadian light. J. Circadian Rhythms **8**, 2–3 (2010)
3. Chung, J., Kim, I.-J., Schmandt, C.: Guiding light: Navigation assistance system using projection based augmented reality. In: Paper presentation at International conference of consumer electronics, Las Vegas (2011)
4. Moffat, S.D.: Aging and Spatial Navigation: What Do We Know and Where Do We Go? Neuropsychol. Rev. **9**, 478–489 (2009)
5. Crowely, K.: Sleep and sleep disorders in older adults. Neuropsychol. Rev. **21**, 41–53 (2011)
6. Stripling, A.M.: Ambient light and sleep in community dwelling older adults. Zeitschrift für Gerontologie und Geriatrie **33/3**, 155–168 (2008)
7. Kempter, G., Maier, E.: Intelligent lighting. Eur Union Public Serv Rev **16**, 431–438 (2008)
8. Iachini, T., Ruggiero, G., Ruotolo, F.: The effect of age on egocentric and allocentric spatial frames of reference. Cogn Process **10/2**, 222–224 (2009)
9. Chang, Y.-J., Peng, S.-M., Wang, T.-Y., Chen, S.-F., Chen, A.-R., Chen, H.-C.: Autonomous indoor way finding for individuals with cognitive impairments. J. Neuro Eng. Rehabil. **7/45**, 2–13 (2010)
10. Roters-Möller, S.: Den Ruhestand gestalten lernen. In: Böllert, K. (Hrsg.) Soziale Arbeit als Wohlfahrtsproduktion, S, pp. 47–69. VS Verlag, Wiesbaden (2011)
11. Dühring, A.: Macht das „Setting"den Unterschied? Der Beitrag der verschiedenen Formen der stationären Altenhilfe zur subjektiven und objektiven Lebenszufriedenheit dementiell erkrankter Menschen. PhD Thesis at University of Kassel (2006)
12. Alpini, D., Cesarani, A., Pugnetti, L., Mendozzi, L., Cardini, R., Kohen-Raz, R., Hahan, A., Sambataro, G.: Project to prevent mobility-related accidents in elderly and disables. In: Proceedings of 3rd international conference about disability, virtual reality and associated technologies, Alghero, Italy (2000)
13. Münch, M., Knoblauch, V., Blatter, K., Schröder, C., Schnitzler, C., Kräuchi, K., Wirz-Justice, A., Cajochen, C.: Age-related attenuation of the evening circadian arousal signal in humans. Neurobiol. Aging **26**, 1307–1319 (2005)

Vibroacoustic Monitoring: Techniques for Human Gait Analysis in Smart Homes

Klaus Dobbler, Moritz Fišer, Maria Fellner and Bernhard Rettenbacher

Abstract The aim of this research project is to automatically analyse human behaviour in indoor environments using vibration sensors attached to the floor. Vibration sensors are used in an absolute passive manner so the monitored persons do not have to wear any sensors. Furthermore this technology preserves the private sphere of monitored persons. In order to extract geometric gait features like motion trajectories out of the vibrational signals, the main part of this work focuses on methods for localization of seismic sources on the two-dimensional floor surface. Starting from a conventional TDOA (Time Difference of Arrival) based technique using uniaxial acceleration sensors we show how to minimize installation cost by reducing the number of sensor entities via tri-axial sensor technology. We describe the main challenges of wave propagation in solids, namely dispersion and multi-path propagation and discuss their implications on robustness and accuracy of localization results. To conclude our work we introduce potential application scenarios.

1 Introduction

Since walking is one of the most frequent and important human activities, much attention has been paid to the analysis of gait patterns [1] in the context of developing smart Ambient Assisted Living solutions [2]. The investigation of gait parameters as well as motion sequences reveals valuable information about the individual mobility, the daily expenditure level and health stability of a person. Furthermore several specific diseases like diabetes, depression and peripheral neuropathy are correlated with symptoms characterized by gait instability or

---

K. Dobbler (✉) · M. Fišer · M. Fellner · B. Rettenbacher
Joanneum Research Forschungsgesellschaft mbH, DIGITAL—Institute for Information and Communication Technologies, Steyrergasse 17, A-8010 Graz, Austria
e-mail: klaus.dobbler@joanneum.at

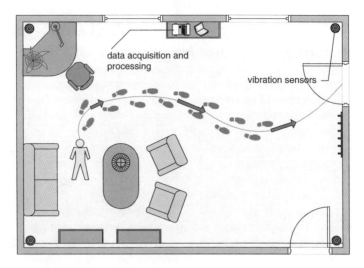

Fig. 1 Principle of vibroacoustic monitoring

unsteadiness, which are indicated by a shorter step length [2]. However, much of the work concerning the observation of gait properties focuses on wearable devices. Marschollek et al. [3], try to assess the risk of a fall for an elderly person using an acceleration sensor attached to the waist by a custom belt. In contrast, our work focuses on pure passively operating sensor systems, and therefore provides a non-invasive permanent monitoring process of a person's well-being. The principle idea is illustrated in Fig. 1.

2 Principles of Vibroacoustic Monitoring

We are interested in extracting two relevant pieces of information from the sensor signals as described below. First, we would like to determine the cause of the vibration signal which allows us to discriminate between signals of interest like a footstep of a person and undesired noise sources like washing machines or walking frames. This requires the implementation of a classification algorithm. Second, the exact location of a seismic event is of great interest. Under the constraint of a purely passive operation mode there are two basic methods of determining the position of a seismic event.

The first method (Fig. 2) is an extension of the conventional TDOA (Time Difference of Arrival) approach, which is well known in the processing of airborne signals. However, this method suffers from diverging waveforms caused by dispersion and therefore only produces robust results in non-dispersive media like air. By taking the dispersion characteristics of the underlying medium into account, we are estimating the so called range differences between every pair of sensors in a

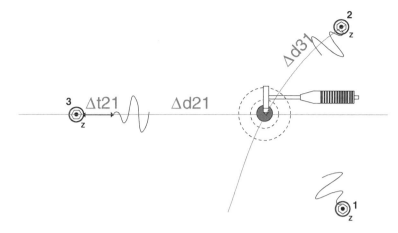

Fig. 2 Illustration of TDOA-approach

direct way and subsequently formulate the nonlinear system of equations that describes the localization problem in a mathematical way [4]. It is worth noting that this approach makes use of uniaxial sensors i.e. it uses the vertical acceleration of floor-surface-particles (z-axis).

In contrast, the second approach (Fig. 3), which is called AOA (Angle of Arrival)-approach, uses tri-axial sensors and consequently processes the acquired surface wave (Rayleigh Wave) as a vector-wave represented by the acceleration of the particles into x-, y- and z-direction of the Cartesian sensor coordinate system. The interpretation of these three components reveals the angle at which a surface wave impinges at the sensor [5]. We compute the exact location of the seismic event by triangulation using two or more tri-axial sensors.

After this short review of existing localization procedures, we present a novel idea on the development of a localization-scheme making use of only one tri-axial sensor station (Fig. 4). The key idea behind this scheme, besides the usual

Fig. 3 Illustration of AOA-approach

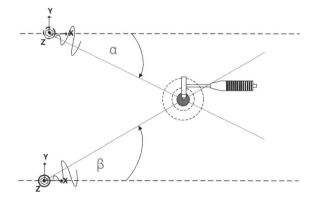

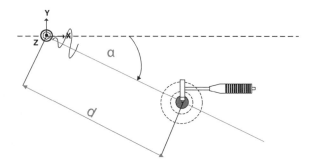

Fig. 4 Illustration of single-station approach

AOA-estimation process, is to exploit the effect of dispersion in order to estimate the distance to the seismic source. To accomplish this, we use advanced time–frequency transformations [6], which allow a frequency-dependent time of arrival estimation. Knowing the dispersion curve it is possible to translate this information into an estimated distance and in combination with the determined AOA into the position of the seismic source.

Localization in a continuous manner becomes a problem of tracking which offers e.g. the extraction of motion patterns. Another processing of the location-information allows the distinction between more than one entities of the same class (e.g. the distinction between footsteps stemming from different persons) by using probabilistic models representing the transition probabilities as a function of space.

3 Challenges

3.1 Multipath Propagation

A robust operation of the developed algorithms assumes the signal to be noiseless and stemming from a single-mode Rayleigh wave. In a real indoor environment however, we expect disturbances in the form of body wave reflections stemming from the bottom side of the floor as well as Rayleigh wave reflections from bordering walls (Fig. 5).

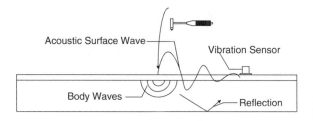

Fig. 5 Illustration of wave propagation

3.2 Dispersion

In vertically layered media, Rayleigh wave propagation is characterized by dispersion, meaning the propagation speed is frequency dependent. This leads to diverging waveforms (Fig. 6) and makes conventional localization methods from air borne sound signal processing inapplicable because they generally do not include the effect of dispersion in their signal model. Therefore we intend to reliably estimate the dispersion curve of the floor in order to improve localization results.

3.3 Space Variance

In general we have to consider a floor as an inhomogeneous anisotropic propagation medium because parts of the floor construction, room furniture or other objects related to floor properties do affect Rayleigh-wave propagation. This implicates dispersion behaviour which is dependent on the excitation point and differs for propagation into different directions. Consequently, consideration of the dispersion effect into algorithm will require a manageable model of the floors space variant properties.

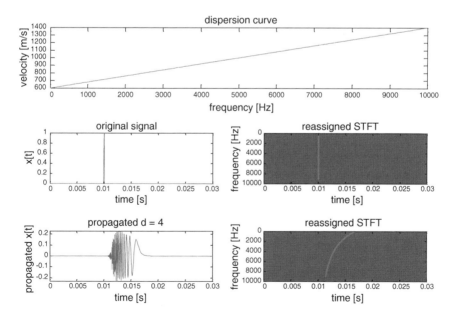

Fig. 6 Illustration of dispersion

3.4 Noise

We not only expect noise sources like road noise but also interactions with objects such as doors or kitchen furniture cause disturbing vibrations that need to be ignored in the localization process. Moreover, sound pressure originating from TV, radio or other participants cause vibrations at the sensor that influence signal evaluation [2].

4 Results

Figure 7 shows the measurement setup for evaluating the localization performance of the TDOA-approach. A typical parquet floor of size 6.9 × 5.57 m was excited by an impulse hammer at ten different positions, five times each. Floor vibrations where gathered by four IMI Sensors with a sensitivity of 100 mV/g and a resolution of 350 µg. Table 1 contains the results we gathered by applying a conventional adaptive threshold based TDOA-algorithm. We assumed the velocity of the fastest occurring frequency to be 1,150 m/s. Localization results vary considerably for different excitation points. Moreover these results show the performance without taking dispersion into account.

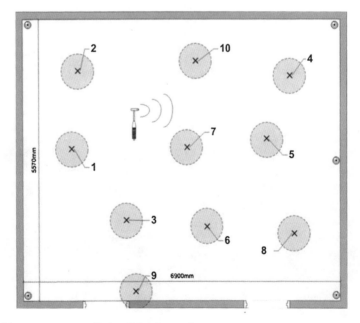

Fig. 7 Measurement setup for localizing impact hammers

Table 1 Results of impact hammer localization TDOA-approach

Point	Mean error (m)	Point	Mean error (m)
1	0.31	6	0.40
2	0.60	7	0.67
3	0.19	8	0.47
4	0.72	9	1.46
5	0.39	10	0.72

We tried to deploy the TDOA based localization algorithm in order to localize real footstep signals. Figure 8 shows a screenshot of a demo video, which should demonstrate the algorithms operational functionality. In the left part of the picture you can see the original sequence—a female person walking over the floor. The right part illustrates the estimated footstep-locations as a walking path. As can be seen in Fig. 8 we were able to successfully follow the path of the person. However, reasonable localization results are limited to steady walking people with appropriate footwear.

Further results refer to the determination of the AOA approach a wave impinging at a tri-axial sensor station. Figure 9 shows the measurement setup, where a PCB 356A17 sensor was placed on a concrete floor, not too close to any floor borders. Every test-angle was excited five times with an impulse hammer to compute an average mean error of the estimated angles, which are presented in the left part of Table 2a. Next, we placed the sensor directly a few centimeters to a wall and repeated the experiment. The results can be seen in the right part of Table 2b. Apparently, this sensor position leads to a much higher mean error of the estimated angles. We conclude that there are reflections from the wall, sumperimposing with the direct signal and degrading bearing-estimation results.

These promising results led us to the decision to further follow the approach using tri-axial sensor technology and AOA-estimation by testing it on real footstep data. Figure 10 illustrates the experiment's setup, where again a PCB 356A17 sensor station was stimulated by 15 barefooted heel strikes of medium to low energy at each angle. Corresponding AOA-estimation results can be seen in Table 3 and show reasonable accuracy.

Fig. 8 Screenshot of demo-video

Fig. 9 Measurement setup for determining angle of impinging wave

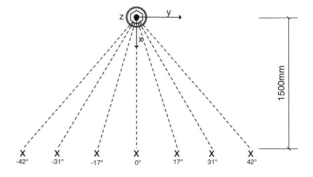

Table 2 Results of impact hammer localization AOA-approach

a		b	
Angle (°)	Mean error (°)	Angle (°)	Mean error (°)
0	3	0	2
17	2	17	11
31	3	31	8
42	1	42	7
−17	2	−17	4
−31	3	−31	13
−42	2	−42	15
Total	2.3		8.6

Fig. 10 Measurement setup for the evaluation of AOA-algorithm on real footstep (*heel strike*) signals

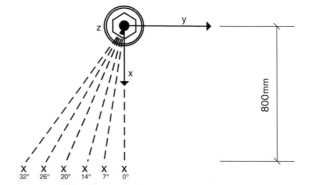

4.1 Real-Time Footstep Detection and Localization System

In order to further test the developed algorithms in real life scenarios, we are implementing a real-time footstep detection and localization system. Its overall structure is depicted in Fig. 11.

First we have to make sure to detect what we are interested in, namely footsteps. Figure 12 shows sample footstep signals stemming from a person walking

Table 3 Results of heel strike localization AOA-approach

Angle (°)	Mean (°)	Median (°)	Standard deviation (°)
0	05.58	6.08	3.15
7	11.08	10.72	1.03
14	14.50	14.93	3.36
20	22.38	22.65	2.24
26	26.17	26.28	1.63
32	29.82	29.82	2.51

barefooted nearby the sensor on a laminate floor in a steady motion. The steps are characterized by broadband high frequency contents stemming from toes and tangential interaction with the floor-surface as well as low frequency content produced by the heel-strike. Due to the strong damping of high frequency components, we focused on detecting the low frequency content, propagating over longer distances. Therefore we decided to implement a heel-strike classifier in order to detect footsteps. To this end we compute Mel Frequency Cepstral Coefficients (MFCCs) in addition to a general low frequency content energy feature. Even though MFCCs are known to have been developed for speech processing algorithms and may not be the best choice for this classification task, they are able to characterize the relatively short heel-strike signal rather compactly. After further reducing the dimensionality of feature space by Principal Component Analysis (PCA), a Support Vector Machine (SVM) classifier decides whether the observed signal block contains the heel-strike of a footstep signal.

The detected footstep signals are processed by the localization stage, which consists of AOA-estimation for every tri-axial sensor station and finally the determination of source location by triangulation. To further increase robustness, the whole estimation is carried out three times with slightly shifted windows and filtered through a median function to accommodate for classification inaccuracies. During the development of the AOA-estimation algorithm based on Zhang [5], we tried different implementations in order to achieve the most robust and reliable result on real footstep signals. The most promising one is described as follows.

First of all the extracted heel strike windows of the three axes are transformed into frequency domain by a conventional Fast Fourier Transform.

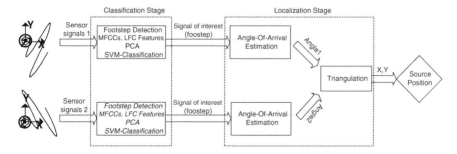

Fig. 11 Overall structure of real time vibroacoustic monitoring system

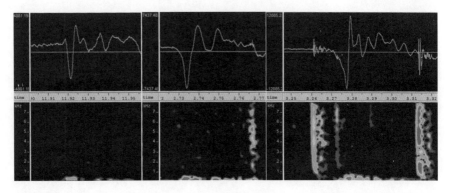

Fig. 12 Exemplary footstep signals

$$A_z = fft(a_z) \quad (1)$$
$$A_y = fft(a_y)$$
$$A_x = fft(a_x)$$

Because of the elliptical motion of floor-surface particles we perform a 90 degree phase shift on the z-axis of the acceleration signal which can be obtained by using the imaginary part of a Hilbert Transform

$$A_{zs} = A_z e^{-j\frac{\pi}{2}} \quad (2)$$

In order to compute the frequency dependent cross correlation of the z-axis and the y-axis respectively x-axis to

$$r_{zx} = A_x * conj(A_{zs}) \quad (3)$$
$$r_{zy} = A_y * conj(A_{zs})$$

By applying the arcus tangent function to these signals, which represent the on the y-axis and x-axis projected compressional part of the surface wave we obtain the frequency dependent angle function.

$$\phi = \arctan(\frac{r_{zx}}{r_{zy}}) \quad (4)$$

Zhang [5] developed a weighted averaging method in order to compute the estimated angle out of the obtained function in a way that weights frequency components according to their energy. However, this approach did not work very well in our conditions so we developed an alternative approach for the final angle computation.

Due to the upper limit on reasonable phase accuracy of the used sensors we restricted the analysed frequency content to the range between 20 and 2,000 Hz. Within this band, the frequency dependent angle function is then windowed using overlapping rectangular windows. For each window the variance is calculated and the ones with the three lowest variance values are selected. Only frequency components exceeding an adaptive energy threshold are considered. For each selected window the median is calculated and eventually the median out of these three values leads to our final angle estimation. Should either the energy level of the signal be too low or the variance of estimations too high, a reject option can be implemented to avoid non-robust results.

First experiments with the system showed that reasonable results strongly rely on a sufficient SNR of the footstep signals.

5 Applications

We assess that the introduced technology has great potential to be deployed in Ambient Assisted Living and Surveillance solutions.

- **Activity-Recognition/Quantification:** How mobile is a monitored person? Ratio Activity/Passivity?
- **Abnormality-Recognition:** Falls, falling objects
- **Determination of gait parameters** as indications for pathological gait patterns and/or fall risk assessment
- **Localization/Tracking:** Analysis of motion patterns, Multi-Person-Recognition

In addition to these functions, the idea of monitored persons consciously interacting with the system reveals other remarkable application scenarios in which the system is used as a tangible acoustics interface (TAI) [7]. In this way, elderly people could interact with any solid object to communicate their needs for example by furnishing the table of the living room with our system. By discriminating different types of excitation activities even more interactive applications with input specific responses are imaginable. First experiments exploring this point of view were limited to the localization of finger tips on table surfaces. They showed general operational functionality of localization but a severe degradation of achievable accuracy caused by reflections of table boundaries. First ideas to overcome this problem include the application of windowing techniques as well as the construction of appropriate table shapes.

6 Future Work

Future developments will concentrate on the evaluation of the whole classification and localization system on real footstep data. Furthermore, because of the transient character of the signals we would like to use more appropriate features like the coefficients of the discrete wavelet transform for the heel strike classification task. We also want to further explore the single station localization approach. Therefore we will thoroughly study a STFT-based approach offered by Chun [4] in order to estimate the dispersion curve of floors and to introduce the obtained dispersion curve into our localization algorithms. Finally, we will carry out further experiments on tables in order to investigate the disturbing reflections from table boundaries currently leading to worse localization accuracy.

7 Conclusion

We introduced the concept of vibroacoustic monitoring by means of classification and localization of seismic events. Focusing on the localization part, after reviewing existing localization methods, we presented a novel idea using only one sensor station. The proposed method minimizes installation cost without significantly increasing the financial cost. However, more work has to be done in order to be able to apply the theoretical idea in a real application scenario. Besides that, AOA-estimation of two sensor stations followed by triangulation seems to be the most promising approach. A robust localization result depends on how well we are able to account for multipath propagation as well as the inhomogeneity and anisotropy of floors as an underlying medium. Furthermore the deployment of the described technology in a real world application will raise the need to account for external noise sources in order to ensure operational localization functionality. The purely passive functionality as well as the low installation cost makes the introduced non-invasive technology easily deployable in several potential application scenarios and therefore interesting for further research.

Acknowledgments Funded by the Austrian Ministry for Transport, Innovation and Technology within the funding agreement 2011–2012.

References

1. Ekimov, A., Sabatier, J.M.: Vibration and sound signatures of human footsteps in buildings. J. Acoust. Soc. Am. **120**(2) (2006)
2. Lee, H., Helal, A.S.: Estimation of indoor physical activity level based on footstep vibration signal measured by MEMS accelerometer for personal health care under smart home environments. Control Instrum. **5801**(5) (2009)
3. Marschollek, M., et al.: Sensor-based fall risk assessment–an expert 'to go'. Methods Inf. Med **50**(5) (2011)
4. Chun, L.K.: Development of Source Localization Algorithms in Dispersive Medium, City University of Hong Kong (2010)
5. Zhang, J.: Bearing Estimation of Seismic Sources from a 3-Axis Seismometer—A Study of Seismic Footstep Detection. Lambert Academic Publishing (2011)
6. Roueff, A., Mars, J.I., Chanussot, J., Pedersen, H.: Dispersion estimation from linear array data in the time-frequency plane. **53**(10) (2005)
7. Fabiani, M.: Development of a tangible human-machine interface exploiting in-solid vibrational signals acquired by multiple sensors. KTH Stockholm (2006)

Part II
Activities of Daily Living

Unobtrusive Respiratory Rate Detection Within Homecare Scenarios

Bjoern-Helge Busch and Ralph Welge

Abstract Against the background of Ambient Assisted Living, this article proposes a new kind of unobtrusive, non-stigmatizing and continuous acquisition of vital signs as respiratory rate and related features on the basis of ultra-wide-band radar sensing. Through the runtime analysis of the reflection data, surrogating mechanical signals e.g. the excursion of the thorax are detected and linked with physiological values like the breathing rate. After a brief introduction to the application context including an explanation of specific user demands and restrictions of current solutions of the telemedicine, physical fundamentals of measurement and the utilized electronics, the applied principles of spatial and temporal data mining are described. Finally, experiments including real measurements with the subsequent discussion of the measurement results provide an outlook to the capabilities of our approach and grant information about open issues and the steps in research.

1 Introduction

Due to the demographic trend in most of the highly industrialized countries, in particular in Germany, the scientific area of *Ambient Assisted Living*, abbreviated by the acronym AAL, gained much more importance in recent years and was expedited by a couple of research projects funded by the BMBF, a German governmental institution for research and education. AAL comprehends technical

---

B.-H. Busch (✉) · R. Welge
Leuphana University Lueneburg, Institute VauST,
Volgershall 1, 21339 Lueneburg, Germany
e-mail: bhbusch@leuphana.de

R. Welge
e-mail: welge@leuphana.de

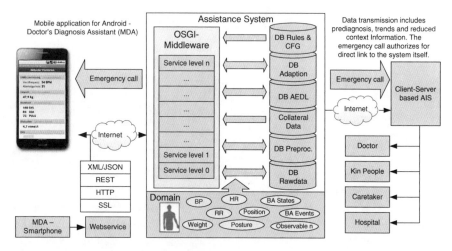

Fig. 1 Concept of an AAL-infrastructure [1]

solutions and concepts for the unobtrusive support of elderly people in their familiar environment. Obviously, this approach implies the usage of ambient intelligence technologies for the collection of user data and the initiation of domain interventions (e.g. emergency detection or domain regulations). Many AAL-research projects deal with the development of human centered assistance systems, including the development of agents for early emergency detection, preventive diagnosis, energy awareness, compliance control and medicine management (refer to Fig. 1). The context aware analysis of certain living patterns covering snap shots of health parameters requires a robust and reliable acquisition of these values—the main objective of further considerations.

1.1 Limitations of the Telemedicine for AAL-Domains

Usually, in home care domains, established components of the telemedicine are applied. In our approach for an assistance system, we utilize telemedical solutions for the acquisition of weight, heart rate, blood oxygen, blood glucose, respiratory rate through nasal prongs et cetera equipped with a Bluetooth interface. However, these devices reveal a lot of disadvantages considering usability, significance and validity of gathered raw data and questions of ethics as listed below.

- The patient has to use telemedical devices autonomously on his own. In accordance to anamnesis and latest therapy recommendations, this has to be done at determined instants of time. Otherwise, the raw data is not trustworthy or out of sync with the aim of acquisition. The extracted information may be diagnostically less conclusive.

- The usage of telemedical devices is often inconvenient. Further, many body attached sensor networks (BSN) restrict the user in their autonomy, agility and mobility. Chest straps are impractical for long-term surveillance.
- The usage of BSNs is not only a question of comfort but also a factor of stigmatization. Therefore, visible measurement of vital signs should be avoided in home care scenarios.
- Imaging techniques based on the evaluation of cameras for the identification of position, posture, activity, agility and emotional shape are inappropriate. The user rejects, or often only accepts them with reservations because this approach procures an impression of observation.

Regarding to these mentioned deficits, it is recommended to avoid camera-based systems or BSNs. Therefore, we prefer the use of a UWB-sensor and concentrate first of all on the most significant vital parameter we can observe: the respiration rate.

1.2 Main Target: Identification of Respiratory Rate

Two main aspects are in the focus of interest—the determination of the physical shape of a human and the extraction of convincing parameters for diagnosis purpose itself (e.g. for the early detection of deteriorating states of health). Considering the physical shape, the vital capacity VC (depicted in Fig. 2) is a useful score and common in spirometry physicals for the determination of the pulmonary function; an indicator for the physiological performance capability. To interpret the values for the vital capacity VC, depending on the standard breathing volume, the maximum inhaled volume and the maximum exhaled volume, the following Eq. (1) can be used to determine the typical set point for the comparison with measured values. The reference value for the residual volume (remaining air in the lungs) drops from 3–4 L at an age of 20 years to 2 L at an age of 65 years.

$$VC = \frac{height^3}{k} \cdot \left(1.03 - \frac{age - 25}{100} \cdot 0.75\right) \quad (1)$$

with gender factor k

$$k(g) = \begin{pmatrix} 1.0, & g = male \\ 1.1, & g = female \end{pmatrix} \quad (2)$$

	Standard breathing volume	~0.5 L
Vital capacity	Inspiration reserve	~1.5-2 L
	Expiration reserve	~1-1.5 L
	Residual volume	~3-4 L / ~2 L

Fig. 2 Vital capacity

For the continuous approximation of the constituent parts of the vital capacity VC, a previous anthropometric survey including test measurements for system calibration is necessary. This has to be done in accordance to a concomitant body plethysmography. Therefore, features like the thorax and abdomen excursion need to be detected and analyzed in the context of reference measurements. Features of interest are respiratory rate RR, the variability of RR, the amplitude and it's variability, significant linear and sometimes exponential trends. These trends can be associated with changes in the residual volume (short-time). That means, the person is breathing erratically. For diagnosis effort, there are also additional features to detect. Most common is the examination of breathing patterns gathered by polysomnographic surveys. Usually, these surveys cover apnoea, hyperpneas or characteristic patterns like Cheyne-Stokes respiration (CSR), Biot's respiration (sometimes called ataxic respiration), Kussmaul breathing and other forms of labored breathing. These phenomena can also be linked to distinct excursion patterns.

Current developments for the detection of breathing parameters can be classified by three main categories: direct measurement of air flow, the measurement of different gas concentrations in the expired air [2] or the blood and finally, the observation of the movement of the thorax or the abdomen or alterations in the tissue impedance. One approach based on the analysis of tri-axial accelerometer data for the detection of breathing rate and flow waveform estimation was proposed by [3]. The extension of this approach deals with the simultaneous monitoring of the activity and the respiration [4]. Also focused on the examination of mechanical signals evoked by the upheaval of the breast, we integrated a sensor component within an armchair and in a bed. Thereby, it is aimed to monitor the patient while he is watching TV, reading a book or executing other activities which can be done in a sitting position. Additionally, the patient can be supervised during the night while he is sleeping.

2 Methodology

The basic concept deals with the investigation of alterations in the UWB-reflection runtime. The concerned reflection is caused by a solid object. In this case, the object is the human body and changes in the body's position or the body's appearance induce also changes in reflection runtime (refer to Fig. 3).

2.1 Related Work

UWB-systems with various types of signal modulation are propagated and tested for diverse applications. In order to discover hidden objects or, in particular humans through massive walls the usage of a *m-sequence radar* was proposed by [5]. Another approach of through-wall radar measurement was introduced by [6]. The assignment of localisation tasks in the context of automobile parking system to

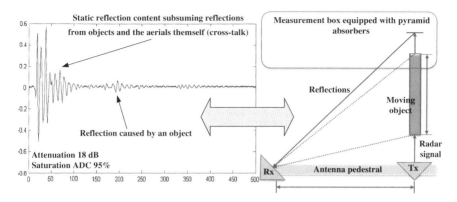

Fig. 3 Working principle

UWB-solutions was proposed by [7]. Much more focusing on aspects of hardware design, [8] presents an on-chip solution with integrated electronics and on-board data processing for UWB radar systems for the detection of humans. Recapitulatory, the elaborate survey about existing through-wall imaging technologies from [9] grants a fruitful overview about the state of the art in this area. Addressing the field of wireless communication, UWB-systems seem to be a promising concept for diverse challenges. Thus, short range communication based on UWB is discussed by [10]. A survey about existing MAC-protocols for ultra-wide-band communication was drafted by [11]. The main objective was the identification of development potentials for an optimal MAC-layer considering the demand for high data rates. [12] proposes a novel methodology for the signal acquisition in UWB-based transmit-only wireless sensor networks. Thereby, the addressed autonomous sensor nodes implement a single code approach to avoid concurrent transmissions and collisions. [13] focusses on the design of filter-antennas for UWB-based communication networks with a bandwidth from 2.65 to 8.52 GHz. An approach for short range communication is proposed by [14] who examines and optimizes the usability of UWB-communication for portable devices. The benefit of UWB-components for BAN-communication in the domain of telemedicine is part of the work of [15]. Concerning our focus of work, UWB-radar was also used to detect vital parameters. In order to correct and calibrate MRI-systems, [16] introduced ultra-wide-band sensing for the detection of the patients exact position regarding imaged body tissue, in particular the chest. The examination of different organic materials in accordance to their content of water is the topic of [17].

2.2 System Setup and Data Processing

For the acquisition of vital signs, an ultra-wide-band radar system, consisting of one transmitting channel and two receiving channels, was embedded within a resting furniture. The connected antennas (type Vivaldi) are installed within the backside of an armchair and aligned to the abdomen of the sitting person. The armchair provides the advantage that the person sitting inside is tied to a fixed position. This fact is essential; motions of the body exert a disturbing influence on the quality of the reflection pattern because these occurrences are belonging to the same or similar dynamic range as the addressed vital signs. In order to cover a larger body area (possibly to detect heart rate), two sensor systems can be coupled to increase the number of receiving channels up to four. The UWB-system itself modulates a m-sequence of 9th order with a length of 511 samples—the signal energy is equally distributed over a bandwidth $B = f_{max} - f_{min}$ of 3.9 GHz from DC (cut-off -10 dB and a master clock rate of 10.09 GHz. Each sample corresponds to a certain distance resp. expressed by the elapsed propagation time of the signal response. The current sample rate (for a complete frame/scan) is about 0.01496 s $\leftrightarrow$ 66.8449 Hz which is sufficient for the addressed target. For more details about the utilized hardware and the m-sequence modulation refer to [6]. The general procedure for the treatment of the n-dimensional reflection data is depicted in Fig. 4 and starts naturally with the parameterization of the sensor device and the reading of the raw data. A single scan (refer to Fig. 5) leads to a vector

$$X_n = (x_{ij})_{\substack{i=1,\ldots,511 \\ j=1}} = (Sample_1 \ldots Sample_{511})' \tag{3}$$

with $x_{ij} \in \mathbb{R}$. Unfortunately, due to hardware properties, measurement frames are sometimes mismatching. Hence, it is important to identify a characteristic signal content for adjustment which is present in every single measurement vector of X_n. Referring to the signal curve drawn in Fig. 5, there is an eye-catching signal content marked by the dotted rectangle one. This large polarization marks the reflection caused by the cross-talk of the receiving and transmitting antenna. Knowing the position of both antennas, a point of origin can be determined in order to span a coordinate system. Each vector of index n has to be scaled/shifted to this origin (depends on the reference vector of index r) by the shift factor c_{s_n}.

$$c_{s_n} = |\operatorname*{argmax}_{i=1}^{i=511}(|(x_{i1})_n|) - \operatorname*{argmax}_{i=1}^{i=511}(|(x_{i1})_r|)| \tag{4}$$

$$X_n^* = (x_{i-c_{s_n}j})_n \tag{5}$$

Now, in accordance to the number of samples Δi between the cross-talk polarization and the object (e.g. marked by the dotted rectangle two), and the spatial resolution $Res_{.s}$ of a single sample, the distance of the object can easily be determined via $d = \Delta i \cdot c \cdot Res_{.s}$. Completing a measurement campaign with the length m (scan) with consecutive scaling, the result is represented by

$$X_c^* = (x_{ij})_{\substack{i=1,\ldots,511 \\ j\ldots m}} \tag{6}$$

A graphical representation (of the signal power over time mTs) is illustrated in Fig. 6. Obviously, there are two noticeable regions in the plot—certain, suspicious

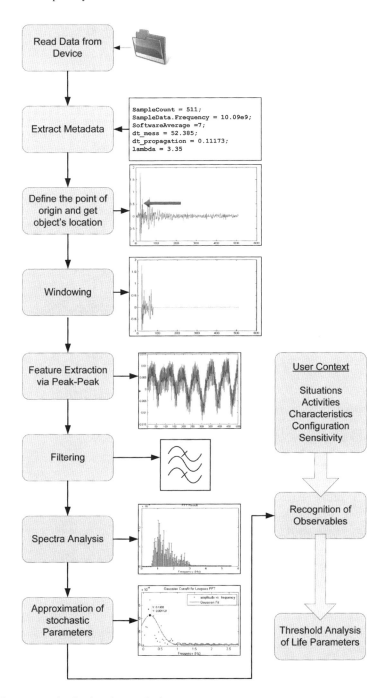

Fig. 4 Process of reflection data analysis

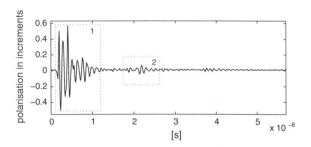

Fig. 5 A single scan with 511 samples

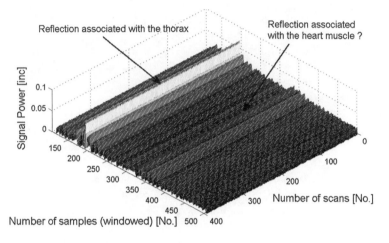

Fig. 6 Temporal behavior of reflections

reflections. These signals may be associated with the cyclic excursion of the thorax and the contraction of the heart muscle. Addressing dynamic processes as the respiratory rate, it is important to consider the static content in the reflection data representing the walls or the furniture in the environment. Therefore, it is important to execute some preliminary measurements X_s^* with a scan length p for the removal of the background. Thus, the underlying dynamic process can be isolated. Due to noise reduction, it is an appropriate method to use the mean vector of $X_s^* = p^{-1} X_s^* 1$ for the subtraction.

$$X_d^* = \forall_{i,j}(x_{ij}^* - \bar{x}_{i1}) \tag{7}$$

In order to isolate interesting features within the reflection pattern, data frames X_d^* are parsed for maxima of signal power. Tracking these significant points in data allows the determination of the vertical window size $|w|$.

$$c_m = \operatorname*{argmax}_{i=1}^{i=511}(x_{d_{im}}^*) \tag{8}$$

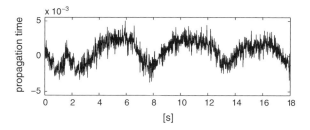

Fig. 7 Extracted movement signal of the first reflection

$$|w| = \max((c_m)) - \min((c_m)) \tag{9}$$

The initial values for the observation window were set in accordance to average amplitude values of breathing. After analyzing the observation window for the first dielectric transient, a timeseries $Y(nTs)$ represented in Fig. 7 is the result.

$$Y(nTs) = T(nTs) + Z(nTs) + S(nTs) + R(nTs) \tag{10}$$
$$n \in \mathbb{N}^+, Ts = 0.0149s$$

$T(nTs)$ covers the trend in the data, $R(nTs)$ grasps all the noise, and $Z(nTs)$ and $S(nTs)$ represent periodic short- and long-term developments in the signal. A typical reason for an occurring $T(nTs)$ may be a slowly decreasing amplitude in breathing due to emerging edema in the lungs—reciprocal, the frequency might increase. In order to reduce the high content of noise in the signal, filtering with e.g. a moving average is the most common solution. Right at the start, an IIR-filter following the transfer function

$$S(z) = \frac{Y(z)}{X(z)} = \frac{\sum_{k=0}^{P} b_k z^{-k}}{\sum_{l=0}^{Q} a_l z^{-l}} \tag{11}$$

$$S(z) = \frac{b_0 + b_1 z^{-1} + b_2 z^{-2} + \ldots b_m z^{-m}}{a_0 + a_1 z^{-1} + a_2 z^{-2} \ldots a_n z^{-n}} \tag{12}$$

was selected to flatten the noise (refer to the filter output drawn in Fig. 8 of the filter input depicted in Fig. 7). Filter coefficients are computed with the aid of

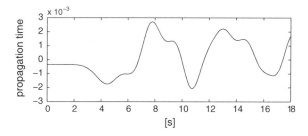

Fig. 8 Response of the IRR-lowpass filter

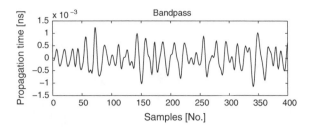

Fig. 9 Response of the IIR-Bandpass filter (heart contraction)

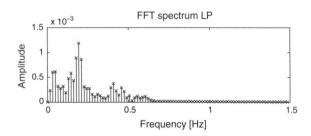

Fig. 10 Spectra of LP-response

Chebyshev-polynomials fitting the demand for an appropriate attenuation regarding a cut-off frequency which corresponds with highest breathing rates. For the isolation of signals with a smaller amplitude and much higher frequencies, a pertinent band-pass filter was designed and implemented (refer to Fig. 9). As expected, filtering the data through an IIR-filter results in a non-linear phase shift of the signal and in an alteration of the underlying curve characteristic. If respiratory rate is the only object of interest, this does not matter; the consecutive windowed FFT-Fast Fourier Transform provides reliable results (refer to Fig. 10). But if it is necessary to identify and localize variances in signal amplitudes or frequencies, and in addition, particular patterns, this filtering procedures deliver timeseries lacking important details of information. Instead of using sinus functions for the decomposition of the timeseries, alternatives of functions only defined over a small interval provide the chance to localize in the time and the frequency domain. Therefore, curve approximation by the superposition of wavelets, an established approach for the compression of data e.g. for pictures, seems to be the best solution to reduce the noise and to preserve important and significant parameters for the feature extraction process. The fundamental idea deals with the approximation of an unknown function f through the superposition

$$f = \sum_k c_k \Psi_k \qquad (13)$$

through the basis functions of Ψ_k and the coherent coefficients c_k. The basis functions Ψ_k belong to the orthonormal set of functions $(\Psi_k)_{k \in J}$ with $J \subseteq \mathbb{N}^+$ under the following condition:

$$<\Psi_i, \Psi_k> = \delta_{i,k} := \begin{pmatrix} 1, & \text{if } i = k \\ 0, & \text{if } i \neq k \end{pmatrix} \quad (14)$$

That means, the scalar product vanishes. Considering the preferred wavelets, there exist many different types with varying properties. In general, the basis function of Ψ_k is described by

$$\Psi_{c_1,c_2}(t) \frac{1}{\sqrt{c_1}} \Psi\left(\frac{t-c_2}{c_1}\right), \quad c_1 \in \mathfrak{R}^+, \; c_2 \in \mathfrak{R} \quad (15)$$

and called mother wavelet. By varying the parameters c_1, c_2 for the shift and scaling of the wavelet, best fitting coefficients for the approximation of the function f can be derived. Using an recursive procedure, coefficients for a couple of single wavelets for the linear combination can be computed. Thinking about continuous functions of f, the transform is done by

$$CWT(c_1, c_2) = \frac{1}{\sqrt{|a|}} \int_{-\infty}^{+\infty} f(t) \Psi^*\left(\frac{t-c_2}{c_1}\right) \quad (16)$$

For discrete values $f(nTs)$, the Fast Wavelet Transform is used. In order to remove the noise content $R(nTs)$ in the timeseries $Y(nTs)$, the signal was decomposed into the wavelet representations using symlets (modified Daubechies wavelets). After computing the coefficients c_k for the signal reconstruction, the gathered values are modified in accordance to the threshold τ as explained below and used for signal estimation (refer to Fig. 11).

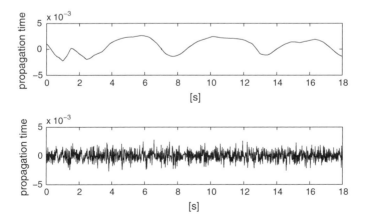

Fig. 11 Denoised signal and separated white noise process

$$c_k = \begin{cases} 0, & c_k \leq \tau \\ c_k, & c_k > \tau \end{cases} \quad (17)$$

This procedure is known as thresholding and realized in two different manners—hard and soft thresholding. Hard thresholding is expressed by the Eq. (17) and means the strict suppression of small coefficients. The alternative soft thresholding deals with shrinkage of the coefficients $c_k \to \tilde{c}_k$ by the value of τ if c_k is greater than τ.

After retransformation, we get an estimation

$$\tilde{f} = \sum_{k \in J} = \tilde{c}_k \Psi_k \quad (18)$$

of the underlying unknown signal of respiratory. Or in other words

$$\tilde{Y}(nTs) = \tilde{Z}(nTs) + \tilde{T}(nTs) + \tilde{S}(nTs) \quad (19)$$

with the separated noise

$$\tilde{R}(nTs) = Y(nTs) - \tilde{Y}(nTs). \quad (20)$$

Figure 12 demonstrates the effect of both techniques. Better results for further considerations are provided by the usage of soft thresholding because the reconstructed signal is much smoother and more often free of any disturbing peaks resp. outliers (refer to the dotted ellipse in Fig. 12). After the removal of the white noise content, the detection of interesting points/areas in the curves was the next issue. Analyzing a couple of reconstructed signals, the detection of extrema sometimes fails due to the movement of the body in the armchair or increasing residual volume—at every intake of breath some additional air remains in the lungs. Assuming,

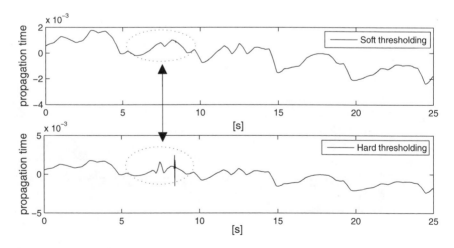

Fig. 12 Hard and soft thresholding of RR-signal

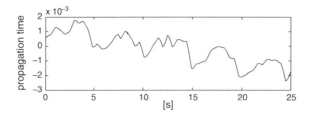

Fig. 13 Breath signal with a decreasing offset

that the reconstructed signal consists of a composition of an indistinguishable long-term seasonal influence ($\tilde{Z}(nTs) \to 0$), some variously weighted trend functions

$$\tilde{T}(nTs) = \sum_{i=1}^{i=\infty} f_i(nTs), \tag{21}$$

and the sought-after hidden process $\tilde{S}(nTs)$, covering the short-time periodic behavior, it is useful to remove disturbing components for the segmentation of the series in the time domain. Using a model reduced to a linear trend and, in accordance to previous observations, some exponential or power behavior, we obtain

$$\tilde{Y}(nTs) = \tilde{S}(nTs) + \tilde{T}(nTs) \tag{22}$$

with a set of basic functions after the removal of the steady component $c_1 nTs + c_2$ in the signal (refer to the example in Fig. 13).

$$f_{T1}(nTs) = c_1 e^{c_2 nTs} \tag{23}$$

$$f_{T2}(nTs) = c_3 e^{c_4 nTs} + c_5 e^{c_6 nTs} \tag{24}$$

$$f_{T3} = c_7 e^{-\frac{nTs - c_8}{c_9}} \tag{25}$$

$$f_{T4} = c_{10} nTs^{c_{11}} + c_{12} \tag{26}$$

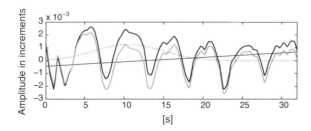

Fig. 14 Decomposition of the extracted RR-signal

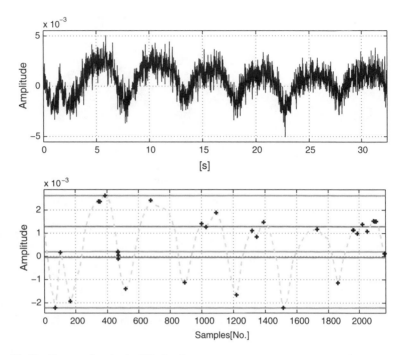

Fig. 15 Significant points in the RR-signal

For the approximation of the influence induced by the movement or motion of the body, coefficients c_i for the mentioned relationships are interpolated. The best fitting function f_{Tr} for the trend removal is selected due to

$$f_{Tr}, r = \operatorname*{argmin}_{i=1}^{i=4} (SSE_i). \tag{27}$$

After the abatement of these components (refer to Fig. 14 with the different trend components of the extracted RR-signal), significant points as extrema are ascertained via interval evaluation. Examining the rectified series, significant extrema are detected. Of course, there are accumulations of these points in relation to the degree of remaining noise or peaks and obviously, to the characteristics of the breathing process itself (ref to Fig. 15). Due to their nearby unique occurrence (in comparison to the high points), the low points of the curve are helpful for the partition of the data to smaller evaluation intervals. Assuming that the largest value in each interval marks the end of each single breathing process (end of inhalation), these points are selected for the computation of frequency, and after rescaling (adding previously removed non-linear trend) of the timeseries, the breathing amplitude. Additionally, the low points are checked due to their Euclidean distance; if it seems that they belong together, a surrogate low point is computed and used for the interval limitation. Furthermore, each identified extrema is weighted

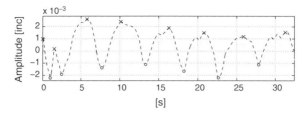

Fig. 16 Selected extrema in respiratory pattern

due to it's location within the σ-bounds of the distribution describing the general stochastic signal properties considering a sensitivity factor c_s. In summary, the process ends with the provision of RR, RRV, AR, ARV and in addition, with the localization of leaps and existing non-linearity within the data. These observations need to be matched to distinct patterns as mentioned in section 2.B (Fig. 16).

3 Discussion of Results

In the early stages of research, preliminary measurements with deterministic signal sources were executed. A programmable linear motor was configured to perform motions with specified values for acceleration, velocity and distance. The maximum periodic displacement of the armature of 45 cm was detected with an exact match of 99.97 %. The lowest possible motion of the drive of 1 mm was detected with an accuracy of 99.6 %; a fine result for further considerations. That means, it was not possible to determine the best resolution because the drive itself is limited to 1 mm movement range.

Considering respiratory rate and variances in frequency and amplitude, the main target of unobtrusive vital sign acquisition via UWB was accomplished. After improving the hardware setup through auxiliary shielding in the backside of the armchair, an increased robustness against disturbing backscattering was achieved. In relation to our reference measurement device, a pulse oximeter with nasal prongs (of course, observed signals are phase shifted, but have the same properties) a sufficient accuracy was reached (refer to Table 1). In addition, the accuracy and

Table 1 Results of RR-measurement-FFT based

Test person	Accuracy (respiratory rate %)	Accuracy respiratory rate (shielded %)
Male, 1.84 m, 74 kg	86	91.5
Male, 1.72 m, 82 kg	87	92.1
Male, 1.92 m, 79 kg	84	92.4
Male, 1.81 m, 76 kg	87	91.6
Female, 1.67 m, 59 kg	71	91.2
Average values	83	91.7

Table 2 Results of RR-measurement—FWT-based

Test person	Accuracy (respiratory rate %)	Accuracy respiratory rate variability (%)
Male, 1.84 m, 74 kg	95.4	67.5
Male, 1.72 m, 82 kg	96.3	72.1
Male, 1.92 m, 79 kg	95.2	69.4
Male, 1.81 m, 76 kg	97.2	69.6
Female, 1.67 m, 59 kg	94.7	68.2
Average values	95.7	69.3

robustness for long-term measurement was improved by the evaluation of different antenna topologies and their recalibration (direction and position). The accuracy of RR-detection indexed by a was determined by measurement campaigns without the mentioned hardware improvement. For the feature extraction in both campaigns, a FFT with a consecutive spectrum distribution analysis was executed. Obviously, the measurement accuracy depends on the constitution of the test person. Thinking about heart rate detection, which is also part of our work, there are still open issues. Accuracy for the detection of signals in this range drops to 43 % in comparison to the accuracy of the RR-detection. The denoising of signals by wavelets including a consecutive analysis in the time domain arose as a convenient alternative; the accuracy of measurement of RR and it's features rose significantly (refer to Table 2). The amplitude of the respiratory signal was measured with an accuracy of 86.4 %. Possibly, an optimization of trend approximation can improve this result. The variance of this feature lacks accuracy and stays at 56 % in comparison to the reference system.

In summary, the results in Tables 1 and 2 show the benefit of analysis of the extracted signals in the time domain. FWT-based denoising is the better approach for filtering, because interesting features in the signal are preserved. In addition, suspicious patterns are easier to recognize which is an important insight for the next steps.

4 Conclusion and Outlook

As shown in the previous sections, the current state of research demonstrates the potential benefit for the unobtrusive acquisition of vital signs which are incidental to mechanical measures. Especially in the domain of geriatric care, according to specific user demands, this approach seems to be worthwhile for practical application. Not only restricted to home care scenarios or applications of Ambient Assisted Living, this technical solution can, embedded within information processing infrastructures, improve the accuracy of situation recognition systems and additionally raise the diagnostic efficiency of such approaches. Thinking about stand-alone solutions, this work can be the springboard for the development of a new type of telemedical devices—the fusion of furniture with ambient intelligence focusing on health monitoring. At the present time, the work concentrates on the detection and collating of respiratory patterns. At the beginning of September 2012, clinical trials with different objectives will start. After examining and improving the robustness and accuracy of proposed approach in daily use for the detection of respiratory rate and related features (patterns also), it is aimed to detect the heart rate and arrhythmias in a reliable manner. In addition, it is planned to aggregate ECG-recordings of atrial fibrillation with reflection patterns in order to identify correlations between mechanical and bioelectrical observations assigned to the same signal source. Furthermore, it is scheduled to perform test campaigns with patients with congestive heart failure in order to detect edema in the lungs by the evaluation of the thorax impedance. Current test measurements in our laboratory with a mock-up prove the meaningfulness of the implemented general functional principle but it is necessary to confirm it for practice by examining persons with altering fluid retentions considering calibrated reference measuring. And finally, derived vital sign must be assessed in the light of the user situation or user activity.

References

1. Busch, B.H., Kujath, A., Witthoeft, H., Welge, R.: Preventive emergency detection based on the probabilistic evaluation of distributed, embedded sensor networks. In: Ambient Assisted Living. Springer, Berlin (2011)
2. Folke, M., Cernerud, L., Ekstroem, M., Hoek, B.: Critical review of non-invasive respiratory monitoring in medical care. Med. Biol. Eng. Comput. **41**, 377–383 (2003)
3. Bates, A., Ling, M., Mann, J., Arvind, D.K.: Respiratory rate and flow waveform estimation from tri-axial accelerometer data. In: Proceedings of International Body Sensor Networks (BSN) Conference, pp. 144–150 (2010)
4. Mann, J., Rabinovich, R., Bates, A., Giavedoni, S., MacNee, W., Arvind, D.K.: Simultaneous activity and respiratory monitoring using an accelerometer. In: Proceedings of the International Body Sensor Networks (BSN) Conference, pp. 139–143 (2011)
5. Sachs, J., Friedrich, J., Zetik, R., Peyerl, P., Klukas, R., Crabbe, S.: Through-wall radar. In: Proceedings of the IRS (2005)
6. Judson Braga, A., Camillo, G.: An ultra-wideband radar system for through-the-wall imaging using a mobile robot. In: Proceedings of the ICC'09—IEEE International Conference on Communications (2009)
7. Mary, g.i., Prithiviraj, v.: Improved UWB localization technique for precision automobile parking system. In: Proceedings of the TENCON 2008—2008 IEEE Region 10 Conference (2008)
8. Chang, S., Chu, T.S., Roderick, J., Du, C., Mercer, T., Burdick, J.W., Hashemi, H.: UWB human detection radar system: A RF CMOS chip and algorithm integrated sensor. In: Proceedings of the IEEE International Ultra-Wideband (ICUWB) Conference, pp. 355–359 (2011)
9. Michal, A.: Through wall imaging with UWB radar system. Dissertation, Faculty of Electrical Engineering and Informatics, Department of Electronics and Multimedia Communications (2009)

10. Ahmadian, Z., Shenouda, M.B., Lampe, L.: Design of pre-rake DS-UWB downlink with pre-equalization. IEEE J. Commun. **40**, 400–410 (2012)
11. Zin, M. S. I. M., Hope, M.: A review of UWB MAC protocols. In: Proceedings of 6th Advanced International Telecommunications (AICT) Conference, pp. 526–534 (2010)
12. Li, Z., Gielen, G.: UWB signal acquisition in transmit-only networks. In: Proceedings of IEEE International Ultra-Wideband (ICUWB) Conference, pp. 126–129 (2011)
13. Panda, J. R., Kakumanu, P.. Kshetrimayum, R.S.: A wide-band monopole antenna in combination with a UWB microwave band-pass filter for application in UWB communication system. In: Proceedings of Annual IEEE India Conference (INDICON), pp. 1–4 (2010)
14. Xiao, Z., Ge, N., Pei, Y., Jin, D.: SC-UWB: a low-complexity UWB technology for portable devices. In: Proceedings of IEEE International Signal Processing, Communications and Computing (ICSPCC) Conference, pp. 1–6 (2011)
15. Hernandez, M., Kohno, R.: UWB systems for body area networks in IEEE 802.15.6. In: Proceedings of IEEE International Ultra-Wideband (ICUWB) Conference, pp. 235–239 (2011)
16. Thiel, F., Hein, M., Schwarz, U., Sachs, J., Seifert, F.: Combining magnetic resonance imaging and ultra-wideband radar: a new concept for multimodal biomedical imaging. In: Rev Sci Inst **80**, 014302 (2009)
17. Marko H., Jrgen S., Ulrich S., Schaefer, M.: Ultrabreitband-Sensorik in der medizinischen Diagnostik, In: 41. Jahrestagung der Deutschen Gesellschaft fuer Biomedizinische Technik BMT, Aachen, Germany (2007)

Context-Enriched Personal Health Monitoring

Barbara Franz, Mario Buchmayr, Andreas Schuler
and Werner Kurschl

Abstract Vital data measurement solely represents a snapshot of the patient's condition. In case of deviant results health professionals can put them in relation to the daily constitution as well as the measurement circumstances and therefore interpret them accurately. Personal health telemonitoring actually provides no possibility for health professionals who interpret the recorded vital data to put them in relation with the context of the measurement and a patient's daily activities. To tackle this contextual knowledge gap, we propose a personal health monitoring approach, which allows enriching measured vital data with contextual information from an *Ambient Assisted Living* (*AAL*) system. Therefore, it is possible to provide additional information, like the person's activity before or during the vital data measurement. Patients and caregivers, which process monitored health data, gain advantages and can improve care giving process. To demonstrate the benefits we use the two use cases (*1*) cardiac rehabilitation and (*2*) mobile nursing care.

B. Franz (✉) · M. Buchmayr · A. Schuler · W. Kurschl
University of Applied Sciences Upper Austria, Research Center Hagenberg, Hagenberg, Austria
e-mail: Barbara.Franz@fh-hagenberg.at

M. Buchmayr
e-mail: Mario.Buchmayr@fh-hagenberg.at

A. Schuler
e-mail: Andreas.Schuler@fh-hagenberg.at

W. Kurschl
e-mail: Werner.Kurschl@fh-hagenberg.at

1 Introduction

Elderly and chronically ill people often have to regularly measure their vital signs, like blood pressure, blood sugar or weight for medical monitoring. Personal health telemonitoring devices, which allow people to measure vital data themselves at home and automatically transfer the measurement results to a healthcare facility, provide support for this repetitious task.

Nevertheless, in comparison to traditional vital data measurement, which is done under controlled conditions by healthcare professionals, telemonitoring still has drawbacks. Incorrect measurement results are more prone to happen due to inappropriate device usage or measurement conditions. Furthermore, since vital data measurement, except for continuous vital data monitoring, represents only a snapshot, the data has to be considered in relation to the current physical constitution of the patient and the measurement circumstances. When health professionals perform vital data measurements they may ask the patient or repeat the measurements in case that uncommon data are retrieved. As example think of a blood pressure measurement which is done after the patient had some physical exercise (walking up the stairs). In case that the vital data measurement is done by professionals they can put the measured data in relation to the context or repeat the measurement under controlled conditions. In case of personal health telemonitoring, this is a problem, because there is actually no possibility for the doctor who interprets the recorded vital data to get feedback from the patient.

Recent developments motivated by the 'quantified-self' movement tackle this issue by providing web portals for self tracking devices, like fitbit ultra [1], which allow to query the measured data as well as to manually add additional (contextual) information. Since this approach requires additional efforts and discipline to manually track all activities, it is not feasible for everyone. Especially, elderly people only accept support devices which are intuitive and easy to use [2].

To tackle the contextual knowledge gap between personal health telemonitoring and traditional vital data measurement, we propose a personal health monitoring approach which allows enriching measured data with contextual information from an *Ambient Assisted Living* (*AAL*) home system. Therefore, relevant and additional information, like the person's activity, can be provided.

2 Enhancement of Personal Health Telemonitoring Data

We will use the following two use cases to illustrate our approach: (1) personal health telemonitoring support of cardiac rehabilitation patients, and (2) personal health telemonitoring in mobile nursing care.

Patients who undergo a cardiac rehabilitation program, have to regularly exercise at home and can benefit from cooperating with their physician by sharing their monitored vital data (see Fig. 1).

Fig. 1 Personal health monitoring

In consultation with health professionals patients have to achieve weekly goals during the rehabilitation period. Such goals can be losing a specified amount of weight or burning an additional quantity of calories each week. The achievement is typically measured at home by each patient using a telemonitoring service.

Although such telemonitoring services support the health professionals in order to keep track of their patients during a rehabilitation phase, the measured data are always just a snapshot of the patient's condition at the time of measurement. For example, if a patient has just finished a workout on an ergometer, his/her blood pressure will be increased. When using a telemonitoring system, there is typically no way to combine this contextual information with the measurement. A health professional examining the measured data will only see the increased value for blood pressure with no additional details about the circumstances.

Mobile home care is a further use case where the usage of a telemonitoring system on its own is insufficient. Mobile caretakers are often confronted with elderly people, who are confined to bed. As supplement to vital data measurements, which are conducted at each visit, mobile health professionals often take handwritten notes about the patient's current condition. These handwritten notes, in conjunction with the patient's vital signs form the foundation for further care planning. The challenge, when using a telemonitoring system is, at which point in the information processing chain should the hand written notes be combined with the measured data?

In order to overcome the described deficiencies this paper introduces a way to enrich data measured by a standard-compliant monitoring system using contextual information. This information can either be provided through a home automation system or by hand, as distinguished in our second use case. Either way the information provided builds the basis to perceive activities for enriching vital data measurements with context, hence providing a way to increase the expressiveness of recorded data stored in a patient's *Personal Health Monitoring Record* (*PHMR*) [3].

2.1 Standard-Compliant Monitoring System

Considering the use cases described in Sect. 2 you can identify different interest groups, like patients, doctors, home care nurses or other health professionals. Thus, a lot of heterogeneous systems are involved in processing a patient's data. Heterogeneous information systems are characteristically within the healthcare domain. To guarantee interoperability and data exchange among the different systems the usage of defined standards is inevitable.

The *Continua Health Alliance* (*CHA*) [4] offers guidelines which describe the implementation of a standard-compliant monitoring solution based on different profiles of *Integrating the Healthcare Enterprise (IHE)* [5], *Health Level 7 (HL7)* version 2 and HL7 version 3 *Clinical Documents Architecture (CDA)* [6]. The described solution of the CHA, as shown in Fig. 2, enables the transmission of data from monitoring devices to a patient's *Electronic Health Record (EHR)*.

From a monitoring device, for example a blood pressure meter, the data are sent to an *Aggregation Manager* and forwarded to a *Telehealth Service Centre (THC)* which stores the data in an EHR. Once the measured data are stored in form of a *PHMR* in an IHE compliant EHR, they can be accessed with respect to the proper security permissions. An EHR manages all documents concerning a person's health, including PHMR documents as well as additional notes from caregivers or information from 3rd party systems. Our approach uses this possibility to enrich the PHMR vital data record with activity information queried from an ADL system (see Sect. 3).

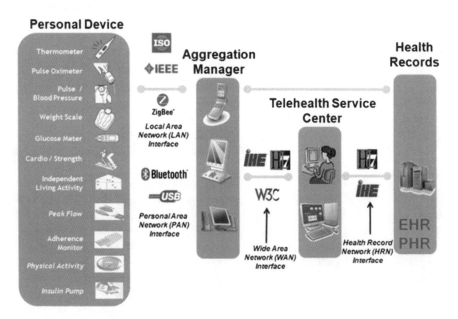

Fig. 2 Standard-compliant monitoring CHA [4]

2.2 Detection of Activities of Daily Living

Motivated by the AAL community the detection of *Activities of Daily Living* (*ADL*) via (unobtrusive) sensor-based systems is an essential foundation for the *aging at home principle* [7]. The reason for ADL detection within AAL is on one hand the detection of critical or even life-threatening situations (e.g. detect deviations from behavioural patterns in a daily lifecycle) and on the other hand the autonomy enhancement of elderly or handicapped people [8]. Furthermore, a recent study [9] showed that the analysis of ADLs can in addition provide relevant medical information.

An ADL can be defined as a simple activity, like getting up in the morning or entering a room, but can as well be more complex, like cooking or personal hygiene. Especially, when activities are overlapping with or are interrupted by other activities, a proper detection becomes a technically challenging task. The project *BehaviourScope* [10] from the Yale University, for example, addresses this issue by describing activities using high-level grammars and refining them during further processing.

Concerning the detection of ADLs there are basically two approaches which are motivated by the available sensor devices and types of activities which should be detected: (1) knowledge-driven ADL detection and (2) utilizing supervised learning techniques to detect activities (patterns) in data/sensor streams. Both approaches have their benefits and drawbacks, always depending on the application purpose and available infrastructure. ADL detection systems reach from sophisticated systems which come along with high investment costs (e.g. expensive TOF cameras) to low cost systems using existing home infrastructure [11] and cheap retrofittable sensor devices (e. g. contact switches, motion sensors). No matter which approach is used, usually the result is a notification that a given activity was detected.

In our work we decided to use a knowledge-driven ADL approach, for the following reasons: (1) knowledge-driven ADL systems store the perceived information in a database (knowledge base) which can be queried for multiple purposes (for instance, which activity was happening during a given time span? Where was the person during a given time span?), (2) knowledge-driven systems allow to gain information from the database without any necessary training, (3) knowledge-driven systems are easier to adapt (in case of new/additional sensor devices), and (4) a knowledge-based AAL framework [12] as well as a simulation environment for AAL homes [13] are available at our institute. Besides, the sensed data in our AAL framework is based on an ontological model which eases the integration into the monitoring system.

2.3 Integrating Contextual Information

To demonstrate how monitored vital data is enhanced with context information from an AAL system, we use the following process illustrated in Fig. 3. A caretaker or the patient regularly monitors the vital signs at home and transmits the data to the *Telehealth Service Centre* (*THC*). During the transmission an *Aggregation Manager* is used to properly interpret and transform the data. According to CHA guidelines, the THC retrieves the patient's latest document containing health monitoring data from the *EHR* and appends the measured vital signs.

The constraints, which contextual knowledge (e. g. activities) should be added to the measured vital data are defined in the THC. This is done via predefined enrichment rules. Each type of vital sign is associated with a set of enrichment rules. These rules are used to query defined contextual information from the ADL system.

When the THC receives vital data, it queries the *Activities of Daily Living* (*ADL*) *system* according to the required information specified in the enrichment rules. For example, which activities happened in the minutes preceding the vital sign monitoring? It would be possible to query all relevant information directly from the ADL system, but for a better abstraction we decided to introduce *enrichment rules* in the THC. This rules abstract 'high-level' activities relevant for doctors and nursing personnel from the specific activities provided by the ADL system. This has the advantage that the person who interprets the data does not have to be familiar with the specific activities defined in the ADL system.

For our prototype we simply decided to distinguish between two high level activity types: (1) *high activity* (doing some exercise, walking around high frequently, cooking, walking in the staircase) and (2) *low activity* (watching TV, sleeping). In addition to retrieving the latest health monitoring document, the THC

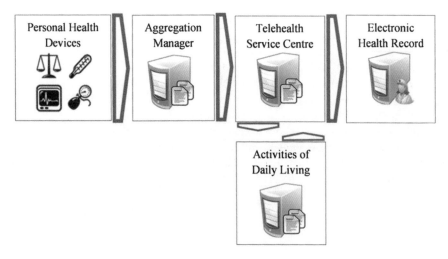

Fig. 3 System overview

queries the patient's latest activity information from the ADL system, appends the obtained data and stores the extended document in the *EHR*.

Afterwards, healthcare professionals can query a patient's health monitoring data from the EHR and set it into context with the patient's activities which happened before the respective measurements. Since activities are stored separately from the monitoring data, for both an adequate model is provided which allows to link activities and measurements using model transformation. Thus, various combinations of monitoring data and activities may be queried, depending on which context is required.

3 System Architecture

In the following section we describe the architecture for the standard-compliant monitoring service based on the CHA guidelines [4]. In addition we give a short introduction into the AAL System we use for ADL detection and context enrichment.

Figure 4 shows the components of our architecture and will be explained in more detail in the following subsections. Each component provides/expects interfaces conforming to the CHA guidelines. For more information on how to implement a standard–compliant monitoring system based on the CHA guidelines please refer to [14, 15].

3.1 Aggregation Manager

Data measured with a personal health device, e.g. a blood pressure meter, is transferred to the aggregation manager. According to CHA guidelines, the

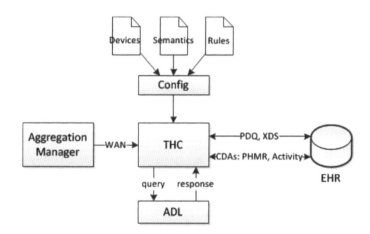

Fig. 4 System architecture

```
MSH|^~\&|||||20111215114509.682+0100||ORU^R01^ORU_R01|88c3|P|2.6.1|||NE|AL|||||
PID|||005.13226488954651^^^&1.2.40.0.10.1.6.1.0.1.100.1.1&ISO||Testpatient^Hage
nberg|||19310320000000+0100|M|||
OBR|
OBX||NM|150021^MDC_PRESS_BLD_NONINV_SYS^MDC|1.0.1.1|125|266016^MDC_DIM_MMHG^MCD
|||||F|||||||||20111215114505.824+0100
OBX||NM|150022^MDC_PRESS_BLD_NONINV_DIA^MDC|1.0.1.2|85|266016^MDC_DIM_MMHG^MCD|
||||F|||||||||20111215114505.824+0100
```

Fig. 5 HL7 message with blood pressure measurement

aggregation manager forwards the measured data as HL7 messages over the *Wide Area Network (WAN)-Interface* to the THC. Supplemental the monitoring data is transformed into HL7 messages,[1] using the HAPI framework [16]. Each measurement is coded as an observation/result (OBX) segment in the message. For example: a blood pressure meter sends the result 125/85 with two values (systolic/diastolic), which are coded in two separate OBX segments shown in Fig. 5.

3.2 Telehealth Service Centre

The THC—the main component of the system—is responsible for associating measured data with activities based on predefined rules. According to the CHA guidelines the THC receives the HL7 v2 messages and transforms the observations into a persistent HL7 v3 CDA PHMR. Using the IHE profile Cross Document Sharing provided by Open Health Tools (OHT, [17]), the component queries the EHR for existing PHMRs based on a patient identifier (PID), which can be retrieved using a radio-frequency-identification as proposed in [4]. The measured vital signs are appended to the latest PHMR. In case that no PHMR exists for this patient or the size of the PHMR exceeds a predefined limit, a new PHMR is created. The PHMR itself is a HL7 CDA document and is based on the *Reference Information Model (RIM* [6]).

The THC additionally queries the *ADL Component* (see Fig. 6) in order to retrieve activities associated with the current measurement. A configuration interface (*Config,* illustrated in Fig. 4) offers the required functionality to define which activity information is required.

The response of the *ADL Component* is used to extend an activity CDA document for the patient. Analogous to the PHMR, the THC component queries the EHR for existing activity documents for the patient, retrieves the latest one and appends the activities which were inferred by the ADL. In case that no activity document exists for this patient or the size of the document exceeds a predefined limit, a new activity CDA document is created. Both documents, the PHMR and activity CDA document, are transferred to the EHR via the *Health Record Network (HRN)* interface. For the transformation and document transfer a communication server [18] is used.

---

[1] HL7 v 2.6 ORU^R01 standard message.

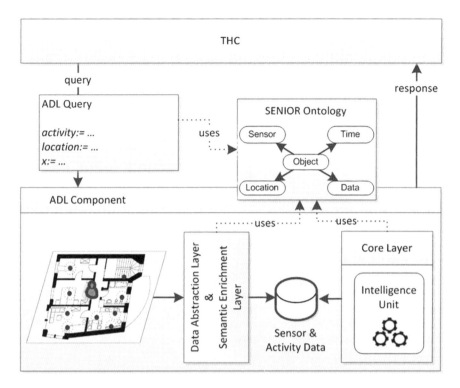

Fig. 6 ADL component

3.3 AAL System for ADL Detection

The AAL framework named SENIOR [12] was originally developed with the purpose to provide an open, OSGi based infrastructure for different AAL components. Various sensor components can be plugged into the framework and the perceived data are transformed into a common data model, enriched with semantic information and stored in a database. The data model is based on the SENIOR ontology which defines the *Objects* and *Relations* within the data store. Objects defined in this ontology are created in the data store and can be queried via rules.

On top of this data model a reasoning component allows the verification and validation of detected situations and activities by checking for defined constraints. For example, if the fall detection unit of an elderly person which lives in an AAL equipped home is signaling a critical situation, SENIOR can be used to verify this alarm and provide additional information on demand. This is done by merging contextual information from installed in-home sensors, like motion sensors, light barriers, switches etc. The additional provided information could be, for example, the location of the person, if an indoor localization system is deployed.

Sensed information is continuously transformed into a model representation and stored in a database. Furthermore we support data specific processing rules. These rules can either be used to query information (e.g. the location of a person at a given time) or to verify if a given constraint can be fulfilled or not (was the person moving before the fall detection unit signaled an alarm). Figure 6 illustrates how the *ADL Component,* which is part of the SENIOR framework, is integrated into the monitoring component (*THC*).

3.4 Accessing Context Enriched Health Data

Since measured vital signs and activities are stored for each patient in the EHR, health professionals are able to query the EHR of a specific patient to retrieve a PHMR and in addition the activities that were associated with past measurements.

A transfer of knowledge between patient and healthcare professionals as well as among different health service providers, demands a clearly structured language. For example, high activity in a nursing home has a different meaning than high activity in a rehabilitation centre. To distinguish these domain specific properties, it is necessary to develop a semantic model based on different domain data models.

Properties with different domain specific semantics, like *high activity*, are represented using a meta model, which in our case is an abstraction of a domain specific model [19]. Particular terminology from health service environments (domains), like rehabilitation centres or nursing homes, is mapped via domain mapping to the meta model.

The PHMR and activity documents are HL7 CDA documents. Thus, the RIM, which CDA is based on, can be used to develop an extended model for the context enhanced personal health monitoring data. RIM guarantees that data is structured in an easily interpretable way. To support semantic interoperability concerning health data processing as well as interoperability among interdisciplinary institutions, the model is extended using domain specific terminology. This requires standardized structures for documents as well as a common language between healthcare service levels. The semantic model has to be developed in cooperation with experts, in our case with healthcare personnel. During the development different terminologies as well as the semantic configuration files from the THC, developed by health professionals, must be taken into account. Therefore, it can be guaranteed that the domain model covers the varying needs of several system partners and services and to assure the necessary coverage. Since activity documents and PHMRs are structured CDA documents, the transformation of context enhanced personal health monitoring models can be done using XSLT.

Thus, the data can be exchanged electronically using the defined model, and can efficiently be integrated and interpreted in context. Furthermore, the model allows the easy integration of further terminologies and domains [20].

4 Evaluation

The test setting used for the evaluation of the proposed concept comprises a productive telemonitoring system as well as a simulated home automation system.

4.1 Telemonitoring System

Using pedometer, scale and blood pressure meter, cardiac patients provide monitored training data from their homes to the EHR and thus to their caregiver. Using a simulated data basis the ADL Component is capable of delivering different activity data to the THC which enriches the PHMR. The enriched data is visualized in the vital data view of the monitoring component (see Fig. 7). Concerning the feasibility of the displayed measurement context (high activity, low activity) an evaluation with healthcare professionals is ongoing.

4.2 Simulation of ADLs

To evaluate our work we decided to use a hybrid evaluation approach. Therefore, we used a simulator [13] which is capable of creating and recording user interaction with different sensor devices within a virtual AAL home. Figure 8 shows a screenshot of our simulator environment.

For our evaluation we used a default blueprint of an apartment and equipped it with different home automation and AAL sensor devices (motion sensors, switch sensors, pressure pads). This allows us to record a predefined set of activities and sensor interactions, like triggering the motion sensor in the staircase or performing

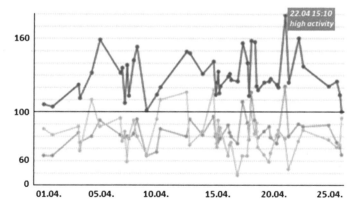

Fig. 7 Blood pressure overview combined with activity data

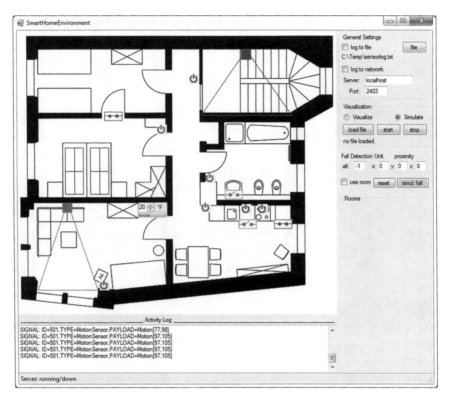

Fig. 8 Simulator for ADL data

some personal hygiene tasks in the bathroom. The recorded user interaction can be sent to the SENIOR framework via a debugging port. The SENIOR framework processes the incoming data as if it was real sensor input, processes the information and stores the data in its database.

Afterwards some vital data measurement using real measurement devices, like a blood pressure meter, was done. The measured vital data is sent to the *THC* which queries the *ADL Component* for additional activity information which is displayed in the vital data view (see Fig. 7).

Our approach to simulate the sensor interaction allows us to properly test if the enrichment done by the THC component corresponds to the definition in the rules. Potential error sources, like faulty sensor input or inaccurate activity detection can therefore be excluded. The evaluation of ADL detection was not purpose of our work. Nevertheless, a test under real world conditions is inevitable.

5 Conclusion and Outlook

After surgery, cardiac patients may participate in an outpatient cardiac rehab program, where their risk factors such as body weight and blood pressure are evaluated and goals for those risk factors are established. During the attendance of the outpatient training the patient is closely monitored, but during the patient's home exercises rehab caretakers have no chance to control execution and outcome. The telemonitoring system allows the patient to transfer vital signs measured during home training to the EHR. By combining telemonitoring with ADL detection, health professionals are able to query health monitoring data from the EHR and properly interpret them considering contextual information. This supports patients and caregivers in correctly interpreting monitored data and improves the care giving process.

The feasibility of our approach was shown by evaluating the two use cases: (1) using cardiac rehabilitation and (2) mobile nursing care. We evaluated our approach using a hybrid process, where we used a productive telemonitoring system in combination with simulated AAL data (as substitution for data from an AAL home system).

A user interface, which allows an easy definition of the enrichment rules and the ontological model by health professionals is outstanding. Furthermore, an evaluation of the activity monitoring with healthcare professionals is planned.

Acknowledgments This research is partially funded by the Austrian Research Promotion Agency (FFG), the European Regional Development Fund (ERDF) in cooperation with the Upper Austrian state government (REGIO 13, Innovative Upper Austria, Health Cluster). Any opinions, findings and conclusions or recommendations in this paper are those of the authors and do not necessarily represent the views of the research sponsors.

References

1. Fitbit ultra, Fitbit Inc.: Available from http://www.fitbit.com/de/product (2012). Cited Sept 2012
2. Demiris, G.: Electronic home healthcare: concepts and challenges. Int. J. Electron. Healthc. **1**, 4–16 (2004)
3. Alschuler, L., Beebe, C., Boone, K.W.: Implementation Guide for CDA Release 2.0 Personal Healthcare Monitoring Report (PHMR), Draft Standard for Trial Use, Release 1.1. (2010)
4. Continua Health Alliance. Continua Health Alliance: Available from http://www.continuaalliance.org (2012). Cited Sept 2012
5. IHE Int. Integrating the Healthcare Enterprise: Available from http://www.ihe.net (2012). Cited Sept 2012
6. Health Level Seven. Health level 7, Available from http://www.hl7.org (2012). Cited Sept 2012
7. Rantz, M., Aud, M., Alexander et al.: Tiger Place: An Innovative Educational and Research Environment, AAAI in Eldercare: New Solutions to Old Problems, Washington DC, USA (2008)

8. Becker, M.: Software architecture trends and promising technology for ambient assisted living systems, In Proceedings of Assisted Living Systems—Models, Architectures and Engineering Approaches, Dagstuhl, Germany, pp. 18 (2008)
9. Rantz, M.J., Marek, K.D., Aud, M. et al.: A technology and nursing collaboration to help older adults age in place. Nurs. Outlook 40–45 (2005)
10. Bamis, A., Lymberopoulos, D., Teixeira, T et al.: The behaviorscope framework for enabling ambient assisted living. Spec. Issue Int. J. Pers. Ubiquit. Comput. (2009)
11. Fogarty, J., Au, C., Hudson, S.E.: Sensing from the basement: a feasibility study of unobtrusive and low-cost home activity recognition. In Proceedings of the 19th annual ACM symposium on User interface software and technology (UIST '06), pp. 91–100, New York, USA, (2006)
12. Kurschl, W., Franz, B., Buchmayr, M., et al.: Situation-aware ambient assisted living and ambient intelligence data integration for efficient eldercare. In: Watfa, M.K. (ed.) E-Healthcare Systems and Wireless Communications: Current and Future Challenges (Contributions to Book: Part/Chapter/Section 15), IGI Global, pp. 315–348, (2011)
13. Buchmayr, M., Kurschl, W., Küng, J.: A simulator for generating and visualizing sensor data for ambient intelligence environments. In Proceedings of the 2nd International Conference on Ambient Systems, Networks and Technologies (ANT-2011), pp. 90–98 Niagara Falls, Canada, (2011)
14. Strasser, M., Helm, E., Schuler, A., Fuschlberger, M., Altendorfer, B.: Mobile access to healthcare monitoring data for patients and medical personnel. In: Proceedings of the XXIV Conference of the European Federation for Medical Informatics, Quality of Life through Quality of Information, MIE 2012, Pisa, Italy, 26–29 Aug 2012
15. Strasser, M., Helm, E., Franz, B., Mayr, H.: Mobile health solutions for empowered, health-conscious individuals and patients, In Proceedings of the 10th International Conference on Information Communication Technologies in Health, ICICTH 2012, Samos, Greece, 12–14 July 2012
16. Sourceforge. Hapi: the open source HL7 API for Java, 2012 [cited 2012 Jan]. Available from: http://hl7api.sourceforge.net/index.html
17. Open Health Tools Inc. Improving the world's health and well-being by unleashing health IT innovation. Available from http://openhealthtools.org (2012). Cited Jan 2012
18. Mirth Corporation. Mirth Connect. Available from http://www.mirthcorp.com/products/mirth-connect (2012). Cited Jan 2012
19. Kühne, T.: Matters of (Meta-)Modeling. J. Softw. Syst. Model. 5(4), 369–385 (2006) (Springer, Berlin)
20. Franz, B., Mayr, H.: Providing technical and semantic interoperability for integrated healthcare using an IHE-compliant system. In: Proceedings of the IADIS International Conference e-Health 2011, Rome, Italy, 20–22 July 2011

Mneme: Telemonitoring for Medical Treatment-Support in Dementia

Torben Wallbaum, Melina Frenken, Jochen Meyer, Andreas Hein and Carsten Giehoff

Abstract This paper presents a system for the evaluation of the physical and mental state of patients suffering from mild to moderate dementia. In addition to the development of a tele-medical monitoring system, a new business model is developed which enables a wide dissemination and a funding concept that supports the patients with financing. The aim of the project is to realize a seamless and cost-effective integration of monitoring at a patients home. Furthermore medical professionals involved in the treatment are enabled to see the monitored data via web-services. The technical realization allows the measurement of parameter like activity, quality of sleep, weight, lean and fat mass as well as total body water.

1 Introduction

The rise of patients suffering from dementia due to the demographic change poses many challenges. One of the most important demands from a patient's point of view is to live at the own home as long as possible. In order to assess the need for

---

T. Wallbaum (✉) · M. Frenken · J. Meyer · A. Hein
OFFIS—Institute for Information Technology, University of Oldenburg,
Oldenburg, Germany
e-mail: torben.wallbaum@offis.de

M. Frenken
e-mail: melina.brell@offis.de

J. Meyer
e-mail: meyer@offis.de

A. Hein
e-mail: andreas.hein@informatik.uni-oldenburg.de

C. Giehoff
Corantis-Kliniken GmbH, Vechta, Germany
e-mail: giehoff@corantis.de

support of the patient, medical experts nowadays work with questionnaires for patients and relatives. Therefore, geriatric assessments—like the Barthel Index, the Mini-Mental State Examination or the clock drawing test—were developed [1]. With these tools, the assessment of a person's health state is limited to a short time period while often more detailed data would be much more meaningful. A long term and seamless monitoring after the diagnosis, applied at the home of the patients could extend the possibility for a longer independent living.

The project Mneme aims the development of an integrated system that allows the continuous and unobtrusive monitoring of the health state for dementia patients. Medical staff and caretakers are enabled to review the recorded data and conclude to the patient's state.

1.1 Medical Motivation

A typical course of dementia disease can be subdivided in three phases differentiable by the remaining cognitive abilities (ICD-10) [2]. The transitions between the phases are fluent and are hard to determine. Furthermore the patient's state of health within one phase can be subject to fluctuations (Fig. 1).

It is questionable and difficult to predict how long a patient stays within a certain phase and how much of his cognitive and physical abilities are remaining.

To assess cognitive abilities, standardized manual assessments are used. Within these tests, questions are used to calculate a score to estimate the remaining abilities. Besides others, important test in the field of dementia diagnosis are:

1.1.1 Mini-Mental State Examination [4]

Developed in 1975 by Folstein et al., the MMSE is a widely used screening test to detect cognitive deficits. It consists of nine task-groups, covering questions

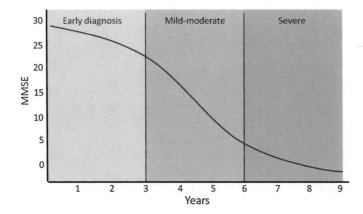

Fig. 1 Typical course of dementia progression [3]

concerning orientation, remembrance capabilities, speech and speech comprehension as well as reading, writing, sketching and calculating. The execution of the test takes about 10–15 min.

1.1.2 Barthel-Scale [5]

The Barthel-scale is used to measure performance in activities of daily life (ADL) and is a systematic way of detecting autonomy or dependencies within daily living. Florence I. Mahoney and Dorothea W. Barthel presented the test in 1965. For each task, a given number of points are assigned. The test results arrange from a minimum of 0 (complete dependency) to a maximum of 100 points (independency).

1.1.3 Clock-Drawing Test [6]

Developed in 1993 by Shulman, the clock-drawing test is a quick screening test to identify problems, which occur in early indications of Alzheimer or other forms of dementia diseases. The patient is asked to draw a clock and a specific time of day. Afterwards a score is determined using the deviations of the drawing.

Utilizing the above tests for diagnosis of dementia disease is a common approach. It is also used to observe the course of the disease. A major disadvantage of this procedure is that these tests can only be performed with presence of a physician. Being in a test situation, under supervision by medical staff, the patient's behavior and answers may differ compared to a normal situation. Furthermore, the execution of questionnaires may not allow a complete observation of all involved symptoms, because the tests are carried out infrequently. Monitoring more symptoms would enable an earlier and better detection of changes within the patient's health state and therefore facilitate improvements in home care.

Adding more detailed data by monitoring patients at their usual environment over a long period of time to support medical decisions regarding the health state of patients as well as needed medication and additional care at home, is the goal of the presented system.

2 Related Work

Ambient Assisted Living (AAL) services are often focusing on monitoring of persons at home. Mainly two kinds of data are interesting: (a) vital signs and (b) behavior patterns. While tele-medical monitoring systems are still a topic of research, there are lots of practical approaches to collect different data.

In Pigadas et al. [7] a continuous monitoring of patient's that are suffering from diseases like dementia, is realized. This is achieved by utilizing a smartphone and further wearable sensors. An algorithm is developed to navigate lost persons to predefined locations. The research presented in [8] focuses on agitation in dementia. A six-DOF accelerometer is used to monitor movement of the person, analyze the data and reason about the behavior. Both of the presented projects have the disadvantage of using inertial sensors that have to be worn on the body, which might be a problem with patients suffering from an advanced dementia disease.

Monitoring persons by utilizing camera-systems to identify activities of daily living within the patient home are used in [9]. Applying camera-based systems for monitoring a patient's behaviors and activities has the advantage of direct visualization of the activity performed. By using current computer vision algorithms like motion detection, motion tracking and object classification, the activity performed can be estimated.

The main drawback of such optical systems is their low acceptance. Participants at research studies often criticize optical methods because of the critical invasion in their privacy. Another optical approach is realized using laser range scanner measurements without utilizing video cameras. In Brell et al. [10] gait velocities are computed by using laser range scanner measurements. This allows an unobtrusive assessment of health related parameters like quality of gait and fall detection.

The active detection of failures in dressing activities is addressed in [11]. The system is able to recognize dressing activities as well as evaluate their quality by using RFID Sensors. Utilizing a layered hidden markov model, the systems high accuracy results without multiple tagging of clothes. RFID sensors have the advantage of being small and relatively low cost and are also used in tagging clothes to enable theft-protection.

In Franco et al. [12] a system is build to detect disruptions in the circadian circle of elderly at home. Therefore, a position-detection is implemented at the patient's flats. To detect the position, passive infrared sensors are used. Using a variant of the Hamming distance enables the measurement of dissimilarity between sequences of activities. The breakdown of activity cycles could be a useful marker for the loss of autonomy of elderly person, especially for patients suffering from dementia disease.

Nevertheless the systems built for chronic diseases like dementia are mainly focus on technical development. To enable a broad use of such systems, the parallel development of a business plan would be necessary. Also most of the systems are only covering one symptom or parameter for measuring. An integrated system that combines multiple sensors could allow a more detailed assessment of the patient's health state.

3 System Design

Subsequently the design of the system is described. Medical requirements have been defined by a consortium of doctors, nurses, caregivers and computer scientist within three workshops. The objectives of the workshops were:

- Determine symptoms, problems within care for dementia patients.
- Define parameters that will be measured at the patient's home.
- Define a meaningful system structure that supports medical stuff in decision making, and assists nursing stuff and relatives of the patients in care.

3.1 Medical Assessments

Table 1 shows the identified disorders that arise within a dementia disease. The symptoms are classified in three categories: (a) conductive disorders, (b) affective disorders and (c) cognitive disorders.

The recognition of one or more of these symptoms may indicate a light dementia disease or a change within the patient's health state.

Conventional tests, as mentioned in Sect. 1.1, are not able to cover all symptoms and aspects of a dementia disease. These tests are often designed to enable a quick and simple first method of diagnosis. They are not designed to monitor the changes of a person's health state and cannot detect some of the symptoms like laziness, changes in daily schedule, etc. without having more detailed data showing the patient's behavior at home.

To expand the data about the patient's behavior in situations of daily life and therefore allow a more detailed view of parameters concerning the health state of affected people, the measurements of the following parameters have been chosen by the medical experts:

Table 1 Identified disorders in dementia diseases

Conductive disorders	
	• Aggression
	• Changes in daily schedule
	• Agitation
	• Laziness
	• Uncleanly
	• Anorexia
	• Dehydration
Affective disorders	
	• Irritability
	• Uncertainty
	• Depression
Cognitive disorders	
	• Disorientation
	• Forgetfulness
	• Memory disorders
	• Lack of criticism

- Weight measurement
- Body fat/lean mass
- Total body water (TBW)
- Position tracking inside the apartment
- Activity measurement
- Sleep quality
- Electricity consumption.

By monitoring these parameters of a person over a longer period of time, the detection of changes in daily life are expected. This allows the involved medical experts to assess the patients conditions and to arrange extra care if needed. Furthermore it allows a more detailed assessment and therefore supports medical decision-making.

3.2 System Architecture

Since the project has an economic aspect, the system design and internal technical structure shall fit to a system-as-service approach. The system consists of four main parts (Fig. 2):

3.2.1 Home Monitoring

For each patient, a set of different sensors is installed at their home. The set will be individually chosen for each patient and will be adjusted to his or her needs.

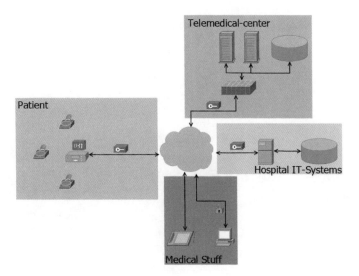

Fig. 2 Four main-parts of the system

Each configuration needs a base-station that locally receives data of the sensor-nodes and transmits the data via a secured Internet connection to the tele-medical center. Making the used sensors exchangeable and to allow a relatively fast and simple integration of new sensors, the OSGi-Framework is used [13]. This framework allows a plug-in-based structure for Java software and is available for free for non-commercial usage. It is easy to integrate new sensors by adding new plug-ins (bundles) into the system. These bundles can be added to the system at runtime, allowing a flexible software development process.

3.2.2 Tele-Medical Center

The tele-medical center is the core component of the system. It is responsible for receiving, processing and storing each patient's collected data. Moreover, the tele-medical center provides a web-service to allow a web-based presentation of data as well as querying data from other services via an API. The implementation of the interface enables a machine-to-machine communication and therefore an easy way to integrate the system in existing software solution like patient records.

To review the patient's data, the involved medical staff needs to log into the system. Therefore, a role system will be implemented that allows showing different views regarding the user's authorities.

Initially an automatic interpretation or processing of the data is not planned. This could be easily achieved by extending the existing server-side software.

3.2.3 Hospital IT-System

Integration of the system for reviewing patient's data in existing hospital IT-systems is also planned for the future. Because of the large variety of heterogeneous software-systems present in hospitals the integration is not straight forward.

Hospital IT-systems are not designed to allow a comprehensive presentation of current and past health related data. To represent a patient's health state over a longer period of time and for different cases, a patient record is needed that can be accessed by multiple systems. Possible examples are electronic health card or electronic patient record systems.

However, the combination with existing medical records is a major advantage and necessity.

3.2.4 Medical Staff

Medical staff is enabled to review the patient's data via a website. To evaluate the patient's health state, doctors as well as caretakers have the possibility to review the collected information in the form of trends and detailed data of certain time-periods.

4 Technical Approach

In this section an initial set of sensors is presented that are able to measure the parameters listed above. Though not all sensors have been implemented yet, this current state of work will be completed in the further development process.

4.1 FS-20

To determine the patient's activity context by assessing mobility and an active lifestyle, home-automation sensors will be installed within the patient's home. To simplify installation and allow a quick development process for the project, existing hardware modules are used. The FS-20 sensors are developed and distributed by ELV, Germany [14].

A variety of different systems exist for home-automation. Since the technology will be installed in inhabited flats, it shall communicate wirelessly and needs to be removable without residues. Besides FS-20, other systems are available e.g. Homeatic [15] or EnOcean [16]. However, the FS-20 system has the advantage of many available types of sensors. Furthermore a Java implementation for receiving sensor-data is available from previous projects [17]. To measure activity in home environments, light barriers, motion detectors and door contacts are used. Events are collected by a base station and transferred to the tele-medical center. The detected events will be processed and presented as abstracted data showing trends of activities rather than raw data.

4.2 Pedometer

A "FitBit Ultra" step counter is used to allow activity measuring outside the flat. It also enables the measurement of the overall distance walked. The sensor is body worn and transmits collected data to an USB-powered adapter wirelessly. The data is transferred to the provider's web-service and is associated with a specific user profile. To access the data and integrate them in the project's web service for reviewing, a Python script is used that collects data each night for the previous day. In addition a job-scheduler is used that is implemented on the tele-medical center servers. This script is receiving all pedometer related data. Besides the step counter this involves also the measurements of the impedance scale described below.

4.3 Impedance Scale

Detecting dehydration or changes in food intake behaviors is important for an extensive assessment of dementia patients. To measure parameters like weight, lean and fat mass as well as total body water (tbw), the FitBit Aria impedance scale is used. This scale utilizes bioelectrical impedance analysis (BIA) to estimate body compositions. Data is transmitted via wireless network to the manufacturer's web-service. Like the step counter, the data is also associated with a specific user profile. The scale itself does not estimate total body water. To calculate the percentage of total body water the Watson formula [18, 19] is used (TBW-W). To calculate the tbw estimation, the formula includes age, height and weight information. It also depends on the gender of the patient, because different algorithms are used for women and men. The formulas are shown below:

Male:

$$TBW - W = 2.447 - (0.09156 * age) + (0.1074 * height) + (0.3362 * weight)$$

Female:

$$TBW - W = -2.097 + (0.1069 * height) + (0.2466 * weight)$$

Other formulas, like the Hume term [18] may be evaluated during the evaluation of the system to compare different approaches.

4.4 Current Sensor

Measuring electricity consumption also allows inferring the activity of patients in their home environment. Furthermore it allows concluding about the context the user is currently in. Small and easy to install sensors are used, developed by the company Plugwise [20], allowing an installation that is easily reversible and can be changed or extended. The identification of activities and user-context is achieved by adding device profiles to the used software system and detect known patterns of the devices electricity consumption. The system has the ability to identify different devices like the toaster, the coffee machine or a water boiler. Together with information about the patient's location and contextual information like time of day etc., the patient's activities, e.g. eating, resting, can be estimated.

4.5 Bed Sensor

To identify parameters regarding the patients sleep quality and enable the medical professionals to determine possible sleep disorders like day-and-night change or how often a person wakes up at night, bed-sensors are deployed. Again, different systems are available, capable to analyze a patient's sleep quality.

Within the project PAALiativ [21], beds are equipped with force sensors to detect cough and behaviors like toss and turn. Since such sensors deliver accurate force measurements, the sensors can also be used to weigh the patient during sleep. This can help to detect a disorder in food intake or a possible dehydration. The disadvantages are costs and a complex installation process due to the sensors type of construction.

A much cheaper solution the recognize context in beds is by using capacitive sensors. In Braun and Heggen [22] small stripes of copper foil are used to measure deformation of the springboard. This approach is limited to detect movement and simple poses. Further, is it not possible to assess the patient's weight. The advantages of these systems are (a) low costs and (b) relatively simple installation that can easily be removed. Within this project different solutions for force measurements in beds will be tested and compared.

To present the collected data to the involved medical staff, a website will be developed. To enable a platform-independent visualization of the data using desktop computers, notebooks or mobile devices like tablets or smart phones, modern web technologies like HTML 5 and JavaScript are used. Figure 3 shows a first prototype of the interface, showing collected activity data.

The server-side is designed and implemented as an API. This allows an easy data access through different consumers like the website created for the medical professionals as well as possible external applications like hospital IT-systems.

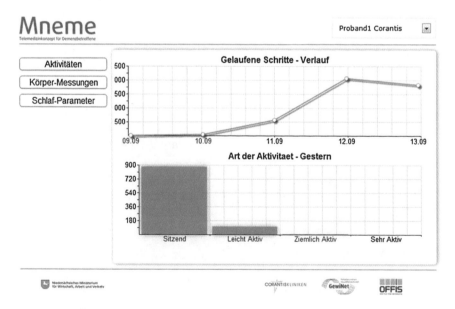

Fig. 3 Prototype for the data presentation

5 Evaluation

To evaluate acceptance, usability and accuracy of some of the involved sensors, a pre-test is carried out at four of the involved hospitals and care facilities. Doctors, nurses and caretakers are enabled to use the sensors by themselves and also with their patients. Involved sensors are a bio-impedance scale[1] and a body-worn pedometer[2] both developed by the FitBit Inc. The web-service that will be developed in this project is not yet used within the pre-test. Instead the manufactures web-service is used to validate the measurements for both systems. Especially the acceptance of the body-worn pedometer is interesting, as this is the most intrusive of the system's components.

Four care facilities have been equipped with both sensors. After 7–14 days of usage at 11 dementia patients, first results are available. Patients are between 66 and 91 years old and 70 % where female. The patients are using the sensors mostly the whole day, without removing it. Also during night-sleep, the sensor is worn. This may change if the patient isn't in a good mood. Some of the patients are feeling responsible for the devices and are taking extra care for them. Some of the patients experienced problems with standing on the weight for a longer time. The caretakers expressed mostly positive about the sensors. The pedometer has the disadvantage, that it's not recognizing very slow walking.

Activities are measured as steps taken. Within the pre-test between 4 and 4,768 steps are measured. Figure 4 shows the steps taken of all patients within the test phase. The distance walked is between 0.02 and 3.56 km. Even on active days, the patient spends most of the time sedentary (746–896 min).

Six nights of sleep could be measured. The patients are wearing the sensor the whole night. The time of sleep is 783 min on average with a minimum of 659 min and maximum of 961 min. To fall asleep, in the sense, that the patient is lying quiet after going to bed, the patients needed 12 min on average. The sensor has counted 20 awakenings during the nights on average.

Figure 4 shows an exemplary course of the movements at night. The measurements are taken every minute and are normalized to a range of 1–3, with 1 equals sleeping and 3 means awake. Longer periods of activity may indicate that a patient is restless at night. Shorter periods could be interpreted as rolling from one side to the other.

Figure 5 shows such long and short movements at night. The first peak starts at 22:45 and ends at 22:51. The second and third peaks are 1 min long and occur at 03:31 and 03:35 (Fig. 6).

The first period of movement can be interpreted as awake, while the second and third are seem to be a short toss and turn behavior.

---

[1] http://www.fitbit.com/de/product/aria.

[2] http://www.fitbit.com/de/product.

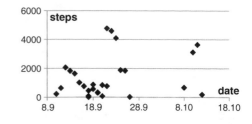

Fig. 4 Steps taken during the pre-test

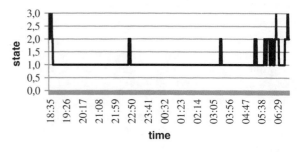

Fig. 5 Measurements of a patient's movement at night

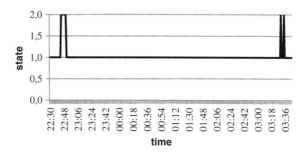

Fig. 6 Longer and short activities at night

The scale has been used eight times within test-period. The measurements range from 75 to 90.5 kg. The body fat varied around 45–46 %. A BMI is calculated between 23 and 31.

Additionally questionnaires were presented to collect data about usability, wearability, etc. on a scale from 1(completely agree) to 6(completely disagree). Some of the data are presented in Fig. 7.

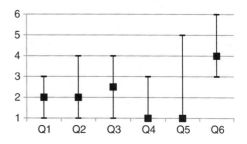

Fig. 7 Results of the questionnaire

Table 2 Criterions for survey participation

Statistical population	Subjects with a diagnosis of mild to moderate dementia
Criterion for inclusion	The subject is… • Able to walk and get up by oneself • Able to communicate verbally • Able to obey instructions regarding the surveys execution • Living alone in his/her own home. Relatives or care takers are visiting the subject regularly
Criterion for exclusion	• Immobility • Serious affective or cognitive disorders. • Serious diseases, that prevent a participation • Presence of domestic animals that move freely in the home

All patients have a positive first impression of the devices (Q1). Handling as well as usability of both devices is rated with a median of 2 and 2.5 (Q2, Q3). The significance of the measured sensor values is rated with a median of 1 and a maximum of 3 (Q4). The acceptance of the device by the patients during the test was positive. Only two of the patients had problems with the scale, because they had problems standing on the scale long enough (Q5). The autonomous use of the devices at the person's home would be impossible for most of the patients (Q6).

After execution of the pre-test, an extensive test will be conducted within a later stage of development of the system. This survey will cover a technical function-test of the whole system as well as further acceptance and usability tests. The participants for the survey are acquired by the involved hospitals and care facilities. Participants need to comply with the given requirements listed in Table 2.

The test subjects are suffering from a light to mild dementia disease and returning from the hospital to their own homes after diagnosis.

The system will be installed within the homes of the test subjects for a time period of 1–2 months. Within this period all measured data are transferred to the tele-medical center, where they are stored and processed. Physicians have access to their patient's data via the web-service developed. The data is presented as trends as well as detailed data for each day. Besides testing the technical functionality of the systems the survey serves as a proof of concept for such an application of monitoring systems. The main goals of the test are a validation of the collected data by medical professionals as well as the usability and user experience of the developed web-service.

The test subjects will be interviewed with reference to the installed sensors at their homes. The focus of the questionnaires is (a) unobtrusiveness and (b) the usability of the body-worn sensor. Further the test subjects will be asked about possible problems that arise during the survey e.g. distracting wires, noises etc. Using standardized usability questionnaires allows a comparable evaluation of the system. Therefore, the IBM Computer Usability Satisfaction Questionnaires [23] or the System Usability Scale by Brooke [24] could be possible solutions.

The data obtained is analyzed with the goal of concluding whether an additional monitoring of patients, suffering from dementia diseases can be recommended.

6 Conclusion and Future Work

This paper has presented a system for unobtrusive and long-term monitoring of patients suffering from dementia disease. The system shall enable medical staff to assess the patient's behavior e.g. changes in activities of daily living, and supports reasoning about the current health state, medications or other medical related decisions. The current state of the art as well as typical assessment test of today has been presented. Furthermore related work in technical development has been shown. Afterwards symptoms of dementia diseases and possible parameters for monitoring the patient at home have been highlighted. Based on these results, a system design has been presented, that covers the involved modules like telemedical-center, hospital IT-systems as well as medical staff. The involved technical solutions are shown in Sect. 4. Sensors used for the system are presented and their possible integration into the system has been illustrated. Furthermore various possible solutions regarding the bed-sensors have been discussed; these sensors are still in test-phase. The envisaged way of evaluation has been illustrated and will start within the later phase of the project. Besides the further development of the system and the integration of FS-20 sensors and current-sensors, the first evaluation will take place as a next step. Also a fourth workshop will be organized to plan and design an appropriate business model for the project. Therefore, the consortium will discuss various possible methods of funding the service for related patients. Also a consultation with health insurances will take place, allowing an early integration of possible financiers.

Acknowledgments The project is sponsored by the Lower Saxony Ministry of Economic Affairs, Employment and Transport in the funding programme "Zukunft und Innovation Niedersachsen—Förderbereich A—Innovationen für ein gesundes Leben".

References

1. Böhmer, F., Füsgen, I.: Geriatrie: Der ältere Patient mit seinen Besonderheiten. Boehlau Verlag, Vienna (2008)
2. Förstl, H.: Demenzen in Theorie Und Praxis. Springer, Berlin Heidelberg (2001)
3. Gauthier, S.: Alzheimer's disease in primary care: Pocketbook: Pocketbook. Taylor & Francis, United Kingdom (1999)
4. Folstein, M.F., Folstein, S.E., McHugh, P.R.: Mini-mental state': a practical method for grading the cognitive state of patients for the clinician. J. Psychiatr. Res. **12**(3), 189–198 (1975
5. Mahoney, Fl.I., Barthel, D.W.: Functional evaluation: the Barthel Index. Maryland State Med. J. **14**, 56–61 (1965)
6. Shulman, K.I., Pushkar Gold, D., Cohen, C.A., Zucchero, C.A.: Clock-drawing and dementia in the community: a longitudinal study. Int. J. Geriatr. Psychiatry **8**(6), 487–496 (1993)
7. Pigadas, V., Doukas, C., Plagianakos, V.P., Maglogiannis, I.: Enabling constant monitoring of chronic patient using Android smart phones. Proceedings of the 4th International Conference on PErvasive Technologies Related to Assistive Environments—PETRA'11, p. 1 (2011)

8. Bankole, A., Anderson, M., Knight, A., Oh, K., Smith-Jackson, T., Hanson, M.A., Barth, A.T., Lach, J.: Continuous, non-invasive assessment of agitation in dementia using inertial body sensors. Proceedings of the 2nd Conference on Wireless Health—WH '11, p. 1 (2011)
9. Mégret, R., Dovgalecs, V., Wannous, H., Karaman, S., Benois-pineau, J., Khoury, E.E., Pinquier, J., Joly, P., André-obrecht, R., Gaëstel, Y., Dartigues, J.: The IMMED project→: wearable video monitoring of people with age dementia. In: Proceedings of the international conference on Multimedia, pp. 1299–1302 (2010)
10. Brell, M., Meyer, J., Frenken, T., Hein, A.: A mobile robot for self-selected gait velocity assessments in assistive environments: A robotic driven approach to bring assistive technologies into established homes
11. Kalimeri, K., Matic, A., and Cappelletti, A., RFID: Recognizing failures in dressing activity. Proceedings of the 4th International ICST Conference on Pervasive Computing Technologies for Healthcare (2010)
12. Franco, C., Demongeot, J., Villemazet, C., Vuillerme, N.: Behavioral telemonitoring of the elderly at home: detection of nycthemeral rhythms drifts from location data. 2010 IEEE 24th International Conference on Advanced Information Networking and Applications Workshops, pp. 759–766 (2010)
13. OSGI Alliance: The OSGi Architecture, (Online). Available http://www.osgi.org/Technology/WhatIsOSGi (2012). Accessed 07 Sep 2012
14. ELV FS20-Funkschaltsystem: (Online). Available http://www.elv.de/fs20-funkschaltsystem.html/refid/SEM_30004/gclid/CIv6l_X8rLICFUJf3godGGMA1A (2012). Accessed 11 Sep 2012
15. EQ-3 Homeatic, (Online). Available http://www.homematic.com/ (2012). Accessed 11 Sep 2012
16. EnOcean: Building Automation, (Online). Available http://www.enocean.com/en/building-automation/ (2012). Accessed 11 Sep 2012
17. Hellrung, N., Ludwig, W., Frenken, T., Lipprandt, M., Steen, E.E., Helmer, A., Veltin,B., Bargen, T., Güvercin, M., Wegel, S., Brell, M., Thoben, W., Steinhagen-Thiessen, E., Haux, R., Hein, A.: Einbettung assistierender Technologien in Gesundheitsnetzwerke—von der Wohnung zum Arzt. In: K.A., Leimeister, J.M. (eds.) Technologiegestüzte Dienstleistungsinnovation in der Gesundheitswirtschaft, pp. 241–262, Gabler Verlag, Shire (2012)
18. Lee, S.W., Song, J. H., Kim, G. A., Lee, K. J., Kim, M.: Assessment of total body water from anthropometry based equations using bioelectrical impedance as reference in Korean adult control and haemodialysis subjects. Nephrol. Dial. Transplant. **16**(1), 91–97 (2001)
19. Watson, P.E., Watson, I.D., Batt, R. D.: Total body water females estimated volumes for adult males and from simple anthropometric. Am. J. Clin. Nutr. **33**, 27–39 (1980)
20. Plugwise: Circle, (Online). Available http://www.plugwise.com/de/idplugtype-f/circle (2012). Accessed 11 Sep 2012
21. Gaefke, C., Baumgartner, H., Brell, M., Simon, S., Hein, A.: System architecture for palliative care in the home environment. In: Wichert, R., Eberhardt, B. (eds.) Ambient Assisted Living, pp. 103–115, Springer, Berlin Heidelberg, (2011)
22. Braun A., Heggen, H.: Context recognition using capacitive sensor arrays in beds. In: Technik für ein selbstbestimmtes Leben (AAL 2012) (2012)
23. Lewis, J.R.: IBM computer usability satisfaction questionnaires: psychometric evaluation and instructions for use. Int. J. Hum. Comput. Interact. **7**(1), 57–78 (1995)
24. Brooke, J.: SUS-A quick and dirty usability scale, Usability Eval. Ind. **189**, 189–194 (1996)

PASSAge: *P*ersonalized Mobility, *A*ssistance and *S*ervice *S*ystems in an *Age*ing Society

Matthias Bähr, Sarah Klein, Stefan Diewald, Claus Haag,
Gebhard Hofstetter, Maher Khoury, Daniel Kurz, Andreas Winkler,
Andrea König, Nadine Holzer, Monika Siegrist, Axel Pressler,
Luis Roalter, Thomas Linner, Matthias Heuberger, Kerstin Wessig,
Matthias Kranz and Thomas Bock

Abstract The demographic change in modern societies has a significant impact on the future planning of self-determined mobility and mobility means. An optimized accessibility of the means of transportation is required, as well as their connection towards buildings and residences. These connections have to be

---

M. Bähr (✉)
Citysax Mobility GmbH, Dresden, Germany
e-mail: info@citysax.com

S. Klein · T. Linner · T. Bock
Technische Universität München, Chair for Building Realization and Robotics, Munich, Germany
e-mail: sarah.klein@br2.ar.tum.de

T. Linner
e-mail: thomas.linner@br2.ar.tum.de

T. Bock
e-mail: thomas.bock@br2.ar.tum.de

S. Diewald · L. Roalter
Technische Universität München, Distributed Multimodal Information Processing Group, Munich, Germany
e-mail: stefan.diewald@tum.de

L. Roalter
e-mail: roalter@tum.de

C. Haag
Haag Rehatechnik GmbH & Co. KG, Kronau, Germany
e-mail: c.haag@haag-rehatechnik.de

G. Hofstetter
Sunrise Medical GmbH & Co. KG, Malsch, Germany
e-mail: g.hofstetter@sunrisemedical.de

M. Khoury
HMM Diagnostics GmbH, Dossenheim, Germany
e-mail: MK@hmm.info

modular and compatible to the mobility means of the users. Barrier-free accessibility according to the existing norms can address these problems only partially. Broader and holistic concepts are needed here. The project *PASSAge* aims at the implementation of seamless mobility chains that smoothly connect private and public space. Mobility shall be ensured by the extension of existing mobility means with user-oriented components. The project follows the approach to complement the barrier-free access and usage of public transportation with mostly electrically powered compact vehicles and micro vehicles. These have to be adapted by physical means and information technology means to residences and building structures. Core of the project is to develop a flexible socio-technical infrastructure with a multitude of mobility means and modular buildings, thus creating synergy effects. An important goal of the approach is the development of business models, which allow for and ensure the allocation and coordination of mobility services. Interfaces will be created for all compact vehicles and micro vehicles that extend their functionality both digitally and physically and thereby enable their ubiquitous connection to the envisioned services.

D. Kurz
metaio GmbH, Munich, Germany
e-mail: daniel.kurz@metaio.com

A. Winkler · A. König · N. Holzer
SOPHIA mit P.S. Südbayern gGmbH, Holzkirchen, Germany
e-mail: andreas.winkler@sophia-suedbayern.de

A. König
e-mail: koenig@sophia-suedbayern.de

N. Holzer
e-mail: nh@nur-holzer.de

M. Siegrist · A. Pressler
Technische Universität München, Department for Prevention, Rehabilitation and Sports Medicine, Munich, Germany
e-mail: siegrist@sport.med.tum.de

A. Pressler
e-mail: pressler@sport.med.tum.de

M. Heuberger · K. Wessig
Ludwig-Maximilians-Universität München, Generation Research Program, Bad Tölz, Germany
e-mail: heuberger@grp.hwz.uni-muenchen.de

K. Wessig
e-mail: wessig@grp.hwz.uni-muenchen.de

M. Kranz
Department of Informatics and Mathematics, Universität Passau, Passau, Germany
e-mail: matthias.kranz@uni-passau.de

1 Motivation and Background

"Everyone wants to grow old, but nobody wants to be old." This saying can be heard quite often. In most cases, it refers to the ailments accompanying the process of growing old. Especially the physical handicaps have a serious negative effect on the mobility of the elderly people. Distances young and healthy people can do in minutes on foot can get an insuperable barrier for the elderly. In rural and small-town contexts, larger distances have often to be traveled to reach, for example, the next clothing shop or specialist. At the same time, public transportation is often underdeveloped in these contexts. That means, for example, the nearest station is too far away for reaching it on foot, or the timetable is very sparse. In order to ensure the self-determined mobility of elderly people seamlessly, it is essential to create seamlessly mobility chains that can be used at any time.

Mobility aids and assistance systems appropriate for the age of the users can heavily contribute to the maintenance of independence of older people. The devices and services can further assist and promote the physical activity [1, 2]. In that way, the physical functions as well as the quality of life and the social participation could be improved. However, these effects can only occur when the aids and systems are accepted and used by the elderly people.

A broad variety of technical products and solutions exists that can contribute towards ensuring the mobility. The examples range from walking aids to electric vehicles. By offering safety, navigation systems and medical alarm systems can also improve the people's mobility situation. However, isolated usage of single aids cannot create a seamless mobility chain. An integral approach is necessary that combines existing aids with information and communication services as well as with health services and mobility services. A special focus has to be on intermodality, i.e. changing between mobility devices has to be ensured. The project "PASSAge—*Personalized Mobility, Assistance and Service Systems in an Ageing Society*" addresses this fact and develops a system for ensuring seamless mobility chains, public transportation, and added value in society as a whole.

2 PASSAge: Intermodality, Mobility and AAL

The project *PASSAge* focuses on ensuring the mobility for elderly people by extending existing mobility means with user-oriented components. Barrier-free access and usage of public transportation is complemented with mostly electrically powered compact vehicles and micro vehicles. Core of the project is to develop an interconnected flexible socio-technical infrastructure with a multitude of mobility means and modular buildings, where the individual elements do not compete but complement each other and thereby create synergy effects. An important goal of the approach is the development of business models, which allow for and ensure the allocation and coordination of mobility services.

2.1 An Innovative Approach

The technical and economic questions are dealt together with medical and nursing users' needs right from the beginning. At the same time, consequences on society, environment and architecture are examined. The inclusion of all related subject areas is an innovative approach in this field.

By involving the target group in design and development of solutions and products, the probability of high demand and user acceptance shall be maximized. The planned modularity and adaptability to the individual user and to environmental conditions allow creating individual mobility supporting solutions. In that way, it can be ensured that each user can maintain her/his self-determined mobility without being limited by unnecessary aids.

2.2 Means of Mobility

Electric mobility is a necessary and (not only) energetically meaningful addition to the existing means of mobility (cf. Nationaler Entwicklungsplan Elektromobilität [3]). The project *PASSAge* wants to exploit this technology for the elderly population in order to strengthen their individual mobility. Due to special mobility equipment, people with disabilities and physical limitations are nowadays able to drive their cars on their own. This special equipment, e.g. manual control units, entrance and loading aids, could partly also be used for supporting elderly individuals with ailments. The ongoing rapid changes in the automobile industry requires constant further development of these mobility aids for vehicles. Especially since the project is focusing on electrically powered compact vehicles and micro vehicles, the aids have to be adapted to the constricted room.

The market demand for mobility aids is constantly increasing, since more and more disabled and elderly people want to maintain their self-determined social life. Especially in rural areas it is important to support the mobility in a way that allows traveling larger distances independently.

In this project, a broad range of mobility means is considered, starting from aids for pedestrians over bicycles to traditional cars. Besides existing models (e.g. walking frames, scooters, electrical bikes), the consideration also includes future devices (e.g. "wearable robots", exoskeletons) which are currently only available as prototypes. However, the focus is on already widely-used mobility means and aids, such as walking frames or wheel chairs.

The rising cost pressure in the field of aids and appliances has intensified the tendency towards developing more modular mobility concepts that can cover a larger range of applications. Interconnectability is another trend on this sector. For example, for electric wheel chairs it is expected that they get more and more connected to the periphery in the future. Especially the usage of mobile devices has revolutionized this area in the last few years. In this project, both concepts, modularity and interconnectability, shall be combined in order to improve the mobility situation for the elderly.

2.3 Information and Communication Technology

For promoting the activity of the users and creating synergy effect in the *PASSAge* mobility chains, the different means of mobility will be interconnected by integrating information and communication technology (ICT) into them. That way, *PASSAge* mobility chain users can not only access a broad range of means of locomotion, but can also access online services, for example, for ordering goods, getting information on medical topics or for leisure activities. Car pooling, or finding and meeting nearby people with similar interests can also be simplified by using ICT.

The development of live-in laboratories [4] as a foundation for later commercialization of intelligent environments and homes for the general public and especially for elderly people allows for seamless and comprehensive ICT support for all individuals. An example is the Fraunhofer inHaus [5]. However, mobility is missing in all these approaches—comprehensive and seamless mobility chains especially outside of large cities, as considered in *PASSAge*, have not been investigated in detail so far. The interconnection is usually realized autonomous, automatically and ad-hoc. It is based on modern communication technology, such as ZigBee, 6LoWPAN, power line communication or dedicated bus systems, such as KNX. The field of application ranges from private homes to large production plants. It is used for automation, ambient assisted living [6], or autonomous production. Today's modern information systems allow presenting location and context based information from various sources that are also connected to the Internet [7]. Parts of this information (e.g. latest news, public transport interruptions, etc.) can also be presented on public displays. Since it is possible to have an Internet connection without the need of wired infrastructure, a broad range of context sensitive applications can be realized. The user is no longer limited to a certain place and new interaction approaches are possible:

- (multi dimensional) bar codes: taking a picture of the bar code allows to get additional information.
- Near field communication (NFC) [8]: bidirectional transmission of information via NFC tags and NFC readers and writers.
- Motion-controlled interaction: sensors in mobile devices create new interaction possibilities.

2.4 Augmented Reality

Augmented reality (AR) is in general a view of the real-world environment augmented with virtual data. As a basic technology, AR can be used wherever three-dimensional and/or additional information could be added.

The diversity and amount of available information as well as the complexity of products and daily processes (e.g. the operation of an ATM) is increasing from day to day. Especially elderly people that are not used to modern devices cannot access information via the Internet and thus cannot benefit from it. Although AR can be seen as a powerful assisting technology, content providers are not yet offering solutions for elderly users. For the establishment and the long-term usability of the *PASSAge* mobility concept it is crucial that the serviceability is suitable for the target group. In order to allow individual shaping of one's life, the user needs support for accessing and comprehending the necessary information. This can, for example, be done by providing a central device that allows accessing the latest information and using the available services. Augmented Reality systems, especially mobile AR applications, show an immense potential for context-sensitive assistance solutions.

AR is, for example, used for furnishing planning. These applications allow the visual evaluation of virtual furniture from different vendors within one's own four walls or in the garden. A similar application could, for example, also be used for planning barrier-free homes. Other AR scenarios that will be examined by the *PASSAge* project team are intuitive AR-supported instruction manuals, AR shopping lists, and an AR navigation system for the elderly users. Virtual information, so called points of interest (POIs), can be integrated in the live camera image and connected to a real reference point. An example for such a system is the AR browser *junaio* [9] developed by *metaio*. This kind of information presentation allows for an intuitive interaction with virtual content and can create a better connection between reality and virtual data.

2.5 HealthPhone

The *HealthPhone*, an all-purpose smartphone, will be the mobile interaction device which allows using the navigation and health system anywhere anytime. In order to help the user monitor her/his health state, it integrates devices and interfaces for measuring vital parameters. The integrated inertial sensors together with the GPS module and the camera will be used by the augmented reality based navigation system. That way, a localization system with a high positioning accuracy can be implemented. Currently available localization systems either need additional devices for reaching a high accuracy, or do not make use of AR for creating an intuitive illustration of the surrounding. This gap shall be closed by the HealthPhone.

An accurate AR-based navigation system allows for reliable orientation, even in unknown environments. In this way, it can reduce the fear of getting lost and can extend the mobility beyond the district. For integrating the different means of mobility in the intelligent mobility chain, they will be extended with digital interfaces. These interfaces can be used, for example, by the HealthPhone's services and apps for acquiring their current state (e.g. battery level) or for controlling them. By using individual independent apps, the HealthPhone can be matched with the user's demands and her/his utilized mobility aids. Advantages of using mass market smartphones over dedicated hardware as hardware basis for the HealthPhone are lower prices, easier integration and many extension possibilities.

2.6 Car Sharing Concept

In contrast to public transport, car sharing can deal with personal needs. The user is not limited to certain lines and stations (except to certain areas the vehicle has to be returned to). It is not to be mixed up with car pooling which is focusing on users sharing the same route where a driver shares her/his car with other passengers while driving to her/his destination.

Car sharing vehicles are normally equipped with a mobile data connection which connects the vehicle with the car sharing provider's infrastructure (e.g. for getting the current location or for reservation and billing). This connection could additionally be used for coupling the system with a central traffic guidance system and a parking lot management system, in order to assist in finding the optimal route and parking space which matches the user's needs. By integrating information from public transportation, e.g. the current timetables and the loading, recommendations for changing the means of transport could be given.

Due to the steadily increasing amount of more and more complex functions in today's cars, human–computer interaction has become a central part of the car development process in the last few years. The integration of mobile devices allows users to control the in-vehicle infotainment systems with their mobile devices' interaction paradigms they are used to [10]. Especially when the means of mobility are changed frequently, as it is the case with car sharing, this can enormously simplify the operation for the user, since the users are more familiar with the handling of their mobile devices than with handling changing systems.

One of the planned *PASSAge* services is a car sharing like system for giving users vehicles equipped with different mobility aids on loan. This can immensely reduce the costs for the individual user and, thus, enable her/him to stay mobile.

3 Methodical Approach

Common aids try to reduce the mental effort for movements and sensory, for example, by the usage of walking frames, fall protectors, or remote controls. In that way, the user of the aid can concentrate on other tasks; the mobility task gets into the background. This can be further optimized by a system that can automatically adapt itself to the user's needs, behavior and current situation.

In order to be able to support a user with such a modern technology, it has to be ensured that the system matches the user's cognitive abilities. This means, among other things, that the operation of an aid must not need more concentration than the user can gain by utilizing the mobility aid. When this is not fulfilled, additional risks can occur. At the same time, the technology may not force the user to completely give up her/his own remaining competences when using the aid. The aids shall only support the user's abilities in order to maintain her/his social life, autonomy and activities of daily living. Hence, the system has to be able to adapt itself to the current situation of the user. In order to create an effective and efficient mobility system, a complementary combination of actuators, sensors, ICT and decision-making components is necessary. The system has to support and/or replace weakened competences that influence the user's mobility negatively.

Four use cases are considered for covering all relevant urban and small-town mobility chains. They refer to different scales: mobility at home, in the district, in the city, and in the surrounding area. An overview of the use cases is depicted in Fig. 1. All four mobility chains are made out by a detailed analysis of existing means of mobility and users' demands. After that, possible solutions for bridging identified gaps are compiled. Three field studies spread over the project's duration will support and ensure the purposeful and target group oriented development. In addition, the experiences of companies working regularly with elderly people will be included. These experiences are also the basis for the development of corresponding business models for the allocation of mobility services and system components. Sound business models are an important element for the success of the project.

In order to establish standardized laboratory conditions, parts of the field studies will be conducted in and around an experimental flat. Intermodality shall not only consider changing between different means of mobility at junctions in public space outside of buildings, but also the mobility in buildings. Especially at the entrance area, a clear discontinuity between indoors and the building's surrounding can be noticed. Besides the physical dimension (e.g. steps, different means of mobility, unloading of goods), there is also a discontinuity at the information technology level (e.g. privacy, display sizes, Internet connection speed). In contrast to currently existent solutions, which are only focusing on single aspects (e.g. Toyota Shopping Car, eTRON Home Delivery Box), the *PASSAge* project is focusing on a comprehensive approach.

Studies have shown that the biggest problem besides traveling longer distances are differences in height. Examples are the common two or three steps in front of

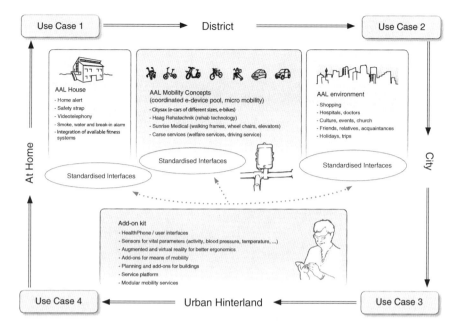

Fig. 1 Four use cases are considered in the *PASSAge* project. They refer to different scales: mobility at home, in the district, and in the surrounding area. Standardized physical and information technology interfaces shall enable a seamless mobility chain for the elderly

buildings, the threshold between rooms, or the different height and gaps between the means of mobility and the pieces of furniture. The mentioned mobility hindrances and many more can be evaluated in and around the experimental flat. An adjacent parking space that can be reached via fliers help in analyzing and optimizing the interface between indoors and the individual means of mobility for longer distances (e.g. electrically driven compact vehicles and micro vehicles).

The group of subjects will be composed of elderly people without physical limitation, elderly people with physical limitations and where applicable people suffering from dementia. Blind individuals will not be part of the group of subjects, but people with limited eyesight can take part. Before taking part in movement experiments the subjects' capabilities will be analyzed (e.g. ability to walk, fall risks, cognitive abilities, etc.). The results from the analysis will be part of the metric for measuring improvements all over the project. For that reason, it would be beneficial when most of the subjects take part in all three field tests. In order to ensure statistical meaningful results, the number of subjects will be chosen high enough so that retirements of subjects can be compensated for. The studies will be conducted with questionnaires, interviews, 3d sensor recording, vital parameter recording, and by observing the subjects in daily situations.

The advantages of the planned solution in comparison to existing approaches can be summarized by the following points:

- An individual device which allows elderly people and users with physical limitations accessing mobility services.
- A platform that provides location and context-sensitive services.
- Technical aids for daily situations.
- Reduced costs due to shared usage of systems, such as adapted means of mobility

The *PASSAge* system's modularity shall allow for adapting the available components to the individual needs of the users. The modularity enables an inexpensive solution which can be smoothly extended in the future.

4 Conclusion and Outlook

The project *PASSAge* aims at safeguarding seamless mobility chains, safeguarding public transportation, as well as safeguarding of the added value related to society as a whole. Mobility shall be ensured by the extension of existing means of mobility with user-oriented components. The project follows the approach to complement the barrier-free access and usage of public transportation with mostly electrically powered compact vehicles and micro vehicles. Core of the project is to develop a flexible socio-technical infrastructure with a multitude of mobility means and modular buildings, thus creating synergy effects. The development of high-tech aids is not only an end in itself, but shall create an aesthetic functional aid that matches the users' needs. An important goal of the approach is the development of business models, which allow for and ensure the allocation and coordination of mobility services.

The multi-functional, interconnected system components are not limited to the main target group of elderly people. They can be used by anyone that needs support in mobility. The whole system is intergenerational and can be upgraded with aids when needed. Based on the comprehensive consideration of the mobility chains, new research fields in the area ambient assisted living (AAL) may be identified which could be a combination of the topics AAL home, AAL city and AAL mobility.

Acknowledgments We gratefully acknowledge the financial support by the German Federal Ministry of Education and Research (BMBF, Förderkennzeichen: 16SV5748) and thank all project partners, their employees and all other contributors. More information on the *PASSAge* project can be found online at http://www.passage-projekt.de/.

References

1. Kranz, M., Spiessl, W., Schmidt, A.: Designing ubiquitous computing systems for sports equipment. In: Proceedings of the 5th International Conference on Pervasive Computing and Communications, PerCom'07, pp. 79–86 (2007). doi:10.1109/PERCOM.2007.12
2. Schutzer, K.A., Graves, B.: Barriers and motivations to exercise in older adults. Prev. Med. **39**(5), 1056–1061 (2004). doi:10.1016/j.ypmed.2004.04.003
3. Bundesregierung, D.: Nationaler Entwicklunsplan Elektromobilität der Bundesregierung (2009)
4. Intille, S.S., Larson, K., Tapia, E.M., Beaudin, J.S., Kaushik, P., Nawyn, J., Rockinson, R.: Using a live-in laboratory for ubiquitous computing research. In: Proceedings of the 4th International Conference on Pervasive Computing, Springer-Verlag, Berlin, Heidelberg, PERVASIVE'06, pp. 349–365, (2006). doi:10.1007/11748625_22
5. Schraft, R., Schaeffer, C., May, T. Care-o-Bot: The concept of a system for assisting elderly or disabled persons in home environments. In: Proceedings of the 24th Annual Conference of the IEEE Industrial Electronics Society, IECON'98, vol. 4, pp. 2476–2481, (1998). doi:10.1109/IECON.1998.724115
6. Roalter, L., Linner, T., Kranz, M., Möller, A., Bock, T.: Robotics for homecare: Auf dem Weg zur Entwicklung maßgeschneiderter Unterstützungssysteme. In: Feuerstein, G., Ritter, W. (eds.) uDay IX—Intelligent Wohnen, Pabst Science Publisher, pp. 70–77 (2011)
7. Kranz, M., Schmidt, A., Holleis, P.: Embedded interaction: Interacting with the internet of things. IEEE Internet Comput. **14**(2), 46–53 (2010). doi:10.1109/MIC.2009.141
8. Broll, G., Siorpaes, S., Rukzio, E., Paolucci, M., Hamard, J., Wagner, M., Schmidt, A.: Supporting mobile service usage through physical mobile interaction. In: Proceedings of the 5th International Conference on Pervasive Computing and Communications, PerCom'07, pp. 262–271 (2007). doi:10.1109/PERCOM.2007.35
9. Madden, L.: Professional augmented reality browsers for smartphones: Programming for Junaio, Layar and Wikitude. Wiley, New York (2010)
10. Diewald, S., Möller, A., Roalter, L., Kranz, M.: Mobile device integration and interaction in the automotive domain. In: AutoNUI: automotive natural user interfaces workshop at the 3rd international conference on automotive user interfaces and interactive vehicular applications (AutomotiveUI 2011)

Part III
Telemonitoring

Facial Movement Dysfunctions: Conceptual Design of a Therapy-Accompanying Training System

Cornelia Dittmar, Joachim Denzler and Horst-Michael Gross

Abstract In this work, we present the scenario of a camera-based training system for patients with dysfunctions of facial muscles. The system is to be deployed accompanying to therapy in a home environment. The aim of the intended application is to support the unsupervised training sessions and to provide feedback. Based on conversations with speech-language therapists and the analysis of existing solutions, we derived a theoretic model that facilitates the conceptual design of such an application. Furthermore, the work is concerned with implementation details, with main focus on the automatisation of the face analysis. We motivate the selection of the features and examine their discriminative power and robustness for the automated recognition of therapeutic facial expressions in a real-world application.

1 Introduction

Facial expressions play an important role in interpersonal communication. Diseases like Parkinson, stroke, or mechanical injury of the facial nerve can lead to a dysfunction of facial muscle movements. The resulting problems are manifold. One consequence of this is that the structure of daily life needs to be adapted to the health impairments. For example, food intake affords more time, if eating and

---

C. Dittmar (✉) · H.-M. Gross
Neuroinformatics and Cognitive Robotics Lab, Ilmenau University of Technology, Ilmenau, Germany
e-mail: cornelia.dittmar@tu-ilmenau.de

H.-M. Gross
e-mail: horst-michael.gross@tu-ilmenau.de

J. Denzler
Computer Vision Group, Friedrich Schiller University, Jena, Germany
e-mail: joachim.denzler@uni-jena.de

swallowing difficulties exist. Patients with impaired eyelid closure need to wear a bandage at night to protect their cornea and the loss of eyelid blink can contribute to drying of the eye. In the long term, this leads to damage of the cornea and may result in blindness. Furthermore, leisure activities like swimming have to be stopped because of the poor corneal protection [1]. Besides implications on daily life and physical abilities, facial muscle dysfunctions can also have negative effects on mental health. Lack of appropriate facial expressions may lead to misunderstandings in face-to-face communication. In combination with impaired appearance of the face caused by imbalance of the facial muscles, low self-confidence and social isolation may be the consequences. In addition to medicinal treatment, the regular practice of therapeutic face exercises under supervision of a speech-language therapist is an important part of rehabilitation. Due to the need for a high practicing frequency, patients need to conduct unsupervised exercises at home—accompanying to therapy. A view in the mirror supports the self-supervised training (Fig. 1). However, the incorrect execution of exercises can impede the training success or even lead to further impairment [2]. The development of technical assistance systems aims to overcome these problems. Such systems can be realized in various forms, e.g., as pure software applications running on a notebook, or as a multifunctional robotic assistance platform. The latter can additionally comprise reminder, communication and training functionalities. Training functionalities aim at improving the patients cognitive [3] and physical [4] state. A therapy-accompanying training system for facial exercises would complete the recent developments of such systems. Against this background, we aim at the development of an automated, therapy-accompanying training system for patients with facial muscle dysfunctions. In this publication, we give an overview of the status of our work with respect to design- and implementation-related tasks. We present a theoretic model, which supports the conceptual design of a training system that is suited to the needs of the target user group. The theoretic model is appropriate for the design of a variety of systems for cognitive and physical stimulation. However, in this paper, we concentrate on the topic of facial exercises. Further emphasis is put on the automation of the training session monitoring. In this context, we motivate the application of the depth features, which we have selected for the specified task. To enable a better understanding for the practical side of this application scenario, we additionally present and examine the features' suitability for a real-world scenario by evaluating their discriminative

Fig. 1 Patients regularly have to conduct unsupervised facial exercises at home in front of a mirror

power and their robustness. A more detailed description and evaluation of the features is given in [5]. The images that are necessary for the analysis of the training sessions are captured using the Kinect from Microsoft (www.xbox.com). Although there are other methods that are suited for the recording of depth information with higher resolution, we decided for the Kinect because of its moderate price and widespread availability.

2 Existing Practical Solutions

In this Section, we give an overview of therapy-accompanying solutions that are already employed for the rehabilitation of facial muscle dysfunctions. We discuss these solutions to identify the main functionalities, which are needed for the design of a comprehensive and automated training system. However, the use of media-technology is slowly evolving in this field. Conventionally, the therapist selects a set of exercises and hands out printed drawings or images as an instruction manual and reminder. The software PhysioTools was developed in order to facilitate and streamline this process (www.theorg.de). It includes a database of various exercises for physical therapy and enables the therapist to compile a set of exercises for a training session. Furthermore, it arranges the images and their associated text in a printer friendly layout. Therapists do not need to search or create descriptive images and to write instructions on their own. However, a video can even be more descriptive because it depicts the process of the exercise execution, instead of the final state only. The software LogoVid comprises demonstrative videos of various exercises that are supplemented by oral instructions (www.logomedien.de). Both mentioned solutions mainly fulfill a tutorial function. The software CoMuZu is supplemented by documentation and feedback functionalities (www.comuzu.de). The target audience are teenagers. As a result, the whole user interface and the story is rather playful in order to give a motivating add-on. The therapist is able to unlock required exercises and in this way design an individual exercise schedule. Instructions for exercise execution are provided in videos. After each training session the teenager is advised to keep a diary about the training with respect to its success and difficulties. Afterwards, the diary can be reviewed by the therapist in order to get an impression of the training performance. However, it is rather impractical and questionable that the patient has to do the evaluation on his own. The three examples show, that current solutions lack an objective and sophisticated feedback function, because the patients have to perform the unsupervised exercises in front of a mirror and evaluate their correctness for themselves. This involves several difficulties. Experience and knowledge of the patient with respect to exercise evaluation may be insufficient, and especially children depend on the support of their parents. In addition, the patient has to concentrate on the execution and evaluation of the exercises simultaneously, which can be very demanding. As a result the patient may lack attention with respect to important details of the exercise. There are studies that indicate that incorrect execution of exercises may lead to an impairment of the facial muscle capabilities [6]. This impairment comprises synkineses that are caused by compensatory motions. Synkineses are involuntary facial movements that accompany voluntary facial movements. For example, a patient may tend to close the eyes to perform an exercise that is physically demanding, e.g., the stretching of the mouth. After a while, this leads to miswiring of nerves and both movements will be involuntarily connected. Further compensatory motions are the raise of the chin, if the patients have to touch the nose with their tongue. A third difficulty is the lack of objective documentation. The evaluation that is made by the patient may be disproportionally optimistic or pessimistic, depending on the current mood. Every person is susceptible to "non-objectiveness", even a therapist. However, patients, who are directly affected by success or failure, might even be more biased in their evaluation because of their mood. More objective feedback is given by biofeedback approaches that employ electromyography to measure the electrical activity of the muscles during practice. This enables the detection of subtle muscle movements that are not visible to the eye. However, the method is more common in earlier states of facial muscle dysfunction, when no movements are visible, and has limited suitability for use in a home environment [1]. Besides the documentation of single practicing sessions it would be helpful to have a solution that enables long-term documentation. The therapist could browse through the exercising history and may identify processes of improvement or impairment, which developed slowly over a larger time span. Other solutions focus on a more playful aspect. The game Mimik Memo is designed for children between 3 and 8 years (www.haba.de). It can be played by two to six children. The game consists of cards that show drawings of animals, which perform facial exercises (e.g., tongue touches the tip of the nose). The task to mimic the exercises is embedded in a game scenario. Concerning the therapy of children with facial dysfunctions, a game scenario adds an important motivational component. Summarizing the above yields four main functionalities which constitute an assistant and comprehensive training system. These functionalities refer to tutorial, feedback, documentation, and motivational aspects that are able to support and enrich exercising. In the next Section, we will discuss these aspects in more detail. Furthermore, we will derive a schematic model that is suited to support future developments of such training systems.

3 A Schematic Model or: What is Lacking in Practical Solutions?

The four aforementioned functionalities roughly coincide with the specifications that we have determined in discussions with speech-language therapists. In the following, we give a detailed description of each functionality and construct a schematic model as a basis for the design and implementation of an automated training system (Fig. 2). The model is suited for various systems of cognitive and

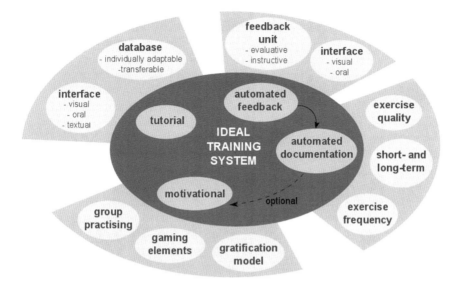

Fig. 2 Schematic model for the conceptual design of an automated training system

physical stimulation, however, we focus on facial exercises. The schematic model facilitates the conceptual work by enabling the identification of beneficial subfunctionalities (outer area of the illustration). The determination of the subfunctionalities is based on the analysis of the presented solutions and the discussions with speech-language therapists.

The design—or exclusion—of each subfunctionality depends on the needs of the target users and the intended price and complexity of the system. The schematic model represents an ideal system. 'Ideal' refers to the inclusion of a comprehensive range of subfunctionalities—a larger range than a real-world application in general may need. The tutorial functionality consists of two elements: a database and an interface. The database provides a collection of therapeutic face exercises, which can be activated by the therapist for each of the patients individually. This allows for the creation of individual training schedules that can be adapted according to the success or failure of preceding training sessions. For each exercise, there is instruction material in form of videos including oral explanations. Important background knowledge can be documented in textual form as well. The interface element of the tutorial functionality visualizes and verbalizes the instructions for the patient. It is important to keep the target users in mind, when designing this interface. While an adult patient may get along with a rather simple video and some textual instructions, a child needs more playful and vivid instructions to keep its attention. Furthermore, some patients may be impaired by additional disease patterns. As mentioned in the introductory Section, possible causes of facial dysfunctions are brain lesions, generated by a stroke. Besides decreased physical abilities, brain lesions can also result in cognitive impairments. One example is the language ability impairment aphasia, which is characterized by difficulties with respect to reading, writing, speech production and speech processing. For persons with decreased speech processing abilities a high amount of visual instructions is essential in order to understand the correct exercise execution. Similar to the tutorial functionality, the feedback functionality consists of two elements as well: a feedback unit and an interface. The feedback unit automatically generates information about mistakes and imprecisions in the exercise execution. Thereby, the mirror is replaced by a camera and the video of the patient doing the exercise is shown on the screen. As mentioned in the introductory Section, exercise evaluation by the patient may be affected by the lack of experience, the emotional state, and the disability to concentrate on the execution and the evaluation at the same time. An automated feedback, however, guarantees results that are more objective and reproducible. The feedback unit provides two sorts of feedback: evaluative and instructive feedback. Evaluative feedback gives a rating of the exercise execution. This rating can vary from a binary rating (good/bad) to a refined scale (0–100 % similarity to the ideal exercise). Additionally, it is possible to realize such a rating for different areas of the face, e.g., mouth or cheeks, separately. The challenge is to find a suitable measure for the assessment of exercise quality. Instructive feedback comprises advice on inaccuracies during practice and gives concrete feedback for improvement ("Puff your left cheek stronger."). Therefore, it is more similar to a real therapist than the evaluative feedback. The interface of the feedback functionality conveys the feedback information in an oral or visual form to the patient. A textual form is less feasible, because patients would have to watch the text and the video of their face simultaneously. Besides the output of evaluative and instructive information, the interface can be used to provide an avatar that synthesizes the face of the patient. The objective of this is twofold: first, an avatar would add a motivational aspect for children, e.g., by enabling children to slip into the role of their favorite comic character. Secondly, a neutral avatar helps patients who are emotionally affected by the impaired appearance of their face and who avoid looking in the mirror. This property of the feedback interface is closely related to the motivational functionality. Detailed information about the conceptual design of the feedback unit will be given in the following Section. The ideal training system additionally comprises a documentation functionality. This functionality is fully automated and focuses on the exercise quality and the exercise frequency. The exercise frequency can be logged to establish a schedule, in which every day of practice is registered, supplemented by the exercise duration. The exercise quality unit comprises the documentation of the exercise success or failure. Retrospectively, the therapist can see which exercises have been performed incorrectly or which have been less difficult for the patient. The automation of the documentation process allows the patient to fully concentrate on the exercise execution during practice. Additionally, no manipulation of the documentation with respect to exercise quality or frequency would be possible. The functionality can be used for short-term documentation, which may comprise information about one single training session or about long-term documentation, which would capture the process over several weeks or even months. The unit that documents the exercise quality needs input from the feedback functionality. Thus, to have a consistent documentation, it is important that the evaluation tool gives objective and reproducible results. The motivational functionality contains elements that motivate the patients to do the practicing sessions with a regular frequency and with certain accuracy. The design of the motivational functionality depends on the target audience. Although one may assume that the inclusion of gaming elements is mainly beneficial for the motivation of children, studies showed a positive impact on the motivation of adults as well, when, e.g., using Wii sports [7, 8] (www.nintendo.co.uk). Furthermore, the integration of the documented training success, e.g. in form of high-score lists, may motivate the patient to practice with a higher frequency in order to exceed earlier performances. Additionally, some extra functionalities may be unlocked, if a patient achieves a further level, which may also enlarge the motivation. Group work may also be more motivating, e.g., as intended with the Mimik Memo game. However, in case of an application that is planned to be highly adaptable to the needs of an individual patient, practicing in groups may enlarge the complexity of system development. Summarizing the above, we think that the feedback functionality plays an essential role because it contributes to the construction of a consistent documentation and the documented success, on the other hand, can be integrated into the motivational functionality. Therefore, in the following Sections, we focus on the embedding of the feedback unit into the training system and discuss and evaluate the automation of the feedback process.

4 Conceptual Design and Details of the Automatic Training System

In this Section, we focus on the conceptual and implementation-related aspects of the training system. First, we provide a schematic overview of the process steps comprised by the system. Additionally, we describe the collection of test images, the selection of features and the choice of the camera type. Finally, we present our preliminary results and status on the way to the solution of this extensive task.

4.1 Overview of the Training System

In the following, we examine the embedding of the feedback unit in the process of automated feedback generation. As shown in Fig. 3, the training system is divided into three layers: the human actions, the interface and the algorithm. The layer on the top comprises the actions of the patient. Via an input interface, such as a camera, the algorithmic layer receives an image or a video of these actions. The algorithmic layer is the basis of the automated feedback and consists of two units:

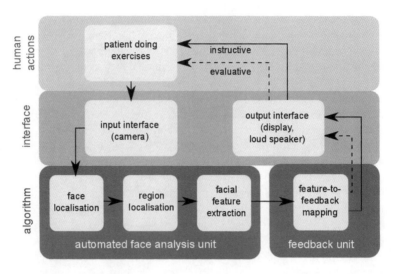

Fig. 3 Embedding of the feedback unit in the process of automated feedback generation

the automated face analysis unit and the feedback unit. The task of these units is to analyze the appearance of the face in order to derive information about the training performance. The properties of the face are captured by the extraction of descriptive features from facial regions. As a result, in each image of the data stream, the face has to be localized and distinctive facial points (e.g., the nose tip) and regions (e.g., the cheeks) have to be detected. The extracted features are analyzed automatically in order to generate evaluative and instructive feedback. The feedback is forwarded to the output interface, e.g., a display or a speech synthesis (or both). The instructive feedback comprises information about necessary changes in exercise execution and, therefore, directly affects the actions of the patient. Evaluative feedback only comprises an assessment of the exercising quality. However, we assume that a negative evaluation of the training will also affect the actions of the patient.

To be more precise, we can say, that the described scenario does not involve a face-appearance-to-feedback mapping but rather a feature-to-feedback mapping. However, features only describe a part of the face properties. Thus, an important question for the selection of the features is, whether they are suited to represent the properties of the face and the quality of the exercise. If the feedback that is given by the training system does not correlate to the feedback of a therapist, then there are two main possibilities: the mapping of the describing features to the feedback is incorrect or the features are not suited to represent the appearance of the face. In order to reduce the probability for the latter, the features need to be examined more closely (left image of Fig. 4). The first question is, whether the features are suited to separate the different exercises. If the features are not able to capture the characteristics that distinguish the different exercises then it is unlikely that the features are able to describe the more detailed differences that are necessary to

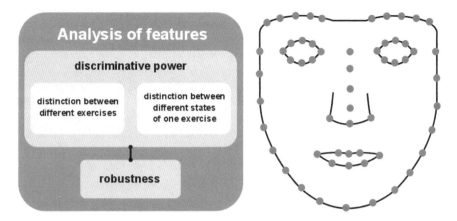

Fig. 4 *Left image* Analysis of features comprises the evaluation of the discriminative power and the robustness. The discriminative power consists of the features abilities to discriminate between the different exercises and the different states of one exercise. *Right image* 58 manually labeled landmarks

characterize a correct or an incorrect exercise execution. The second question is, in how much detail the features are able to describe different states of an exercise: How do the feature values change if a face expression changes from a neutral state to the final state of the exercise? The third question refers to the robustness of the features. In a real-world application it is not feasible to localize the position of the points and regions for feature extraction manually. As a result, they have to be detected automatically. However, automated labeling is less accurate than manual positioning. Thus, we need to evaluate the robustness of the features with respect to varying regions of feature extraction. The performance of the features— extracted from manually labeled points and regions—must be compared to the performance of features extracted from automatically detected areas. The right image of Fig. 4 shows the 58 manually labeled landmarks used in our approach. For the automated labeling of these landmarks, we train an Active Appearance Model (AAM) [9, 10]. AAMs have various applications in the area of object detection. Commonly, they are employed for the classification of facial expressions. In this work, however, we apply them for finding and placing the landmarks. The nose tip is detected robustly by a threshold-based localization algorithm using curvature analysis [11]. This approach is more accurate for the nose tip detection than the solution found by the AAMs, however, it is not suited for landmarks that lie in areas with less characteristic and changing surface shape, as for example the corners of the mouth or points on the cheek. In the following, we motivate the selection of depth features as robust descriptors of the landmarks and evaluate their discriminative power with respect to the distinction between different exercises. Furthermore, we compare the results for manually and automatically labeled regions. Prior to that, we will have a closer look on the exercises to be included into the training system that is currently developed. These exercises are the basis to define, which regions of the face need to be localized and analyzed more closely.

4.2 Therapeutic Face Exercises

In cooperation with speech therapists, we selected a set of nine therapeutic face exercises by certain criteria (Fig. 5).

The first criterion was the ability to transfer the exercises to various disease patterns because speech-language therapy is geared towards people with various facial movement dysfunctions. Facial palsy for example comprises a reduced ability as well as the total inability to move facial muscles [12]. It can be caused by brain lesions or mechanical injury of the facial nerve. Another result of brain lesions can be dysarthria, which results in speech disorders and articulation problems. A further disease pattern is the myofunctional disorder, which is caused by an imbalance of facial muscle strengths, and often affects children [12]. Typical symptoms are a constantly opened mouth and an incorrect swallowing pattern. The exercises that we selected are beneficial for each of these disease patterns. Additionally the exercises should train several face regions: the lips, the cheeks and the tongue. Each exercise has to be retained for around 2 or 3 seconds. The speed of the performance is not important. Therapeutic tools like spoons and spatulas, as well as movements of the head, e.g., moving the chin to the chest,

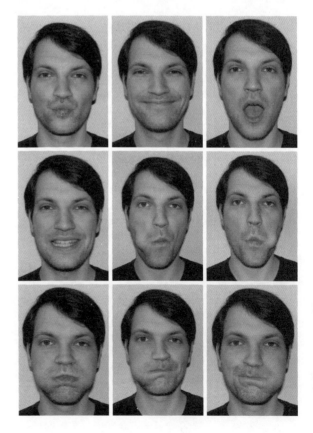

Fig. 5 Exercises that have been selected in cooperation with speech therapists (from *left to right* and *top to bottom*): pursed lips, taut lips, A-shape, I-shape, cheek poking (*right/left side*), cheeks puffed (*both/right/left side*(s)). Exercises are performed by a person without facial movement dysfunctions for better visualization

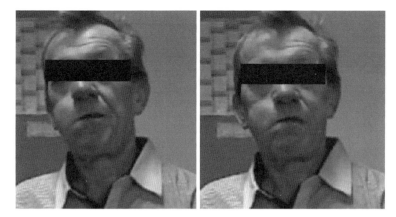

Fig. 6 Patient with facial paresis on his right side. *Left image* The exercise right cheek puffed is conducted correctly, because the bulge of the cheek is a passive process as reaction of a higher air pressure inside the mouth and a contraction of the buccinators on the left facial side. *Right image* The exercise left cheek puffed is conducted incorrectly. The lack of contraction in the right buccinators leads to the bulge of the right cheek

should be avoided in order to prevent occlusions. Occlusions lead to missing information, which would necessitate more cameras for observing the patient. However, we constrain the number of cameras to a frontal one to reduce hardware costs, which is important in order to guarantee widespread use of such a system. Additionally, the complexity of camera calibration is reduced. The selected exercises are easy to practice and build a set of sub-exercises that can be combined to more complex and dynamic series of exercises: As an example, the alteration between pursed and taut lips or pursed lips and a neutral face are possible. Due to the lack of a public database that comprises facial exercises, we collected our own dataset which will be made available as soon as possible. It contains eleven persons conducting the nine exercises. For each exercise, there are around seven images showing different states throughout exercise execution. This amounts in a total size of 696 images in the dataset. For the following tests, we only employ image data that show healthy persons doing the exercises. Because our main focus in this paper is the selection and evaluation of the features, we want to eliminate other sources of error. Thus, we omit data recorded from persons with dysfunction of facial expressions, as we expect their ground-truth to be ill-defined. This is due to the circumstance, that an incorrect execution of an exercise may resemble other exercises (Fig. 6).

4.3 Choice of Suitable Features

Looking at the example images of Fig. 5 reveals that the execution of the exercises has strong and manifold impact on the facial surface. Whereas the exercises pursed lips and A-shape lead to a rather concave cheek surface, the other exercises

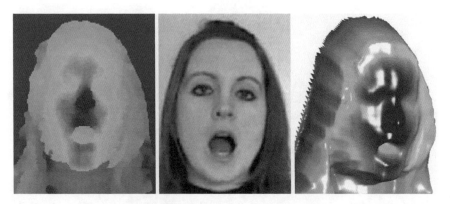

Fig. 7 2.5D image, its corresponding color image and the generated 3D point cloud. For better visualization, the point cloud is shaded using Gouraud's method. The depth information in the 2.5D image is represented by an iterative gray-value scale

produce a convex curvature. But even the convex surfaces are manifold. The cheek boxing exercise results in a rather steep and local bulge, whereas the cheeks puffed exercise causes a more global and smooth bulge. Exercises with a wide mouth, like taut lips and I-shape, produce small wrinkles. However, the magnitude of the surface bulge differs between individuals because it depends on the face type (e.g., full versus slim). Nevertheless, the shape of the face surface is a reasonable property to separate the different appearances of the face as shown in earlier works of [13], and [14]. In total, we use three depth feature types that analyze the shape of the surface: curvature type histograms, point signatures and line profiles. They will be discussed more detailed in the following Sections. To capture depth data, we use a Kinect camera. The camera outputs 2.5D depth images. These are two-dimensional images—similar to a gray-value image—that contain object-to-camera distance information in each pixel instead of intensity information. In addition to the depth image, the Kinect simultaneously captures a color image. Via camera calibration, the intrinsic and extrinsic camera parameters can be determined [15]. Using the information of the 2.5D depth image, these can be employed to generate a 3D point cloud. Figure 7 shows a 2.5D depth image, the corresponding color image and a 3D point cloud.

4.3.1 Curvature Type Histograms

We determine the curvature type for each pixel of the face [11, 16]. The curvature type contains information about the surface that is surrounding the pixel. This information comprises the direction of the surface curvature (convex, concave) and its shape (hyperbolic, cylindric, and elliptic). There are eight different types of curvature. The left image of Fig. 8 shows four examples. In an image, the face is represented by 8.000–13.000 pixels. If—for each pixel and its neighborhood—the

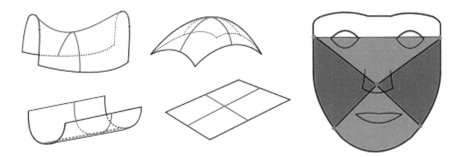

Fig. 8 *Left image* Examples of curvature types. *Top left* hyperbolic convex, *top right* elliptic convex, *bottom left* cylindric concave, *bottom right* planar. *Right image* Four regions that are used for feature extraction

curvature type is determined this results in a feature vector with a length similar to the number of pixels. In order to reduce the dimension of the feature vector, we summarize the curvature values with a histogram. To maintain spatial information, we define several facial regions from which separate histograms are extracted. Here, our approach follows the work of [13], who focus on the classification of six facial expressions. They divide the face into seven regions (e.g., chin, lower cheek, upper cheek) and summarize the curvature types with histograms. In their dataset that is used for testing purposes regions for feature extraction were localized manually by humans. In contrast, we detect the regions automatically, which is less accurate than manual labeling. As a result, we have reduced the number of regions from seven to four in order to increase the size of each region (right image of Fig. 8). This decreases the influence of small variations of the region border locations, but also decreases the accuracy. The borders of each of the four regions are determined by connecting fiducial points of the face. To enable a stable detection of the regions it is important that the fiducial points can be localized easily. Suitable positions lie in distinctive areas of the face that are only slightly influenced by changes of the face surface. This enables a good detection of the same point in different images. In Fig. 9, we show examples for the distribution of curvature types in the left cheek region for two different facial expressions. The curvature types are represented by different colors.

4.3.2 Point Signatures

In [14] point signatures are employed for the recognition of faces. We adapt this approach for our task of therapeutic face exercise classification. Similar to curvatures, the idea of point signatures is to describe the properties of the surface shape. Point signatures capture the slope of a path that runs around a distinctive point in the face to describe the neighborhood of this point. We selected the nose tip as centre point because it can be detected more robustly. The point signature is calculated as follows: A sphere is centered into the nose tip. The intersection of the

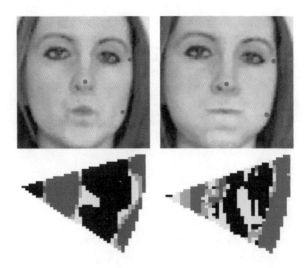

Fig. 9 *Top* Person performing the exercises pursed lips and both cheeks puffed. The points mark the corners of the left cheek region of the person. *Bottom* Detail view for the left cheek area, showing the curvature types represented by gray values (from dark gray to bright gray: elliptic convex, elliptic concave, hyperbolic concave, hyperbolic convex). As expected, the cheek has a large amount of elliptic convex area

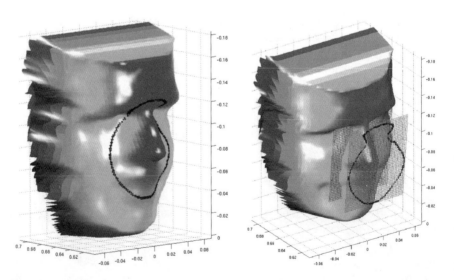

Fig. 10 *Left image* Intersection path of the 3D face point cloud and a sphere. *Right image* Plane fitted in the intersection points and the displaced plane that is used for distance calculation. The curve on the plane is the projection from the intersection curve. The distance between these two curves is sampled in an interval of 15°

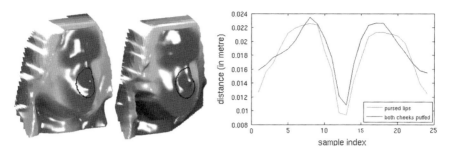

Fig. 11 *Left image* Person doing the exercises pursed lips and both cheeks puffed. Point signature paths (for f = 0.5) are marked on both faces. *Right image* Curves showing the slopes of the paths. Each curve consists of 24 samples (360°:15° = 24). The sample index multiplied with the sampling interval (15°) results in the size of the angle, starting from the point on the path that is intersected by an imaginary connection of the nose tip to a point on the center of the chin. The center and global minimum of the curve represents the root of the nose, which has the smallest distance to the displaced plane. At the beginning and the end of the curves it can be seen that the distance of the lips to the displaced plane is smaller for the exercise pursed lips than for the exercise both cheeks puffed

sphere with the facial surface creates the path (left image of Fig. 10). To capture the slope of the path, the distance information of the points that lie on the path needs to be sampled. Depending on the position of the person to the camera, the absolute distance values vary, although the face may be identical. To obtain a distance measure relative to the position of the person, we fit a plane into the intersection points and displace this plane along its position vector until it goes through the tip of the nose (right image of Fig. 10). The distance of the curve to the displaced plane is now sampled in regular steps of 15°. The slope of the path can be visualized by a coordinate system with the axes 'degree' and 'depth distance' (Fig. 11). The size of the radius is determined by multiplying the distance between the eyes with a factor f. We use the following values for f: 0.4, 0.5, 0.7, 0.8 and 1.0. As a result, we get five point signatures with a length of 24 samples each. To reduce the length of the resulting feature vector, we apply a discrete cosine transform (DCT) on each point signature and retain the first twelve coefficients [17].

4.3.3 Line Profiles

We developed the line profiles on the basis of the point signatures. Whereas a point signature consists of a path that runs radially around a point, a line profile connects two landmark points. We selected line profile paths that comprise the cheeks and the mouth because these regions show characteristic changes if a face performs facial exercises. The paths can be seen in Fig. 12. Seven paths run from the tip of the nose to silhouette landmark points. The two remaining paths connect silhouette points. A path consists of N equidistant points in a three-dimensional space.

Fig. 12 3D face with the marked paths of the 9 line profiles curves

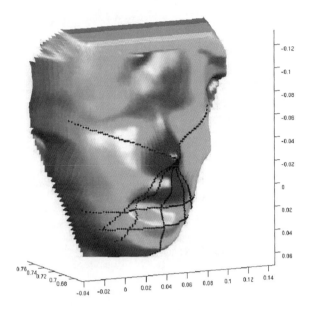

To obtain a representation of the path, which is invariant with respect to translation and rotation operations of the face, we need to extract relative distance values. Therefore, we extract the Euclidean distances between the points. The number of points per path depends on the size of the face and the executed exercise. To get a constant length and to reduce the length of the feature vector, again we apply a discrete cosine transform and retain the first twelve DCT coefficients.

4.4 Feature Evaluation: Results

In the preceding Sections, we introduced three feature types that comprise information about the surface of a face:

- curvature type histograms
- point signatures
- line profiles.

In the following, the features are examined with respect to their ability to discriminate between the executed nine therapeutic exercises. We assess the quality of this ability by the average recognition accuracy, which describes the ratio of the correctly detected exercises to the total number of exercises. According to the number of images in our dataset the total number of exercises is 696. Feature types are evaluated individually and in combination. As mentioned in Sect. 4.1, several steps are necessary for the evaluation of the features' suitability for the

Fig. 13 Results for the single feature types and their combination. The bars in dark gray represent the results for the features that are extracted from manually labeled regions. The bars in light gray show the results for the features extracted from automatically detected regions

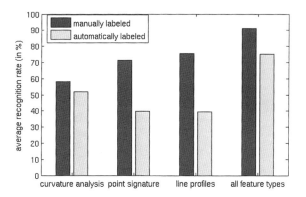

planned scenario. We concentrate on two of these aspects. First, we evaluate the features that were extracted from manually labeled regions in order to exclude other influences like deviating region borders. Second, we evaluate the features that were extracted from automatically determined regions using an AAM and a curvature-based nose tip detection. Training and classification is performed by applying linear Support Vector Machines [18]. The dataset was split up into training and test set using the leave-one-out cross-validation. Additionally, all images of a person that is present in the test image are excluded from the training set. This approach is consistent with the mentioned application scenario in which the images of the test person will not be part of the training data. The number of feature dimensions was reduced from 232 to 8 by using a Linear Discriminant Analysis [19]. If features are extracted from manually labeled regions, line profiles perform better than the other features. However, the best result is obtained by the combination of the three types (Fig. 13). This results in an average recognition accuracy of 91.2 %. The performance of the curvature type histograms is rather low compared to the other two feature types. However, curvature type histograms outperform point signatures and line profiles if automatically detected regions are used. This is due to the fact that curvature features extract information from larger regions than point signatures and line profiles. As a result, the curvature type histogram is more robust against small variations of the region borders. Again, the combination of the three features leads to the best performance and results in an average recognition rate of 75.1 %. This result confirms the suitability of the features for the classification of the presented therapeutic facial exercises, even in an automated scenario. The deviations of the landmarks, determined by the AAM, compared to the position of the manually labeled landmarks were -1.9 pixels (mean value) in x-direction with a standard deviation of 4.7 pixels. In y-direction the mean value of the deviation was 6.0 pixels with a standard deviation of 15.9 pixels. Considering the distances of the persons to the camera, 6 pixels correspond to about 0.95 cm on the face.

5 Conclusion and Future Work

In this publication, we presented our state-of-work for the development of an automated, therapy-accompanying training system. On the basis of existing therapy solutions and conversations with speech-language therapists, we derived a theoretical model that supports the conceptual design and implementation of such a system. Furthermore, we presented nine facial exercises, which were—in cooperation with therapists—determined as beneficial for the therapy of facial dysfunctions. On the basis of the selected exercises, we collected and manually labeled a dataset that comprises 696 depth images with their corresponding color images. This dataset was used to evaluate features that are the fundament for the implementation of an automated face analysis unit. The features were examined in two respects. First, we evaluated their discriminative power concerning the classification of different exercises. Second, we tested their robustness regarding varying locations of feature extraction. The latter is relevant to determine, whether these features are suitable for a real-world application. Future work will be focused on the evaluation of the features' suitability for the separation of different states of an exercise. Furthermore, we will examine the mapping of the feature values to an evaluative and instructive feedback scale.

We would like to thank the m&i Fachklinik Bad Liebenstein (in particular Prof. Dr. med. Gustav Pfeiffer, Eva Schillikowski) and Logopädische Praxis Irina Stangenberger, who supported our work by giving valuable insights into rehabilitation and speech-language therapy requirements and praxis. This work is partially funded by the TMBWK ProExzellenz initiative, Graduate School on Image Processing and Image Interpretation.

References

1. Brach, J., VanSwearingen. J.M.: Physical therapy for facial paralysis: A tailored treatment approach. Phys. Ther.: J. Am. Phys. Ther. Assoc. **79**(4), 397–404 (1999)
2. Wolowski, A.: Fehlregenerationen des Nervus facialis—ein vernachlässigtes Krankheitsbild. Dissertation. Universität Münster (2005)
3. Gross, H.-M., Schroeter, C., Mueller, S., Volkhardt, M., Einhorn, E., Bley, A., Langner, T., Merten, M., Huijnen, C., van den Heuvel, H., van Berlo, A.: Further progress towards a home robot companion for people with mild cognitive impairment. In: Proceedings of the IEEE international conference on systems, man, and cybernetic, Korea, Seoul, 637–644 (2012)
4. Geue, P.-O., Scheidig, A., Kessler, J., Gross, H.-M.: Entwicklung eines robotischen Bewegungsassistenten für den Langzeiteinsatz zur physischen Aktivierung von Senioren. In: Proceedings of Ambient Assisted Living Kongress, p. 5. Berlin, Germany, (2012)
5. Lanz, C., Denzler, J., Gross, H.-M.: Automated classification of therapeutic face exercises using the kinect. In: Proceedings of the International Conference on Computer Vision Theory and Application, Barcelona, Spain, pp. 556–565 (2013)
6. Shiau, J., Segal, B., Danys, I., Freedman, R., Scott, S.: Long-term effects in neuromuscular rehabilitation of chronic facial paralysis. J. Otolaryngol. **24**(4), 217–220 (1995)

7. Halton, J.: Virtual rehabilitation with video games: a new frontier for occupational therapy. Occup. Ther. Now 12–14 (2008)
8. John, M., Häusler, B., Frenzel, M., Klose, S., Ernst, T.: Rehabilitation im häuslichen Umfeld mit der Wii Fit—Eine empirische Studie. In: Proceedings of the Ambient Assisted Living. Berlin, Germany (2009)
9. Cootes, T., Edwards, G., Taylor, C.: Active appearance models. IEEE Trans. Pattern Anal. Mach. Intell. **23**(6), 681–685 (2001)
10. Stricker, R., Martin, C., Gross, H.-M.: Increasing the robustness of 2D active appearance models for real-world application. In: Proceedings of the IEEE international conference on computer vision systems, Liege, Belgium, pp. 364–373 (2009)
11. Colombo, A., Cusano, C., Schettini, R.: 3D face detection using curvature analysis. Pattern Recogn. **39**(3), 444–455 (2006)
12. Gordon-Brannan, M.E.: Clinical Management of Articulatory and Phonologic Disorders. Lippnicot Williams and Wilkins (2007)
13. Wang, J., Yin, L., Wei, X., Sun, Y.: 3D facial expression recognition based on primitive surface feature distribution. In: Proceedings of the international conference on computer vision and pattern recognition, pp. 1399–1406 (2006)
14. Wang, Y., Chua, C.S., Ho, Y.K.: Facial feature detection and face recognition from 2D and 3D images. Pattern Recogn. Lett. **23**, 1191–1202 (2002)
15. Hartley, R., Zisserman, A.: Multiple View Geometry in Computer Vision. Cambridge University Press (2004)
16. Besl, P., Jain, R.: Invariant surface characteristics for 3D object recognition in range images. Comput. Vis., Graph., Image Process. **33**(1), 33–80 (1986)
17. Salomon, D.: Data Compression: The Complete Reference. Springer, New York (2011)
18. Chang, C.-C., Lin, C.-J.: LIBSVM: A library for support vector machines. ACM Trans. Intell. Syst. Technol. 1–27 (2011)
19. Webb, A., Copsey, K., Cawley, G.: Statistical pattern recognition. Wiley, New York (2011)

Detecting Activities of Daily Living with Smart Meters

Jana Clement, Joern Ploennigs and Klaus Kabitzsch

Abstract Smart meters provide us new information to visualize, analyze, and optimize the energy consumption of buildings, to enable demand-response optimizations, and to identify the usage of appliances. They also can be used to help older people to stay longer independent in their homes by detecting their activity and their behavior models to ensure their healthy level. This paper reflects methods that can be used to analyze smart meter data to monitor human behavior in single apartments. Two approaches are explained in detail. The Semi-Markov-Model (SMM) is used to train and detect individual habits by analyzing the SMM to find unique structures representing habits. A distribution of the most possible executed activity (PADL) will be calculated to allow an evaluation of the currently executed activity (ADL) of the inhabitant. The second approach introduces an impulse based method that also allows the detection of ADLs and focuses on temporal analysis of parallel ADLs. Both methods are based on smart meter events describing which home appliance was switched. Thus, this paper will also give an overview of popular strategies to detect switching events on electricity consumption data.

J. Clement (✉) · K. Kabitzsch
Dresden University of Technology, Dresden, Germany
e-mail: jana.clement@tu-dresden.de

K. Kabitzsch
e-mail: klaus.kabitzsch@tu-dresden.de

J. Ploennigs
IBM Research, Belfast, Ireland
e-mail: joern.ploennigs@ie.ibm.com

1 Introduction

Smart meters and smart plugs enable the simple monitoring of a building's energy consumption down to submeters and individual devices. The data can be used to visualize and analyze the energy consumption which saves energy by providing early feedback. They can be used for demand-response management. They enable the identification of individual appliances. But, besides saving energy and costs by feedback and demand-response, smart meters can also help to face another important challenge of the future—the demographical change.

The demographical change all over the world creates new needs and challenges for technical innovations. Older people have a strong desire do stay self-determined in their own well-known homes. Unfortunately, they suffer from normal age-related, physical degeneration and are exposed to a higher fall risk and diseases that reduce their abilities in everyday life. Even if they are not directly affected, elderly and relatives are often burdened by the fear that such things may happen if the elderly is alone and nobody detects it. Technical inventions can help these people and support the home care and increase their quality of life.

Several research projects are working on that particular topic of *Ambient Assisted Living* (*AAL*) to learn and monitor the human behavior inside a residential flat to raise an alarm if health critical situations have been detected. In general the approaches for *Human Activity Detection* (*HAR*) can be classified in detection of inactivity and the detection of Activities of Daily Living. *Inactivity detection* is a very basic approach of identifying cases that people are not active for an unusual time [4, 9]. This addresses the use case of people that fell or lie down with sickness. The detection of *Activities of Daily Living* (*ADL*) needs more advanced approaches that identify human behavior patterns [14, 15]. ADLs are common activities a human is performing in his home, for example: nutrition, grooming, or using the toilet [13]. Detecting ADLs allows doctors and care givers to understand people's diurnal rhythm, it improves inactivity detection, and changes in ADLs indicate development of diseases such as Alzheimer's [17]. But, common approaches require to equip the flats or the persons with multiple, dedicated sensors. This makes the solutions very cost intensive and many elderly are not able to pay for them. The systems also add a stigmatization problem, as the sensors indicate and remind the people and others of problems.

This paper focuses on a low cost approach to monitor human behavior in single households by the usage of smart meters. Smart meters are getting common and their installation are promoted in the US and Europe by laws [25]. In combination with energy and demand-response management they allow new business models for combined ambient assisted living with energy management. This paper will explain the methods that can be used to analyze smart meter data to create a human behavior model. Two promising approaches will be introduced and evaluated on real measurements.

The next section gives an overview of popular monitoring methods based on smart meter consumption. Section 3 introduces two monitoring approaches: the

Semi-Markov-Model and the impulse function using the human forgetting curve. Both methods are compared in Sect. 3.4, which also refers to future work. The paper ends with the Conclusion in Sect. 4.

2 State of Art

Identifying the appliances used is a concept known as *Non-Intrusive Load Monitoring* (*NILM*) developed in the 80th [11]. Since then, smaller, embedded and wireless meter devices developed [12, 22] that are using indirect measurements [23] and improved data mining approaches [6, 18]. This allows even to identify the TV channel watched [10]. We are using a common smart electricity meter from an industrial partner with a modified frequency spectrum analysis based on [20]. It is complemented by a smart water meter which runs a similar algorithm that identifies the usage of basins, tubs, toilets, and washing machines based on the water flow rate and water volume taken.

Any of these smart meter approaches can be used for HAR as long as they provide a labeled event streams about the appliances turned on and off from the smart meters. The detection approaches are also not restricted to smart meters as any sensor information that relates to human activities can be used such as floor sensors, occupancy sensors, audio or video processing. Out of the reasons given above, we are focusing on smart meter.

The simplest form of HAR is *inactivity detection* by calculating the time intervals of incoming events and to raise an alarm if the inter-arrival time is above a threshold. The limitation of these simple approaches lies in the fact that the threshold has to be very high to ignore normal long inter-arrival times and allows no timely reaction to inactivity. Overnight, for example, it is ok if the people are sleeping and no activity is detected for several hours. In contrary, it is an indicator of insomnia, incontinency, and Alzheimer's if people do rise up several times in the night. Therefore, advanced approaches train inactivity profiles for each day and differentiate day and night periods [4, 9] and allow detection times below 3 h. Inactivity diagrams, like the one in Fig. 1, are used to visualize the inactivity detection. They accumulate the time where no events occur caused by the inhabitant's actions. The red line in Fig. 1 shows accumulated inactivity times trained over 196 days. The green and yellow lines are used to automatically detect the day- and night phase. The blue line is a calculated polynomial function based on the red line and is used as threshold. If the current inactivity exceeds the blue line, an alarm message can be sent to care givers.

Labeled event stream from smart meters allows more precise models of the human activity. A general clustering approach for sensor events was demonstrated in [1]. The author of [3] used a Semi-Markov-Model (SMM) to model the human movement based on occupancy sensors placed in each room of the apartment. These approaches correctly model the correlations of events, but are not able to abstract higher semantic meanings in form of ADLs.

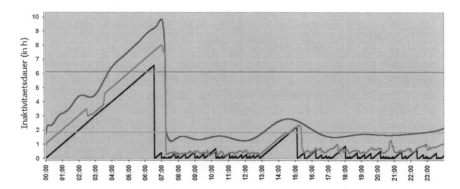

Fig. 1 Resulting inactivity diagram after training phase (*red line*) with constants to automatically detect day- and night period: *green line*—average plus double of the standard deviation, marking the night; *yellow line*—average, marking the day; *blue line*—polynomial function representing the alarm threshold

Activities of Daily Living (ADL) are essential routines every person executes. They were initially defined in care science [13]. We focus on basic ADLs typical for the household such as eating, sleeping, hygienic, sanitation, cooking, household, and media. Each ADL consists of different sets of actions that are typically related to it and may occur in different temporal orders. For example, hygienic relates to the actions: going to the bath room, washing at a basin, taking a shower, or using the tub. Each action now can further be related to appliance(s) such as the lights in the bath room or the tap at the basin, shower, or tub that are assigned respectively. However, each of these relations are n:m associations, where an appliance can implicate different actions that imply different ADLs. The challenge now is to deduce the correct ADL from the appliance detected by the smart meters.

The challenge has two aspects that need to be addressed [19]. First, the *segmentation problem*, which targets the question if a concurrent series of events belong to the same ADL. In its simple form it uses the assumption that ADLs are performed sequential and non-interleaving. The more complex segmentation problem is to consider and detect parallel ADLs with interleaving actions. Second, the *recognition problem* of relating the events to the correct actions and them to the correct ADL. An interesting approach is to mine the probabilities for assigning appliances to actions and ADLs from the how-to web pages [19]. But, this general information is usually known like which appliances are involved in hygienic. It is more promising to individualize the probabilities to the monitored people, as particularly elderly developed very specific behaviors over the years in executing their ADLs [17]. It is to note, that this is an assumption common to all approaches. Younger people have a more flexible daily routine. In this case, the integration of real-time information from smart phones calendars or social web pages open new opportunities to individualization.

Most approaches apply temporal classification models such as Hidden Markov Model (HMM) [2, 16], Dynamic Bayesian Networks (DBN) [21], and Conditional

Random Fields (CRF) [14]. These approaches assume sequential and non-interleaving actions to create individual models for each ADL. However, concurrent activities are common and cannot be well identified by these approaches [14, 19]. Other approaches use machine learning classification algorithms such as Naive Bayes Classifier [24], Artificial Neural Networks (ANN), and Support Vector Machines (SVM) [8]. They require labeled and segmented data for training, which requires manual effort that is often not practicable.

3 ADL Recognition Approach

The goal of the approach presented in this paper is not only to identify ADLs, but also to individualize them to the user and detect changes in their execution. Therefore, the approach divides into three phases: the initialization phase, the training phase and the application phase. In this paper two approaches for ADL identification and training are introduced: an SMM approach and an impulse-based approach.

During the *initialization phase* the available smart meters and home appliances inside each apartment are defined. Furthermore, possible relations to ADLs are analyzed. In the *training phase* individualized ADL models are learned. These characteristics models will be compared to the current consumption data to find abnormalities in the *application phase*. If abnormalities are detected, which may be caused by falls or fainting, an alarm message will be generated to organize help.

3.1 Initialization Phase

During the initial phase general information about available appliances are gathered. This is:

1. Available smart meters and optional sensors.
2. Available appliances that are relevant for ADLs.
3. The events detectable by the smart meters and to which appliances they map.
4. General relevance weights of how appliances relate to ADLs.

Appliances can be classified, in extension to [1], in their involvement of user activities (insignificant, significant), in their temporal phenomenal appearance (temporary, permanent), and in their mobility (static, mobile). The mobility classification is used to assign static appliances to rooms of the apartment, while mobile appliances can be used in any room (e.g. vacuum cleaner). Permanent Background Appliances are on all time and have no significant user interaction (e.g. stand-by power). They produce a constant noise that needs to be initially measured and filtered by the meters. Temporary Background Appliances are also

Fig. 2 Floor plan to show the allocation of home appliances to the rooms

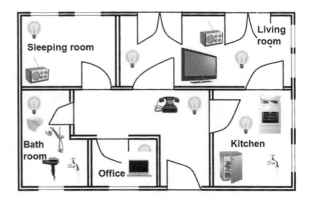

not user controlled and autonomously switch on and off (e.g. fridge). They produce events detectable by smart meters that need to be ignored either by not assigning them or by assigning them as false-positives. Passive Appliances are switched on by users and then remain in this state without their presence (e.g. light). Active Appliances require user presence and are usually of a short term (e.g. vacuum cleaner). Only the last two appliance classes are relevant for HAR while active appliances have the strongest impact.

Figure 2 shows an example of home appliance assignment for a three room apartment. Some characteristic consumptions for home appliances are illustrated in Fig. 3. Our approach does not analyze these particular characteristics, but the resulting event streams from the smart meter. However, Fig. 3 illustrates that the characteristic consumptions for home appliances can be identical such as the light in different rooms. Then the event from the smart meter is mapped to all these appliances as a differentiation is not directly possible. In later steps it will be possible to reason the appliance by analyzing the temporal and spatial correlation of events, appliances, and their room assignment.

Fig. 3 Assignment of consumption to home appliance

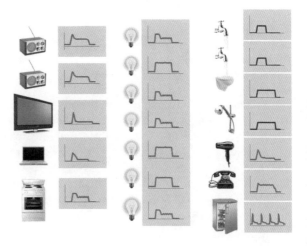

Detecting Activities of Daily Living with Smart Meters   149

Fig. 4 Overview of rooms with assigned appliances and probabilities

Appliances are assigned to one ADL with a relevance weight $G_a(e, a_i)$, which describes how commonly the appliance $e \in E$ is used for the ADL a_i. It will be individualized during training. Figure 4 shows a screenshot of the application that summarizes all these relations between appliances, rooms and ADLs.

3.2 Training Phase: SMM Analyzes

With the information from the initialization phase, the training of ADLs can be started. The first approach presented in this paper is a Semi-Markov-Model (SMM). Each state Z in the SMM represents the set of currently active home appliances ($Z \subseteq E$). This allows us to consider parallel actions and ADLs. A change of states occurs if an active one is turned off or another appliance is turned on. Each state logs the entry and exit times, which are required for ADL comparison. If no appliance is active for a certain time, the SMM returns in an *idle-state*, in which no event is active. In Fig. 5 an excerpt of such SMM is shown. The radio was switched on first (ID 9), followed by the mirror light (ID 1), and then the toilet was flushed (ID 22). Afterwards the light is turned off and sometime later the radio.

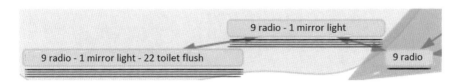

Fig. 5 Excerpt from the behavior model (SMM) in Fig. 6 representing a structure of human habit

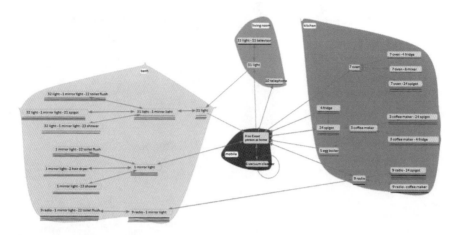

Fig. 6 Semi-Markov-Model illustrating the occupants behavior. The shaded regions group states in the same room. The underlines assign states to ADLs: *orange* sanitation, *blue* household; *red* nutrition; *black* media; *green* telephone; *white* shopping

Figure 6 shows the full SMM created from smart meter data of the three room apartment. The shaded groups represent the assignments of appliances to rooms from the initial phase. Each color represents another room. It is visible that most interactions (transitions) occur between appliances in the same room. This is also often related to a specific ADL, which are indicated by colored underlines of the states. For example, all actions of *sanitation* (underlined orange) take place in the bathroom (shaded light red).

ADLs are detected within this SMM. For this the segmentation and recognition problem needs to be addressed. The segmentation uses temporal and graph structural criteria. One temporal criteria results from the fallback into the idle-state not directly after the last appliance was deactivated, but after a time delay. In result, if another appliance is turned on shortly after the last appliance, it will be recognized as state sequence without passing the idle-stated. The structural criteria detects subgraph patterns in the SMM graph. Such patterns are visible in Fig. 6. The star pattern is dominant in the bathroom, where first the mirror light is turned on and then alternative appliances afterwards. Another relevant pattern is the cycle where several appliances are used in sequence (e.g. spigot, coffee maker, egg boiler). It follows from the sequential and non-interleaving action assumption. These subgraphs represent very dominant habits of the user and are good indicators of an ADL [5].

For the recognition problem each structure is evaluated in their significance $0 \geq G_S \leq 1$ for each ADL a_i in comparison to the full SMM by computing:

Detecting Activities of Daily Living with Smart Meters

Table 1 Sub-criteria to evaluate structural importance

Sub-criterion	Equation part
Similarity of two structures	$\frac{N_{mT}+N_{mZ}}{max(N_{T_1},N_{T_2})+max(N_{Z_1},N_{Z_2})}$
Degree of isolation	$1 - N_{SA}/(N_{TA} + N_{SA})$
Similarity of time intervals	$1 - \sigma_{Zt}$
Similarity of timestamps	$1 - \sigma_t$

N_{mZ} Number of matching states
N_{mT} Number of matching transitions
N_{SA} Number of entry and exit transitions
N_{TA} Number of transitions without match: $N_T - N_{mT}$
N_T Number of transitions
N_Z Number of states
σ_{Zt} Standard deviation time intervals between two states normalized to an interval of $0 \leq t \leq 1$
σ_t Standard deviation of timestamps, normalized to an interval of $0 \leq t \leq 1$

$$G_S = 0.6 \frac{N_{mT} + N_{mZ}}{max(N_{T_1},N_{T_2}) + max(N_{Z_1},N_{Z_2})} \\ + 0.1\left[1 - \frac{N_{SA}}{N_{TA} + N_{SA}}\right] \\ + 0.2[1 - \sigma_{Zt}] + 0.1[1 - \sigma_t]. \quad (1)$$

The significance combines the temporal and structural sub-criteria as summarized in Table 1. The first sub-criterion of G_S is the similarity of two structures. It is calculated as the ratio of identical states and transitions in relation to the sum of the SMM. The matching of identical states is simple as they are distinctly defined by the set of active devices. The transitions are distinctly identified by the tupel of source states and target states. As states and transitions are the base of a SMM, this sub-criterion has major influence and is weighted with 60 % within the G_S. The degree of isolation rates the variation of transitions to leave or enter a structure. It ranges from 0 to 1, while a lower value represents additional transitions to enter or leave. The character of the previously mentioned structures is not only described by states and transitions. To detect these structures the timestamps have to be evaluated as well. Therefore, the similarity of timestamps and their intervals are included in G_S, too. In a pre-processing step the timestamps are normalized to fit between 0 and 1. Human actions are more similar in time intervals than in concrete time dates, as actions take the same duration but can be at different daytime. Thus, the intervals are rated higher.

G_S can be computed by the following algorithm that is executed for each existing ADL a_i evaluating the structure S, its states Z_S and transitions T_S, the ADL a_i, its states Z_i and transitions T_i:

1. Compute the sets Z_i and T_i of a_i.
2. Compute the sets Z_S and T_S as the number of all states and transitions of the structure S.
3. Compute Z_m and T_m, with $Z_m = Z_i \cap Z_S$; $T_m = T_i \cap T_S$.

Table 2 Weighted allocation of appliances from the structure shown in Fig. 5 to the corresponding ADLs

Appliance	ADL	Weight G_a, with $0 \leq G_a \leq 1$
Radio	Media	1.0
	Sanitation, household, nutrition, shopping, telephone	0.0
Mirror light	Sanitation	0.6
	Household	0.2
	Nutrition	0.2
	Shopping, telephone, media	0.0
Toilet	Sanitation	0.70
	Household	0.25
	Nutrition	0.05
	Shopping, telephone, media	0.0

4. Compute N_{SA} and N_{TA} of S.
5. Compute σ_{Zt} and σ_t for all elements in Z_m.
6. Calculate G_S from (1) with: $N_{mT} = |T_m|$; $N_{mZ} = |Z_m|$; $N_{T1} = |T_i|$; $N_{T2} = |T_S|$; $N_{Z1} = |Z_i|$; $N_{Z2} = |Z_S|$.

Figure 5 serves as an example structure to demonstrate the calculations steps to gain G_S. It illustrates quite well the general decision problem in detecting ADLs. Table 2 displays the initial relevance weights G_a of the ADLs for each home appliance. The radio, that stays on the whole time, is assigned to the media ADL with 100 %, but the remaining appliances are most descriptive to the ADL *sanitation*, but also relevant to *household* and *nutrition*. The challenge is to automatically calculate which ADL was actually executed.

Table 3 demonstrates the single values of the parameters of G_S and the final value of G_S calculated for each ADL. Obviously the values of G_S do not differ that much for each ADL, but in comparison with Table 2 the ADL media has a low value of 0.34 where the ADLs *household* and *sanitation* are rated higher with 0.37 and 0.38, respectively.

Obviously the combination of appliances is crucial for the detection of ADLs. Thus, the radio was assigned to media with 100 %, but in combination sanitation and household are weighted higher. These weights are used for the final

Table 3 Calculation of each significance criterion for all possible ADL in reference of the structure in Fig. 5

	N_{Z1}	N_{T1}	N_{mZ}	N_{mT}	N_{SA}	σ_{Zt}	σ_t	G_S
Sanitation	15	16	2	2	0	0.2	0.6	0.38
House-hold	18	17	2	2	0	0.2	0.6	0.37
Nutrition	19	23	2	2	0	0.2	0.6	0.35
Shopping	2	0	0	0	0	0	0	0
Telephone	1	0	0	0	0	0	0	0
Media	6	8	1	0	0	0.2	0.6	0.34

Table 4 Calculation results of the PADL in comparison with averaged values of the structure in Fig. 5

ADL	$\bar{G}_a$	G_{act}
Sanitation	0.43	0.59
Household	0.15	0.21
Nutrition	0.23	0.31
Shopping	0.0	0.0
Telephone	0.0	0.0
Media	0.33	0.44

calculation of the most probable ADL (PADL). Both values of G_a and G_S are relevant for the recognition problem and are finally combined to calculate G_{act} for all ADLs from

$$G_{act}(S, a_i) = \bar{G}_a(S, a_i)(1 + G_S) \qquad (2)$$

with the average $\bar{G}_a$ of the relevance weights of each appliance in each state

$$\bar{G}_a(S, a_i,) = \sum_{Z \in Z_S} \sum_{e \in Z} G_a(e, a_i). \qquad (3)$$

The temporal and structural classificator G_S is used as gain factor to intensify relevance weights that are supported by the temporal and structural criteria. Table 4 displays the final result of G_{act} in comparison to the simple average calculation of $\bar{G}_a$ from Eq. (3). The average $\bar{G}_a$ indicates that *sanitation* and *media* are relatively close, but the values of G_{act} highlight that the *sanitation* ADL is the most dominant.

The ADL with the highest G_{act} value is recognized as the most probable ADL (PADL) that the inhabitant is executing with his action set. As all ADLs are considered during evaluation also parallel ADLs can be analyzed. This is the target of the next approach.

3.3 Training Phase: Impulse Analyzes

The SMM approach has limitations in identifying parallel ADLs. This is related to their temporal classification in individual states and the independence assumption of Markov models. The benefit of the presented SMM approach in contrast to HMM approaches is that it models concurrent appliances and also sequences of them for identifying ADLs. Conditional Random Fields have a similar approach by considering event sequences [14]. However, the focus on sequences and the resulting segmentation problem leads in both cases to a temporal restricted view.

The impulse approach removes the limitation and analyzes the temporal correlation of events. It bases on the assumption that if two events occur in quick succession and relate to the same ADL its significance is increased while longer time intervals have a reduced impact. This is modeled by defining an exponential

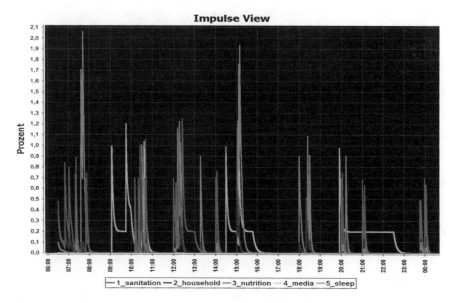

Fig. 7 The ADL significance levels of one day using the impulse approach

decline function for each ADL significance. It basically resembles the human forgetting curve [7] and each ADL gets less important if it is not stimulated. If an appliance is activated the corresponding ADLs are stimulated according to their relevance weight. The stimulus may not drop below 20 % as long as the appliance is active. The stimulus declines after the appliance is deactivated.

The significance $I_{a_i}(t)$ of each ADL a_i computes for the time t from the relevance weight $G_a(f, a_i)$ of each home appliance e depending on its activity via:

$$I_{a_i}(t) = \sum_{e \in E} \begin{pmatrix} G_a(e,a_i)0.2 + 0.8e^{-d(t-t_{on})} & \text{active,} \\ G_a(e,a_i)b_{on}*e^{-d(t-t_{off})} & \text{inactive,} \end{pmatrix} \quad (4)$$

with the decline factor d, the last impulse value from active state b_{on} and the last event times t_{on} and t_{off} of the appliance activation and deactivation, respectively.

This method was evaluated with a set of events over a one day period. Figure 7 shows the plot of the ALD significances of the whole day. Figures 8, 9, 10 and 11 show zoomed details of this plot. A first example is presented in Fig. 8 that shows the impulse of only one active home appliance. This appliance is assigned to the ADLs *sanitation*, *household* and *nutrition* (*nutrition*) with different G_a.

More events occurred in Fig. 9 and the combination of them emphasizes the ADL *nutrition* and the ADL *household* with a lower grade. Figure 11 shows a similar situation in the beginning with the ADLs *household* and *nutrition* active. From 15:06 to 15:16 o'clock the combination of appliances is clearly assigned to the ADL *nutrition*. At around 15:07 o'clock some *media* appliance was turned on. As it is not influencing any other ADL, it is clear that it is executed in parallel.

Detecting Activities of Daily Living with Smart Meters 155

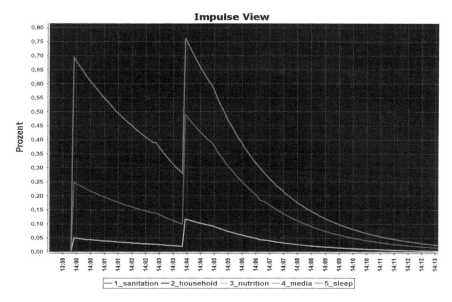

Fig. 8 Zoomed part of Fig. 7 displaying one active home appliances that addresses three ADLs

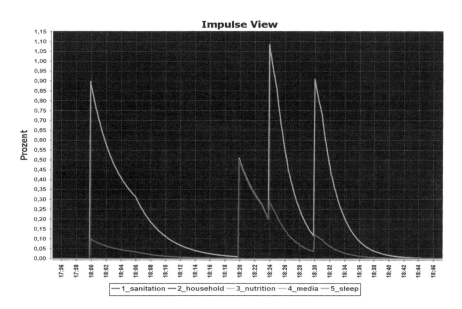

Fig. 9 Zoomed part of Fig. 7 displaying several active home appliances which address two ADLs. The *green* ADL *nutrition* was executed, as it exceeds the *blue* one

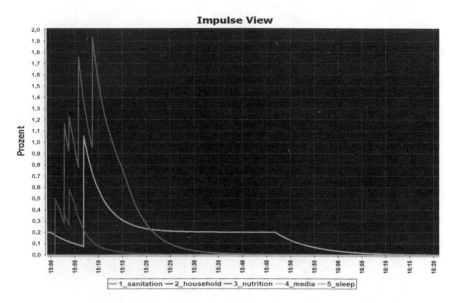

Fig. 10 Zoomed part of Fig. 7 displaying several active home appliances which address three ADLs and two parallel ADLs (*green* and *yellow*)

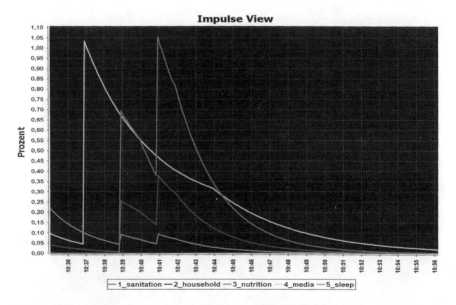

Fig. 11 Zoomed part of Fig. 7 displaying the corresponding impulse to the example in Fig. 5

Figure 11 shows the ADL significances for the example used in Fig. 5. The area of each ADL's significance per minute is summarized in Table 5 and plotted in Fig. 12. The sum of the area of the ADL significance indicates that the *media* and

Table 5 Area calculation for the significance of each ADL

Time [h:min]	Sanitation	Household	Nutrition	Shopping	Telephone	Media
10:36	42	1,210	6,405	0	0	11,752
10:37	31	906	4,796	0	0	56,493
10:38	6,202	2,767	3,397	0	0	42,902
10:39	37,831	13,896	4,760	0	0	37,378
10:40	36,171	13,160	3,879	0	0	30,190
10:41	49,923	17,978	4,363	0	0	23,867
10:42	40,225	14,470	3,474	0	0	23,870
10:43	23,558	8,480	2,036	0	0	18,980
10:44	16,912	6,088	1,462	0	0	17,389
10:45	10,202	3,672	882	0	0	12,713
10:46	6,796	2,444	587	0	0	10,157
10:47	4,476	1,611	387	0	0	8,036
10:48	3,190	1,148	276	0	0	6,925
10:49	1,930	695	167	0	0	5,070
Sum	237,491	88,525	36,869	0	0	305,722

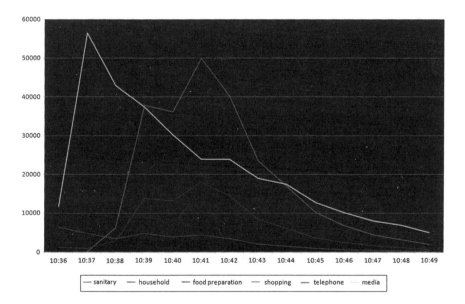

Fig. 12 Display of the impulse area of Table 5 of the Fig. 11

the *sanitation* ADLs are the most important ones. Other ADLs such as *nutrition* and *household* have a significant lower significance value. In addition, the ADLs *media* and *sanitation* are executed in parallel, because their stimulating appliances are disjunct. Figure 11 shows that the *media* device was active first, but over time it fades into the background due to the exponential decline. When the *sanitation* devices are used they dominate the ADL significance and are the currently active

ADL. After their deactivation and around five minutes later the *media* appliance is the only one remaining active and again the dominating ADL. This is the main benefit of the approach as the currently active appliances strongly support their ADLs while previous stimuli remain in effect, but with a lower significance. This allows that ADL that are interrupting other ADLs can be detected as such. An aspect that was not distinguishable in the SMM as both appliances were combined in one state and sequence.

3.4 Comparison and Outlook

The SMM-approach is able to model that appliances are running in parallel by combining them in states. Therefore, it does not require the sequential and non-interleaving assumption of most other approaches. On the other hand, it can compute the PADL only for a sequence of states and cannot well distinguish parallel ADLs. The impulse approach does not base on a state model, but analyzes the temporal correlations of the appliance events using a model that adopts the human memory. The ADL significance is analyzed separately such that parallelism and temporary effects of interleaving ADLs are easy to detect.

Current research targets to combine both models to improve the segmentation in the SMM model by separating disjunct appliances in their own states as well as considering the effect of temporal correlations clearly visible in the impulse-diagrams to detect interleaving ADLs.

4 Conclusion

This paper demonstrates how smart meter data can be used as an affordable and practical data source for human activity recognition and behavior modeling without the need of highly integrated sensors. Two approaches where introduced. Firstly, the SMM-approach allows training and detecting individual human habits by analyzing the relationships between home appliances usage. Secondly, the impulse approach adopts the human memory by using an exponential decline function and the appliances as stimuli. It allows analyzing the relationships of executed parallel and interleaving ADLs. Two important aspects of a humans habits. Particularly, the daily routine of elderly is very stable and therefore qualified for monitoring and modeling. The models allow doctors and care givers to understand peoples diurnal rhythm. It improves inactivity detection and changes in ADLs indicate the development of diseases such as Alzheimer's. In future work both methods will be combined in a more sophisticated model where the SMM-structures, representing human habits, will have an impact on the process of the impulse curves for each ADL. This will improve and individualize the detection of parallel executed ADL in single home apartments.

References

1. Berges, M., Rowe, A.: Poster abstract: Appliance classification and energy management using multi-modal sensing. In: BuildSys—3rd ACM Workshop on Embedded Sensing System for Energy-Efficiency in Buildings (2011)
2. Brdiczka, O., Crowley, J.L., Langet, M., Maisonnasse, J.: Detecting human behavior model from multimodal observation in a smart home. Autom. Sci. Eng. (2008)
3. Bruckner, D., Velik, R.: Behavior learning in dwelling environments with hidden markov models. IEEE Trans. Ind. Informat. **57**(11) (2010)
4. Clement, J., Ploennigs, J., Kabitzsch, K.: Enhanced inactivity diagram to meet elderly needs. AAL-Forum (2011)
5. Clement, J., Ploennigs, J., Kabitzsch, K.: Smart meter: Detect and individualize ADLs. In: Ambient Assisted Living. Advanced Technologies and Societal Change, Springer, pp. 107–122, (2012)
6. Du, L., Yang, Y., He, D., Harley, R.G., Habetler, T.G., Lu, B.: Support vector machine based methods for non-intrusive identification of miscellaneous electric loads. In: IECON—38th Annual Conference of the IEEE Industrial Electronics Society (2012)
7. Ebbinghaus, H.: Über das Gedächtnis. Untersuchungen zur experimentellen Psychologie. Wissenschaftliche Buchgesellschaft Darmstadt ((Nachdr d Ausg 1885) 1992)
8. Fleury, A., Vacher, M., Noury, N.: SVM-based multimodal classification of activities of daily living in health smart homes: Sensors, algorithms, and first experimental results. IEEE Trans. Inf Technol. Biomed. **14**(2), 274–283 (2010)
9. Floeck, M., Litz, L.: Inactivity patterns and alarm generation in senior citizens' houses. In: ECC—European Control Conference (2009)
10. Greveler, U., Justus, B., Loehr, D.: Multimedia content identification through smart meter power usage profiles. In: CPDP—5th International Conference on Computers, Privacy and Data Protection (2012)
11. Hart, G.: Nonintrusive appliance load monitoring. Proc. IEEE **80**(12), 1870–1891 (1992)
12. Jiang, X., Van Ly, M., Taneja, J., Dutta, P., Culler, D.: Experiences with a high-fidelity wireless building energy auditing network. In: SenSys—7th ACM Conference on Embedded Networked Sensor System, pp. 113—126 (2009)
13. Katz, S., Down, T.D., Cash, H.R.: Progress in the development of the index of ADL. The Gerontologist 10(1 Part 1), 20–30 (1970)
14. Kim, E., Helal, S., Cook, D.: Human activity recognition and pattern discovery. IEEE Pervasive Comput. **9**, 48–53 (2010)
15. Liao, L., Fox, D., Krautz, H.: Loction-based activity recognition using relational markov networks. In: IJCAI—19th International Conference on Artificial Intelligence (2005)
16. Matsumoto, T., Shimada, Y., Hiramatsu, Y., Kawaji, S.: Detecting non-habitual life behaviour using probabilistic finite automata behaviour model. In: ICCA—4th International Conference on Control and Automation, pp. 703–707 (2003)
17. Noury, N., Quach, K.A., Berenguer, M., Teyssier, H., Bouzid, M.J., Goldstein, L., Giordani, M.: Remote follow up of health trough the monitoring of electrical activities on the residential power line—preliminary results of an experimentation. In: Healthcom—11th International Conference on e-Health Networking Applications and Services pp. 9–13 (2009)
18. Onoda, T., Murata, H., Ratsch, G., Muller, K.R.: Experimental analysis of support vector machines with different kernels based on non-intrusive monitoring data. IJCNN **3**, 2186–2191 (2002)
19. Palmes, P., Pung, H.K., Gu, T., Xue, W., Chen, S.: Object relevance weight pattern mining for activity recognition and segmentation. Pervasive Mobile Comput **6**(1), 43–57 (2010)
20. Patel, S.N., Robertson, T., Kientz, J.A., Reynolds, M.S., Abowd, G.D.: At the flick of a switch: Detecting and classifying unique electrical events on the residential power line. In: UbiComp—14th International Conference on Ubiquitous Computing (2007)

21. Philipose, M., Fishkin, K., Perkowitz, M., Patterson, D., Fox, D., Kautz, H., Hähnel, D.: Inferring activities from interactions with objects. IEEE Pervasive Comput. **3**(4), 50–57 (2004)
22. Rowe, A., Berges, M.E., Bhatia, G., Goldman, E., Rajkumar, R., Garrett, J.H., Moura, J.M.F., Soibelman, L.: Sensor andrew: Large-scale campus-wide sensing and actuation. IBM J. Res. Dev. **55**(1.2), 6:1–6:14 (2011)
23. Rowe, A., Berges, M., Rajkumar, R.: Contactless sensing of appliance state transitions through variations in electromagnetic fields. In: BuildSys—2nd ACM Workshop on Embedded Sensor System for Energy-Efficiency in Buildings. pp. 19—24. New York, NY, USA (2010)
24. Tapia, E., Intille, S., Larson, K.: Activity recognition in the home using simple and ubiquitous sensors. Pervasive Comput. **3001**, 158–175 (2004)
25. Torriti, J., Hassan, M.G., Leach, M.: Demand response experience in Europe: Policies, programmes and implementation. Energy **35**(4), 1575–1583 (2010)

A Personalized and Context-Aware Mobile Assistance System for Cardiovascular Prevention and Rehabilitation

Alexandra Theobalt, Boris Feodoroff, Dirk Werth and Peter Loos

Abstract Existing approaches for personalized regulation and adaptation of physical activity for cardiovascular disease prevention and rehabilitation are often based on one quantitative parameter, usually heart rate. Other influence factors such as time of day, nutritional condition, outside temperature, exhaustion level are ignored and greatly impede an optimal personalization. In the following an innovative mobile assistance system is presented uses electrocardiogram (ECG) records as parameter for personalization (an electrocardiogram is proven to be better parameter for personalization). Moreover a "connected pedelec", i.e. a bicycles equipped with an auxiliary motor which only assists when the cyclist pedals, is used a exercise machine. Thus, for the first time automated regulation and personalized adaptation based on ECG records are made mobile. In addition, a holistic approach is applied which also considers other parameters for the assistance during the exercise but also for the assistance before and after the exercise. As a result a more holistic approach for personalized regulation and adaptation of physical activity and individualized assistance and motivation is provided before and after exercise.

A. Theobalt (✉) · D. Werth · P. Loos
Institut für Wirtschaftsinformatik im Deutschen Forschungszentrum für Künstliche Intelligenz, Saarbrücken, Germany
e-mail: Alexandra.Theobalt@dfki.de

D. Werth
e-mail: Dirk.Werth@dfki.de

P. Loos
e-mail: Peter.Loos@dfki.de

B. Feodoroff
Zentrum für Gesundheit der Deutschen Sporthochschule, Köln, Germany
e-mail: B.Feodoroff@dshs-koeln.de

1 Introduction

Physical activities as a protection against coronary heart disease was firstly examined in the mid-20th century [1]. Further research proved the relationship between physical activity and reduction of risk of mortality and morbidity from cardiovascular disease [2–4]. Andersen et al. conducted a study which links regular bicycling with a decreased risk of mortality [5]. The authors found out that even after adjustment for other risk factors those who did cycle to work had a 39 % lower risk of mortality. Cycling was identified as one of the solutions to improve health [6], since it has a functional role which does not completely rely on self-motivation and thus can contribute to higher levels of physical activity [7]. In addition to the benefits for the environment, transportation research considers increased walking and cycling (together called active transportation [8]) to improve public fitness, particularly for vulnerable populations such as children or seniors [9]. Furthermore, the goal of the European Union (EU) is to phase out conventionally fuelled cars in urban transport by 2050. This shows that cycling will play a major role as shaping means of future transportation.

The growing public awareness of the health benefits and of the advantages of cycling to solve mobility problems is also reflected in the growing market for electric power assisted cycles (EPAC). EPACs cover two diffcrent concepts of bicycles with an auxiliary elcctric motor [10]:

- Cycles equipped with an auxiliary motor that cannot be exclusively propelled by that motor. Only when the cyclist pedals, does the motor assist. These vehicles are generally called 'pedelecs' short for "pedal electric cycle" and they are today the most popular in the EU.
- Cycles equipped with an auxiliary electric motor that can be exclusively propelled by that motor. The cyclist is not necessarily required to pedal. These vehicles are generally called 'E-bikes'.

In terms of physical activity pedelecs are of greater interest than E-bikes, since motor assistance requires pedaling. There are different levels of motor assistance which can be set by the cyclist. Latest pedelec models provide a mobile app to adjust the assistance level, i.e. a mobile phone instead of the integrated control unit is attached to the rod and sends commands about the assistance level to the motor. In view of the rapid growth of the mobile broadband market worldwide ("Mobile broadband has become the single most dynamic ICT service reaching a 40 % annual subscription growth in 2011." [11]) the integration of a smart phone provides the possibility of a "connected pedelec". Connection on the one hand allows the collection of information about the user and the user's context, since different information sources such as motor, external sensors (e.g. electrocardiogram monitoring device, SO2 sensor, respiratory rate sensor) and smart phone sensors (e.g. GPS or accelerometer) can be connected to the smart phone. On the other hand, the mobile internet connection of the smart phone allows for the transmission of the collected information to a server which is able to process and analyse the collected information.

2 Motivation

Recording of vital signs and their analysis in real-time for personalized regulation and adaptation of exercise for cardiovascular disease prevention and rehabilitation is possible only to a limited extent. Existing approaches in therapy or prevention are often very vague, e.g. recommendation such as "running without heavy breathing" or individual perception of exertion are usually given to a patient. In other approaches personalization mechanisms are based on quantitative parameters such as heart rate. The maximum heart is often used in training or rehabilitation plans as a threshold which must not be exceeded during the exercise. A common approach to determine the maximum heart rate is the equation: $HRmax = 220-$age [12]. However, Robgers [13] showed that origin as well as usage of this equation is very questionable from a point of view of sport medicine and sport science. In general, all of the approaches applied today only consider one parameter to individually regulate and control physical activity for cardiovascular rehabilitation and prevention. As a result, influence factors such as time of day, nutritional condition, prescription drug use, effects of substances such caffeine or beta blocker, outside temperature, exhaustion, etc. are ignored and greatly impede an optimal personalization. There are several studies that show that for individualized regulation and adaptation of physical activities the best parameter is the electrocardiogram (ECG) [14–16]. Currently using ECG for personalization is only possible in inpatient treatment. To achieve a regulated physical activity of a patient, exercise on an ergometric bicycle is medical standard in rehabilitation centers. But even there, the power of the ergometric bicycle is automatically adapted to not exceed a predefined heart rate. The recorded ECG is manually analysed by a physician afterwards and thus not considered in the regulation and control of the physical activity. In this context a connected pedelec which records an ECG and transfers that information and other context information of a user to a backend server for real-time analysis is an approach which neither exists for prevention nor for rehabilitation of cardiovascular diseases. In the following, Sect. 3 looks into existing approaches in industry and research. Section 4 describes our approach named MENTORbike which beside automated adaptation based on ECG data during exercise also looks into assistance and motivation of a user before and after exercise. In this section a use case scenario and the system architecture of MENTOR bike are detailed. The paper finishes with a conclusion and an outlook in Sect. 5

3 Related Work

3.1 Applications of Connected Pedelecs for Health Applications

The German company MTB Cycletec equipped its pedelec e-Jalopy with a new motor called Greenwheel which was developed by the Massachusetts Institute of Technology (MIT). Greenwheel combines for the first time battery, motor and control system in a rear wheel hub motor. Furthermore, a smart phone provides the interface to the Greenwheel motor. It is envisioned that the e-Jalopy will be connected via the smart phone to social networks and with other road users, be located via the GPS integrated in the smart phone and be able to measure pollution and recommend alternate less polluted routes in the future [17]. The Hungarian company Gepida developed an Apple and Android app which enables a cyclist to control the pedelec via his/her smart phone. The app displays information on current, maximum and average speed, daily and total covered distance as well as information on the charge status of the battery and the current assistance level. Via GPS tracking the route can be shown on google maps. Furthermore a heart rate belt can be connected via bluetooth to the smart phone which then can also display the current heart rate and save it. Measured heart rate and other collected information can be sent to medical professionals for analysis afterwards or for real-time monitoring. In addition, the user can specify an upper and lower boundary for the heart rate at which motor assistance is switched on or off automatically [18]. Also the German company Kalkhoff offers automated motor assistance based on an upper and lower heart rate threshold [19].

In summary, pulse control of motor assistance as well as transfer of monitored information to medical professionals are basic assistance mechanisms to support individual health prevention. However, the described approaches are mainly focused on assistance during a ride and do not take an holistic approach for personalized assistance. In the following research approaches for intelligent and mobile assistance systems for personalized prevention and rehabilitation from cardiovascular disease are analysed.

3.2 Mobile Assistance Systems for Cardiovascular Prevention and Rehabilitation

Research on mobile assistance systems for cardiovascular applications can be differentiated into prevention and rehabilitation application.

Ho [20] describes a mobile assistance system for prevention. It generates recommendations during exercise based on the monitored heart rate and exercise intensity. In addition the system can detect anomalies such as incidences of cardiac arrhythmia. Over- or under-exercising is monitored on the mobile handheld. Further evaluations are conducted on a server. Based on collaborative filtering approaches the system determines whether the exercise is suitable or should be adjusted or replaced with another routine.

There are also examples of assistance system for general health prevention and improved life style management or for better life style management such as weight or stress reduction which also contribute to prevention of cardiovascular diseases [21]. However, since they are not particularly focused on cardiovascular topics, they are not further detailed.

Gay et al. [22] describe a mobile assistance system for rehabilitation from cardiovascular diseases. They combine activity and bio signal monitoring (e.g. exercise, ECG, weight, blood pressure, glucose) to generate immediate local feedback to the user without the intervention of a health professional. In addition health professionals can access the user's data and carry out remote monitoring and reporting. Furthermore, the system can detect emergencies such as life-threatening cardiac arrhythmia or a fall and issue an emergency call. Immediate feedback comprises information about over- or under-exercise based on heart rate information. Mainly the system assists with the management of exercise and active (via questionnaires) and passive monitoring (recording of vital signs) of the health status. Deeper analysis of the data is expected from the medical professionals.

The research project HeartCycle develops amongst others personalized mobile assistance system for exercise for users who suffered from heart failure and coronary heart disease. The user is equipped with an exercise shirt that collects the user's vital signs during the exercise. These are transmitted in real-time to a portable station (PDA) which also contains the training plan. The PDA processes the collected data and provides feedback to the patient during the training based on the training plan. At home the PDA is connected to a PC which conducts progress analysis based on the latest information [23].

Similar to Sect. 3.1, the described examples are focused on basic feedback during physical activity, monitoring and data transfer to medical professionals. Feedback during physical activity is even more limited since the user is only notified and no automated adaption is possible due to the fact that the exercise machine is not connected. Thus, a connected pedelec provides additional benefits due to the possibility of automated adaptation.

Improved communication with professionals and peers was a common goal throughout the examples described above. Therefore, social networks for health applications are examined in the following section.

3.3 Social Health Networks for Health Applications

Collaboration, flexibility, a pre-eminence of content creation over content consumption and interactivity are considered the most prominent features of the social web [24]. In this context, Barsky et al. [25] consider amongst others wikis, blogs, and the user comment functionality as social networking enabled technologies relevant in the context of health application. For instance, the website British Medical Journal Rapid Responses [26] and an user comment functionality at the website Patient.co.uk [27] offer the opportunity to record your experience as a patient. At Patient.co.uk a user can even rate the experience entries of others, thus extending the peer rating functionality with a reputation management system. In terms of wikis, wikisurgery [28] is an example of a social networking enabled technology which aims at collaboratively building a surgical encyclopedia for surgeons and patients. Examples of health related blogs are DrugScope DrugData Update blog [29]. In terms of social networking sites, LibraryThing Medicine Group [30] is an interesting example, where users with an interest in books about medicine and medical science share content and ideas.

An example of a social community offering several social web enabled technologies is the mental health social network HealthyPlace [31]. Boulos et al. [24] list several more examples of social networking enabled technologies in terms of health applications for instant messaging and virtual meetings or online social gaming. Eysenbach et al. [32] conducted a review of studies on health related virtual communities and support groups and their effects of online peer to peer interactions. The results lead to the conclusion that no robust evidence exists on the effects on health and social outcomes of computer based peer to peer communities and electronic self support groups. Eysenbach et al. [32] point out that due to the growing number of virtual communities and support groups more quantitative studies are needed in addition to qualitative studies on this topic.

An American study on social media and health conducted by the Pew Internet & American Life Project [33] found out that collaborative filtering mechanisms of the social web are a main feature users look for when using web 2.0 technologies, since they search for "just-in-time-someone-like-me". On the other hand creation of health content is rather low compared to the consumption of health content. The survey furthermore found out that social networking sites are used only sparingly for health queries and updates and that there is a surge of interest in information about exercise and fitness. Overall, social health networks are mainly designed to get information about health topics from professionals. Quality is ensured since information is provided and supervised by professionals. However, information is usually more general and more suitable to get a general understanding of a health issue. Information about a particular health situation or issue is more likely to be found when individual describe their experiences which is mostly via blogs. However, in this context information sources are usually laypersons, so that quality and thus usability of the information is questionable. A combination of information sources consisting of laypersons with similar experiences (i.e. using collaborative experiences) and professionals promises to improve the quality of social health networks.

4 A Personalized and Context-Aware Mobile Assistance System for Cardiovascular Prevention and Rehabilitation

The approaches described in Sect. 2 were mainly focused on assistance during physical activity. However, a holistic approach should support and motivate a user before and after physical activity as well and connect the user with peers and professionals. Therefore, in the following a system is described which comprises a connected pedelec. The system called MENTORbike consists of the following components:

- Connected pedelec
- Wireless body area network
- Sensing system (e.g. mobile electrocardiogram monitoring device, GPS, pedelec motor)
- Central service platform on a backend server.

The smart phone connects the pedelec on the one hand with the wireless body area network and smart phone acts as data collecting and processing device. On the other the smart phone is connected to the mobile internet and transfers the collected data to the service platform on the backend sever.

4.1 Use Case Scenarios

MENTORbike provides great benefits in transitional situations in rehabilitation and for motivational aspects in prevention of cardiovascular diseases.

In rehabilitation the fact that the pedelec is a connected exercise device and thus can be adapted automatically makes it possible to take monitored and regulated exercise outside for the first time. Duration of training, power in Watt and maximum heart rate with which a patient should exercise are indicated in a rehabilitation plan. In order to set a fixed power level usually ergonomic stationary bicycles are used. In addition, the patient's heart rate is monitored during the exercise by a physician who can intervene if the heart rate exceeds the prescribed limit. With the connected pedelec adjustment of power and heart rate monitoring can be automated and taken safely outside. The patient can start exercising with a fixed connected pedelec inside where he/she can get used to the automated adaptation of power, heart rate monitoring and the user interface of the smart phone. When the user feels secure and strong enough, the connected pedelec can be taken outside for rehabilitation exercise. Thus, the patient greatly benefits in the transitional phase from inside to outside training or exercise at home. Moreover, patients with different rehabilitation plans can cycle together with a trainer. Vital parameters of the members of the exercise group are transferred to the trainer's smart phone. The system alerts the trainer based on the individual rehabilitation plans when there are deviations from the plan.

In prevention the transition between stationary and mobile usage greatly benefits the motivation of a user. In case of bad weather training can be conducted and monitored inside. For outside training the connected service platform can provide mobile services based on the user's context in addition to the automated adaptation of the motor. For instance, location-based services such as the next bus station to go home due to the fact that the user over-exercised, or next resting points (e.g. a restaurant or cafe) can be provided based on the monitored user situation. Furthermore, similar to the rehabilitation scenario the pedelec enables cycling in a group with different fitness level. This community aspect is transferred from real to virtual world by offering a MENTORbike social network. Following the conclusions from Sect. 2.3 this social network will provide the opportunity of communication and information exchange with health professionals and peers. A user will be able to give physicians, trainers, etc. access to collected data. Based on for instance the recorded ECG, a physician will be able to conduct long-term monitoring of the training progress and its effects on the user's heart. A trainer will be able to for instance to improve the training plan based on the monitored progress. Furthermore, the user will be able to post the routes he cycled, set up training meetings, etc. Here, the system can recommended for instance suitable training partners based on similar fitness level or or provide a user with suitable cycling routes which match his/her route profile. Also access to suitable services such as diet plans or services about cycling and health topics can be better recommended to a user based a user profile which a user specifies in the beginning but will be also learned by the system from the monitored user interactions.

4.2 System Architecture

Figure 1 depicts the architecture of MENTORbike fun which aims at assistance with prevention of cardiovascular disease. The system consists of a mobile side and a server side. The mobile side contains all external sensors from ECG device and pedelec and additional sensors such as blood pressure and oxygen saturation sensor. With the pedelec sensors the exact amount of power applied by user and the exact amount of power added by the motor can be determined. With the additional information from the ECG device on heart rate and ECG and further information from external sensors much more detailed analysis about the training can be conducted and as a result also an improved personalization is possible.

The smart phone contains internal sensors such as GPS or acceleration and the mobile MENTORbike app. The mobile app controls the pedelec motor based on the collected context information of the user which is partly processed on the smart phone in the application logic component. Furthermore, the smart phone sends the collected information and evaluation results to the MENTORbike server side. The service platform serves as a single entry point and the interaction facade decides what actions to pursue. Via the push notification component on the service platform the smart phone can be addressed from the server side. Also external services

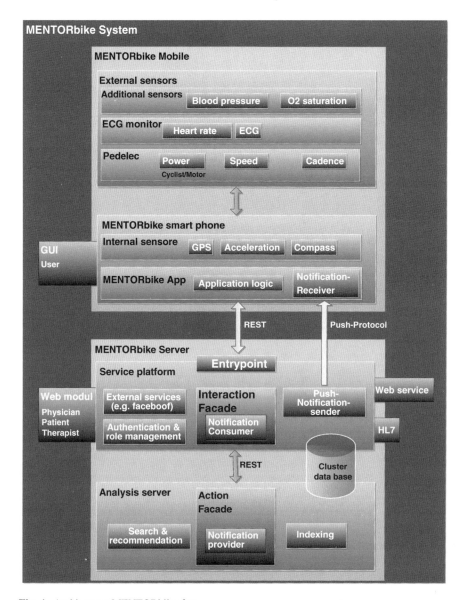

Fig. 1 Architecture MENTORbike fun

such as facebook or google+ are connected to the service platform. Other web services such as bus schedule, location-based services, etc. can be integrated via the web service interface. HL7 interface provides the connection to the medical information system of a hospital or the physician. The web modul enables the user to access his data and the virtual community via a PC, since in MENTORbike cross-modal access to the MENTORbike is envisioned. Also it provides an

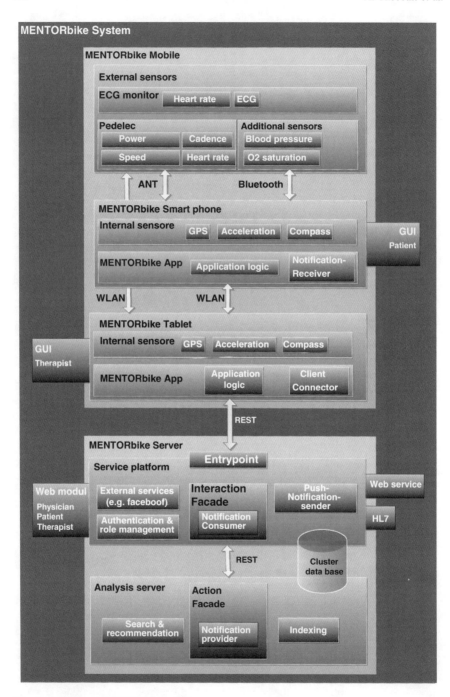

Fig. 2 Architecture MENTORbike Reha

interface for medical professionals to access the collected information of their patients. The authentication and role management component ensure security and privacy of the collected data and data exchange and access. Extensive analysis of the collected data is conducted on the analysis server. Here, the action facade provides the REST interface to the service platform for data exchange and decides how to process the received data. The search and recommendation component contains mechanisms such Bayesian networks for context analysis and content-based and collaborative recommendation mechanisms to detect suitable recommendation objects (e.g. services, training plans, training partners, etc.) for a user. The indexing component contains a semantic knowledge base and a semantic index of recommendation objects. A hybrid search and recommendation approach is envisioned based on traditional and semantic mechanisms. The cluster data base saves and manages the collected information about all users.

Figure 2 depicts the architecture of MENTORbike Reha which aims at assistance with rehabilitation of cardiovascular diseases. The difference to the architecture of the MENTORbike fun is the integration of another mobile device. For the monitored exercise scenario in a group outside with a therapist, the therapists tablet to monitor patients' vital signs is included. As a result, the user interface of the patients' smart phone is much simple than for the MENTORbike fun. Since the user is monitored by a medical professional, only basic information such as heart rate, speed or current position on a map are depicted.

5 Conclusion and Outlook

The paper describes a intelligent and connected pedelec which assist a user with the prevention of cardiovascular disease and the rehabilitation from it. The great advantage of the connected pedelec is the application of electrocardiogram (ECG) for the individualized adaptation instead of the heart rate. Furthermore, the pedelec enables an automated adaptation based on ECG records outside. Also a more detailed monitoring of applied power of the user and added power by the motor compared to stationary ergometric bicycles is possible. The possibility to connect further sensor transforms MENTORbike into a assistance systems which considers much user information than existing systems. The service platform provides the assistance and motivation also before and after exercise and support a user with the management of further activities in addition to exercise for the prevention of or rehabilitation from cardiovascular diseases.

The next steps are first user evaluation via questionnaires and system demonstrators to get feedback about user interface and functionalities at an early development stage of the system.

References

1. Paffenbarger Jr, R.S., Blair, S.N., Lee, I.: A history of physical activity, cardiovascular health and longevity: the scientific contributions of Jeremy N Morris, DSc, DPH, FRCP. Int. J. Epidemiol. **30**(5), 1184–1192 (2001)
2. Blair, S.N., Morris, J.N.: Healthy hearts and the universal benefits of being physically active: physical activity and health. Annals of Epidemiology, **19**(4), 253–256 (2009). http://dx.doi.org/10.1016/j.annepidem.2009.01.019
3. Schnohr, P., et al. 2006.: Long-term physical activity in leisure time and mortality from coronary heart disease, stroke, respiratory diseases, and cancer. The Copenhagen City Heart Study. European J. Cardiovasc. Prev. Rehabil.: official J. European Soc. Cardiology, Working Groups on Epidemiology & Prevention and Cardiac Rehabilitation and Exercise Physiology, **13**(2), 173–179. http://www.ncbi.nlm.nih.gov/pubmed/16575269. Accessed 16 Sept 2012
4. Bull, F.C.: Physical activity. In: Ezzati, M. (ed.) Comparative quantification of health risks: global and regional burden of disease attributable to selected major risk factors, pp. 729–881. World Health Organisation, Geneva (2004)
5. Andersen, L.B., et al.: All-cause mortality associated with physical activity during leisure time, work, sports, and cycling to work. Archiv. Int. Med. **160**(11), 1621–1628 (2000). http://www.ncbi.nlm.nih.gov/pubmed/10847255
6. Hillman, M.: Cycling and the promotion of health. Policy Studies **14**(2), 49–58 (1993)
7. Haines, A., et al.: Fossil fuels, transport, and public health. Br. Med. J. **11**, 1168–1169 (2000)
8. Lawrence, D.F., et al.: Carbonless footprints: Promoting health and climate stabilization through active transportation. Prev. Med. **50**, S99–S105 (2010)
9. Killingsworth, R.E., Nazelle de, A., Bell, R.H.: Building a new paradigm: improving public health through transportation. Institute of Trans. Eng. J. pp. 28–32
10. Anon, App-gefahren: Gepida bringt 2013.: Smartphone-Steuerung für E-Bikes—Infos und Bilder. ElektroBike. http://www.elektrobike-online.com/news/e-bikes-und-pedelecs/app-gefahren-gepida-bringt-2013-smartphone-steuerung-fuer-e-bikes.672513.410636.htm. Accessed 15 Sept 2012a
11. Anon.: British Med. J. Rap. Respon. http://www.bmj.com/comment/rapid-responses. Accessed 15 Sept 2012b
12. Kroidl, R.F., Schwarz, S., Lehnigk, B.: *Kursbuch Spiroergometrie: Technik und Befundung verständlich gemacht* 2nd ed., Stuttgart: Thieme (2009)
13. Robergs, R.A., Landwehr, R.: The suprising history of the "HRmax = 220-age" equation. J. Exer. Physiology Online **1971**(1), 1–10 (2002)
14. Kligfield, P., Lauer, M.S.: Beyond the ST segment beyond the ST segment. Contemporary Reviews Cardiovascular Med. Circulation **114**, 2070–2082 (2006)
15. Bjarnason-Wehrens, B., et al.: Leitlinie körperliche Aktivität zur Sekundärprävention und Therapie kardiovaskulärer Erkrankungen. Clinical Res. Cardiology Suppl. (CRICS) **4**, 1–44 (2009)
16. Trappe, H.J., Löllgen, H.: Leitlinie zur Ergometrie. Zeitschrift für Kardiologie **89**, 976–984 (2000)
17. Anon, Das e-Jalopy von.: MTB Cycletec: Erstes E-Bike mit dem Greenwheel des MIT—Fotos und Video. EletroBike. http://www.elektrobike-online.com/news/e-bikes-und-pedelecs/das-e-jalopy-von-mtb-cycletec-erstes-e-bike-mit-dem-greenwheel-des-mit-fotos-und-video.576213.410636.htm. Accessed 15 Sept 2012c
18. Anon.: DrugScope comment and opinion. http://drugscope.blogspot.de/. Accessed 15 Sept 2012d
19. Anon.: EU Regulations for e-Bikes (Part 1) Type-approval legislation and CEN standards. Bike Europe. http://www.bike-eu.com/Laws-Regulations/Safety-standards/2010/8/EU-Regulations-for-e-Bikes-Part-1-Type-approval-legislation-and-CEN-standards–BIK004232W/. Accessed 15 Sept 2012e

20. Ho, T.C.T. exerTrek—a portable handheld exercise monitoring, tracking and recommendation system. In e-Health Networking, Applications and Services, 2009. 11th International Conference on Healthcom 2009. pp. 84–88 (2009)
21. Ketabdar, H., Lyra, M., 2010. ActivityMonitor: assisted life using mobile phones. In: *IUI'10* Proceedings of the 15th international conference on Intelligent user interfaces. pp. 417–418. http://dl.acm.org/citation.cfm?id=1720050. Accessed 10 Sept 2012
22. Gay, V., Leijdekkers, P., Barin, E. 2009.: A mobile rehabilitation application for the remote monitoring of cardiac patients after a heart attack or a coronary bypass surgery. In: Proceedings of the 2nd International Conference on PErvsive Technologies Related to Assistive Environments—PETRA'09. USA: ACM Press, pp. 1–7. http://portal.acm.org/citation.cfm?doid=1579114.1579135. Accessed 11 Sept 2012
23. Reiter, H.: HeartCylce Deliverable 18.B (2011)
24. Boulos, M.N.K., Wheeler, S.: The emerging Web 2.0 social software: an enabling suite of sociable technologies in health and health care education. Health Inf. Libraries J. **24**(1), 2–23 (2007)
25. Barsky, E., Purdon, M.: Introducing Web 2.0: social networking and social bookmarking for health librarians. J. Canadian Health Library Association, **67**(27), 65–67 (2006)
26. Anon.: Healthy Place America's mental health channel. http://www.healthyplace.com/. Accessed 15 Sept 2012f
27. Anon, Kalkhoff.: Impulse Ergo NuVinci Harmony—E-Bike mit Pulssteuerung. ElektroBike. http://www.elektrobike-online.com/news/e-bikes-und-pedelecs/kalkhoff-impulse-ergo-nuvinci-harmony-e-bike-mit-pulssteuerung.660230.410636.htm. Accessed 15 Sept 2012g
28. Anon.: Key statistical highlights: ITU data release June 2012. ITU World Telecommunication/ICT Indicators Database. http://www.itu.int/ITU-D/ict/statistics/material/pdf/2011 Statistical highlights_June_2012.pdf. Accessed 15 Sept 2012h
29. Anon.: Library thing medicine. http://www.librarything.com/groups/medicine. Accessed 15 Sept 2012i
30. Anon.: Patient.co.uk. http://www.patient.co.uk/. Accessed 15 Sept 2012j
31. Anon.: WikiSurgery The free surgical encyclopedia. http://www.wikisurgery.com/index.php?title=Main_Page. Accessed 15 Sept 2012k
32. Eysenbach, G., et al.: Primary care health related virtual communities and electronic support groups. Br. Med. J. **328**(1166), 1–6 (2004)
33. Fox, S.: The social life of health information, Pew Research Center's Internet and American Life Project (2011)

GlobalSensing: A Supervised Outdoor-Training in Cardiological Secondary Prevention

Tim Janus, Torben Kohlmeier, Viktor Marinov, Janina Marks, Christian Mikosch, Michael Nimbs, Thorsten Panke, Jörn Störling, Oliver Dohndorf, Heiko Krumm, Jan-Dirk Hoffmann, Anke Workowski and Detlev Willemsen

Abstract The GlobalSensing system has the aim to supervise (bicycle-) hiking of a group of patients in a cardiologic rehabilitation by taking advantage of user-friendly components. A smartphone application, the patient component, records the vital data and data of the training by using a sensor-broker. These data is transmitted via internet to the group leader and to the cardiologist. This arrangement offers the possibility to monitor the data in real time, to ensure an optimal individual training and to protect the patient against overloads.

1 Introduction

Cardiovascular diseases are still 41.1 % of all deaths, the leading cause of death in Germany [1]. Especially at older people these diseases often lead to death. A total of 92 % of the dead from the cardiovascular system's disease were 65 years and

---

T. Janus (✉) · T. Kohlmeier · V. Marinov · J. Marks · C. Mikosch · M. Nimbs ·
T. Panke · J. Störling · O. Dohndorf · H. Krumm
TU Dortmund, Lehrstuhl Informatik IV 44221 Dortmund, Germany
e-mail: tim.janus@tu-dortmund.de

T. Panke
e-mail: thorsten.panke@tu-dortmund.de

O. Dohndorf
e-mail: oliver.dohndorf@tu-dortmund.de

J.-D. Hoffmann · A. Workowski · D. Willemsen
Schüchtermann-Schiller'sche Kliniken, 49214 Bad Rothenfelde, Germany
e-mail: JHoffmann@schuechtermann-klinik.de

A. Workowski
e-mail: aworkowski@schuechtermann-klinik.de

D. Willemsen
e-mail: dwillemsen@schuechtermann-klinik.de

older [1]. In the year 2008 these morbidities caused a total of 15 % of medical expenses by 37 billion euros for the prevention, treatment, rehabilitation and care in Germany [2]. Because of this background the meaning of primary and secondary prevention increases.

Morbidities of the cardiovascular system are caused up to 90 % by the cardiovascular risk factors: hypertension, dyslipidemia, diabetes mellitus, obesity, physical inactivity and distress [3]. Cardio respiratory fitness constitutes a health-protective factor. "Fit People", people with a high performance, have a longer life expectancy [4]. Regular physical activity reduces the risk of a cardiac event by positive physiological adaptations [5] and is associated with a reduction in cardiovascular morbidity and mortality [6]. Physical inactivity is a major modifiable risk factor [7], but at the same time an increasing lack of movement in Germany is reported [8].

Therefore the cardiac rehabilitation focuses at the physical activity training of cardiac patients. The efficiency of a traditional rehabilitation after discharge from hospital is largely documented and generally acknowledged. The learned strategies should be continued, however it was found in several studies that the cardiac rehabilitation, as a multidisciplinary intervention, improves the functional capacity and psychosocial resources after a cardiac event. Studies indicate, however, that this success is not a long term effect [9]. Outpatient heart groups are the only offer of cardiology secondary prevention. However, these groups are not represented nation-wide and temporally inflexible, so that only 13–40 % of patients with cardiovascular diseases do participate [10].

Innovative solutions have to be created for this gap. Concepts in the area of Ambient Assisted Living by IT-based systems can fill this gap in the future. Because of these recent developments, the chair 4 of the Technical University of Dortmund and the Schüchtermann-Schiller'sche Kliniken in Bad Rothenfelde developed a hiking application as a part of a two-semester project, which allows patients after curative treatment in a cardiac rehabilitation clinic, to attend supervised outdoor endurance training.

Within this project the approach to motivate patients with cardiovascular diseases for a regular endurance training by cycling or hiking was put into practice. Healthcare professionals and computer science students developed an app for a smartphone, which is connected by a mobile data connection to the terminal of the group leader and the clinical application, so that the data of the participants can be tracked and monitored in real time. The vital signs of patients are detected by an ECG chest strap and transmitted via a Bluetooth connection to the appropriate phone.

The system is designed to enable the cardiologist and the group leaders the possibility to control online training parameters like intensity (speed), time, heart rate, ECG data (heart activity) and body temperature of the patient's during the hike, intervening early in the training process and to prevent overexertion and under exertion and further cardiac events. Thereby heart patients loose their anxiety to exercise and become more safety and body awareness.

Additionally the application features an automated alarm system of the vital data. The cardiologist can enter vital data thresholds. If these were over- or underrun, the physician â€‹â€‹receives an alarm message and he is able to send instructions such as "Go slower" to the patient currently. For the documentation of the train and vital data a report of each session will be generated. These reports were sent to the medical application for further training supervisions.

The project is based on the results of "OSAMI—Open Source Ambient Intelligence Commons", which was funded by the Federal Ministry of Education and Research in Germany (ITEA2) [11].

This paper first introduces related work and afterwards discusses the requirements and possible solutions to the system. The following sections describe the system and its architecture in detail concluding with first test results of the system.

2 Related Work

An important area in the context of IT-based assistance systems is the mobile support for health maintenance. There are numerous solutions that have been created to promote fitness. One example is SportsTracker [12], which provides a smartphone application and an associated community. Users can plan trainings or exchange experiences on the website. The application records the user's pulse in athletic activities if an appropriate pulse belt is available, as well as the distance covered using the GPS signal. The data can be viewed during and after the training session and can be uploaded to the website. There the training sessions can be managed, compared to other ones or shared with other users. Other similar projects are Smartrunner [13] and Runtastic [14], which also focus on sport and social networking, but yet not offer medical assistance.

At least a virtual assistance for training exercises is offered by the Mobile Personal Trainer (MOPET) [15]. This project not only records and analyzes data, but also focuses on motivation and guidance. A virtual trainer provides detailed instructions how to perform a certain exercise or gives acoustic navigation hints during a run workout. The measured pulse data directly influence those instructions. That way the user can be motivated to increase speed at low physical loading.

All these platforms have in common that they are oriented to private sporting activities and meet no stricter medical requirements.

For some time, systems are being developed that allow remote monitoring of patients in their familiar surroundings. On that point, certain vital signs of patients are recorded continuously via wireless medical sensors and passed on a mobile device, on which the data are processed and forwarded to a medical facility. By way of example, the system of JK Pollard et al. is mentionable that allows medical personnel to view live data and compare it with historical data [16]. Such systems are useful if an automatic detection of emergencies is too unreliable or technically still very difficult.

Other systems are able to recognize emergencies independently, such as falls or cardiovascular events, and then to submit the patient's current location with the aid of GPS or other location-based data. An example is the AMON project by Urs Anilker et al. [17], which was specifically designed for cardiac and pulmonary patients with high medical risk. The "Wearable Multiparameter Medical Monitoring and Alert System" combines multiple sensor data in order to increase the detection rate of emergencies on a suitable level. Furthermore, special attention was paid to the unobtrusiveness of the sensors so that they can be worn around the clock on the body without interfering in everyday life.

3 Requirements and Solution Approaches

For the envisaged project GlobalSensing it was necessary to meet various social, technical and medical requirements:

- Social requirements
  - **Easy usage of the patient application**:
    The target group of rehabilitation patients spans a big amount of persons who potentially have little experience in dealing with modern media. To solve this problem, an easy to use smartphone application for patients will be developed. All settings are made remotely by a group leader or a physician and sensors are integrated via Plug and Play functionality.
  - **Building trustfulness**:
    In the sensitive area of medical applications, the mediation of trust is a necessary prerequisite with respect to a high user acceptance. The centralization of the medical supervision can arise, however, the feeling of not being optimally cared. To avoid this, the supervising physician can contact the group leader and each patient for individual feedback during the workout.

- Technical requirements
  - **Achievement of data security and privacy**:
    Sensitive patient data are subjected to high requirements of data security and privacy. To meet these criteria, patient data are transmitted via secure connections to the server and are stored in a database. Access to these data is based on the RBAC principle [18] that ensures access only to authorized persons. A backup system prevents the loss of data and a resulting mistrust.
  - **Dealing with unreliable data connections**:
    Hiking in rural areas increases the risk of unreliable or nonexistent UMTS network coverage. To counteract this problem, the data of a hike are collected on the smartphone and are submitted live to the server while connected. In case of loosing the connection, the live data must be transmitted with priority when reconnected. In the non-supervised period, the data must be analyzed

automatically and patients must be informed about the exceedance of the threshold values.
- **Managing sensors and smartphones**:
For an easy usage of the patient application, newly paired Bluetooth sensors must be registered automatically and unregistered in case of loosing the connection. The unique combinations of smartphones and sensors must be clearly assigned to individual patients to ensure a supervision of different groups and patients.

* Medical requirements
  - **Supervised hiking and cycling training**:
  Core of the GlobalSensing system are training sessions supervised by medical professionals. Groups of patients are led by a group leader, who takes on the on-site first aid. For an easy supervision, patients are ordered within the Group Leader and Cardiologist Applications by their constitution through an automatically analysis of the vital data. In case of critical values, the cardiologist can give advices to the group leader or directly to the patients.
  - **Recording of data**:
  For the acquisition of physical constitution in the course of cardiac rehabilitation and prevention a broad spectrum of medical sensors is supported. These are pulse, ECG and S_pO_2 sensors. GPS sensors are used for location determination, virtual sensors can be used for integrating weather information or for aggregated senors.
  - **Analysis and visualization of vital data**:
  The data recorded in the course of training sessions are automatically analyzed for threshold exceedance. This facilitates the supervision of patients and ensures a safe operation. A detailed evaluation of finished trainings is offered only in the physician application.
  - **Avoid confusion for patients**:
  In the area of eHealth applications, confidence and the sense of security are an important part [19]. For this reason, not all vital signs of a patient during a hike or an ergometer training are displayed. Particularly, the display of the ECG curve is reserved for medical professionals.

4 System Architecture

The system consists of two distributed subsystems, the sensor broker and the application system.

4.1 Sensor Broker

The *Sensor Broker* acts as gateway between sensors and applications. It collects, filters, processes, transforms, multiplexes and forwards sensor data values and data streams to a plurality of applications providing independency between sensor data and application needs: the broker is a mediator between sensors and applications in a multi-domain setting, where sensors, devices, other infrastructures, applications, providers of used services, users and other principals involved can belong to different organizational, security, trust, technical and application domains.

The Sensor Broker consists of distributed software components which reside on server computers and terminals, like intelligent devices and smartphones. As depicted in Fig. 1 the sensors are assigned to *Sensor Hub* components, which manage the sensors and provide access to the sensor data for applications. *Sensor Registry Servers* operate directories of sensor attribute data, sensor type metadata and sensor adaptor software. Moreover, *Authorization Servers* provide an integral user and access privilege management to sensor owners, suppliers of sensor adaptor software and users of applications.

Usually, an application needs a sensor of a certain type and location, e.g., an ECG-device, connected to a certain user. Therefore it retrieves a corresponding sensor from a Sensor Registry Server which in turn verifies the access privileges and informs the application about the managing Sensor Hub, which mediates the sensor data to the application. Figure 2 depicts that process of finding and binding a sensor between an application and a Sensor Hub under participation of a Sensor Registry Server and an Authorization Server.

Communication: The communication between the different components of the Sensor Hub as well as the communication between applications and Sensor Hub components applies the REST paradigm [20] and is based on XML messages. The message types are defined by means of RelaxNG schemes [21]. All messages are

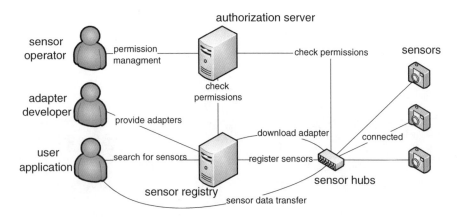

Fig. 1 Sensor broker: components and usage

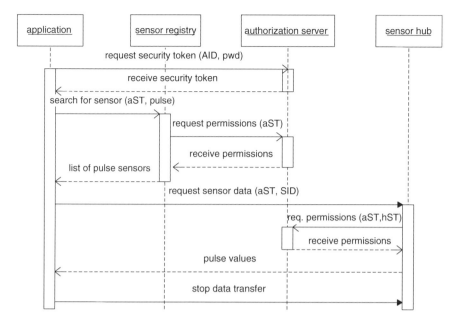

Fig. 2 Process of finding and binding a sensor

exchanged via SSL connections. Sensor Broker components and applications are identified by UUIDs.

Sensors and adaptors: Besides of single sensors, the Sensor Broker supports sensor platforms and virtual sensors. A sensor platform contains a combination of sensors where each can be used as single sensor (e.g., the Zephyr HxM BT heart rate monitor provides speed and distance data in addition to heart rate and R to R interval data). A virtual sensor merges and processes data of several sensors in order to implement a logically defined sensor (e.g., the GPS positions and weather web service retrievals can be linked and provide a virtual local weather sensor).

A *Sensor Adaptor* is a software module which is specific to a sensor type. It controls the basic communication with the sensor, manages the sensor configuration and supports the data conversion between device and broker format. Sensor Adaptors are implemented as OSGi bundles [22] and thus can be loaded and activated on demand at runtime.

Sensor registry servers: The Sensor Registry Servers act as information center of the Sensor Broker system. They cooperate and provide a sensor type metadata repository, a sensor type adaptor implementation repository and a sensor directory. An application which is looking for certain sensor data can request various information about existing sensor types and available sensor instances, particularly about the sensor data semantics and the address of the managing Sensor Hub.

The information is managed under application of the Semantic Sensor Network Ontology (SSN) [23]. Thus complex queries are supported combining different aspects of sensors (e.g., an application can ask for all known weather data sensors of a certain geographic region). Moreover, due to the ontology employment, sensor data is not static and new sensor types and sensor instances can be added during runtime dynamically. In that case, a Sensor Hub which establishes the instantiation of a sensor of a new type retrieves the appropriate adaptor software module from the adaptor implementation repository.

Sensor hubs: A Sensor Hub manages a (possibly dynamically changing) set of assigned sensors which are physically near to the hub. It contains the necessary adaptor modules and like a device driver exclusively manages the configurations and performs the low level access to its sensors. Moreover a hub takes care that all of its sensors are registered in the Sensor Registry Server system.

Towards applications, a Sensor Hub manages bindings between applications and sensors. The bindings can be established and released dynamically on demand of the applications.

Particularly Sensor Hubs which reside on mobile devices temporarily may lose network connectivity. In that case, a Sensor Hub can buffer sensor data and defer its delivery until connectivity is resumed.

Depending on type and location of a sensor, it may be attached either with a mobile device or with a server computer. Therefore we developed an Android embedded service implementation as well as an application server Servlet implementation. Particularly there are vital data sensors which are attached to smartphones via wireless Bluetooth connections.

Authorization server: An authorization server manages the sensor access privileges of system components and applications under fine-grained distinction between several data and configuration access modes.

When a Sensor Hub or a Sensor Registry receives an application request, it in turn requests an application privilege check from the Authorization Server. The Authorization Server verifies the requestor's identity (password authentication), checks the privileges and responds with a corresponding security token. Each security token has limited lifetime. During its lifetime, it asserts the privileges of the bearer and thus reduces the frequency of authorization requests.

Users: In the main, there are three types of Sensor Broker system users: applications, sensor owners and adaptor software suppliers. Applications request temporary bindings with sensors which support the retrieval or the streaming of sensor data to the application. Sensor owners provide sensor equipment and manage their physical connection to a mobile device or a server computer. Moreover they administer the assignment to a Sensor Hub and determine the corresponding access privilege entries of the Authorization Servers. Adaptor software suppliers consider the types, the physical connections and the operating software environment of sensors and provide appropriate adaptor software modules in form of OSGi-bundles.

Scalability: The Sensor Broker is designed as a distributed system of cooperating servers which in principle can support a world-wide and application-independent sensor infrastructure. Registry and authorization servers perform load sharing. Information and load distribution particularly can be based on a hierarchical structuring of geographic regions.

4.2 Application System

The *Application System* (depicted in Fig. 3) consists of a central *Server Application*, and the *Client Applications*: *Cardiologist*, *Group Leader* and *Patient Applications*. The Group Leader Applications reside on the tablet computers of the group coaches, the Patient Applications reside on the smart phones of the patients and the Cardiologist Applications reside on desktop computers. The Server Application is installed on a central server computer.

All application components access sensors via the distributed Sensor Broker system. Moreover, there are REST-based client/server-interactions between the Client Application and the Server Applications.

The communication embodies a star topology, the Client Applications communicate only with the central server but not with each other. Moreover, the client/server-interactions apply the polling paradigm. The clients periodically request information updates from the central server. Thus interactions are initiated by clients only and are addressed to the Application Server. That meets the needs of Network Address Translation (NAT) and dynamic address allocation mechanisms employed by mobile internet providers.

Implementation issues: The Server Application is implemented in C# using the Microsoft.NET Framework and tailored to a Microsoft Windows Server environment. The Cardiologist Application is a C#.NET software, too, but tailored to Windows desktop computers. The desktops computers preferably shall dispose of several monitors in order to facilitate the simultaneous monitoring of multiple patient groups.

The Group Leader Application and the Patient Application are implemented in Java as Android applications and take profit from the Android conceptions of Activities and Background Services.

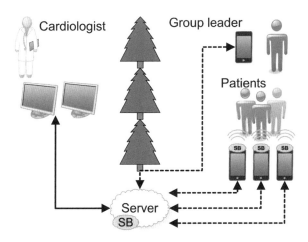

Fig. 3 The application system

Data transfer: The Cardiologist Application (in the modus of online monitoring) as well as the Group Leader Applications periodically poll data updates from the Server Application and vice versa report local status and vital parameter data to it.

The Patient Applications transfer local sensor data according to the LIFO principle, i.e., the youngest data is transferred first. That is due to the changing quality of mobile internet connections (particularly during walks in the woods). There are phases in which the data volume supersedes the available bandwidth and then current data has priority in order to reduce the delay of the newest vital parameter data. The older data, which may be obsolete for online monitoring, however, is buffered and transferred in phases of better connection quality at least after the end of the outdoor training. Thus, the whole training history is available on the server and can be evaluated by the cardiologist later on.

Since the Patient Applications very frequently report vital data to the Server Application, messages, the Server Application sends to Patient Applications need not to be polled but the Server Application piggy-backs them on its responses to the vital data notifications of the clients.

5 GlobalSensing System

The GlobalSensing system is divided into components tailored to the three user groups: The Patient Application, the Group Leader Application and the Cardiologist Application.

5.1 Patient Application

The Patient Application provides the functionality to perform a supervised training. For this purpose, every patient must be equipped with a smartphone containing this application. It performs both the transmission of sensor data to the application server, as well as a monitoring of vital data. With respect to the target group, the installation and administration effort for the patient was reduced to a minimum. The individual steps of a workout with the Patient Application are briefly presented in the following:

When the patient has received his assigned smartphone by the group leader, launching the application is sufficient for starting the training session. At the beginning, all types of mandatory sensors which have to be connected with the smartphone are listed (see Fig. 4). This can be for instance a pulse or an ECG sensor type. Are all sensors connected, the application indicates the group leader its readiness for training start, who finally starts the training.

During the training the Patient Application lists data of the distance covered, the elapsed time and the average speed (see Fig. 5). Furthermore the current pulse

Fig. 4 Patient application: the login screen

Fig. 5 Patient application: the training view

is displayed and a classification to one of the three physical constitution areas (green, yellow or red). In case of the red one, regardless of whether the pulse is too low or too high, an emergency alarm is triggered, which calls the patient to stop and to contact the group leader. Additionally, manually transmitted notifications by the supervising physician are displayed visually and acoustically. When the group leader ends the training, patients are notified visually and acoustically as well. A short summary of the current training data is displayed together with mean

Fig. 6 Patient application: the borg input

Fig. 7 Patient application: the history view

values of the last training's data (see Fig. 7). This helps the patient to assess his today's performance roughly. Finally, the patient is asked to specify a Borg value [24] for the training, which is an important clue for the cardiologist about the subjective intensity (see Fig. 6).

5.2 Group Leader Application

This application enables the group leader to create a training, to supervise it, to receive messages from the cardiologist, to contact the trainees in case of emergency and to end the training.

Fig. 8 Group leader app: training setup

Fig. 9 Group leader app: training view

Fig. 10 Group leader application: a list entry for a single patient

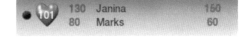

Figure 8 illustrates the various possibilities for the training setup. On the top, you can specify which patient owns which smartphone for the next training session. Due to this, the patient must not log in for identification. Below you can choose if you are going on a hike or start a cycling tour. The drop down menu specifies the level of detail of the vital data displayed in the Patient Application. After pushing the start button, the system waits until the patients have connected all sensors and the training is supervised by a physician before the training starts.

The training view lists all patients who are participating the session (Fig. 9). The heart symbol shows the current pulse of each patient. If it exceeds or falls below the specified thresholds, the background is colored orange or red, depending on whether the pulse is in the hard or soft threshold (see Fig. 10). The emergency button notifies the supervising physician. Choosing one of the listed patients opens the detail view. Here additional personal data of the selected patient are displayed, e.g. the weight, the height and also current values of further sensors. If the patient is wearing an ECG sensor, a special view is available which visualizes the ECG bend.

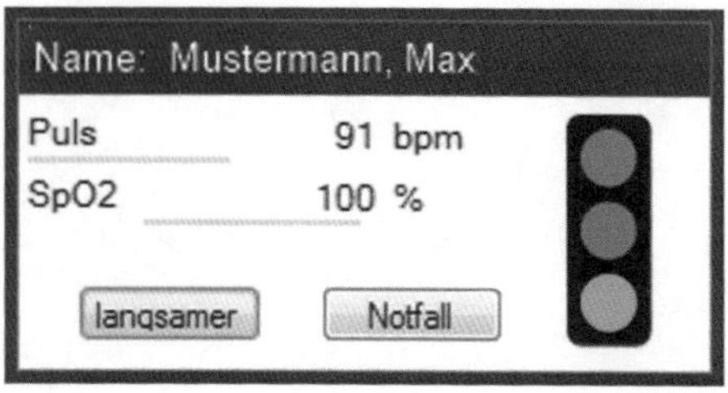

Fig. 11 Cardiologist application: sensor overview for a patient

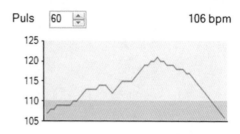

Fig. 12 Cardiologist application: process of a pulse sensor value for 60 s

5.3 Cardiologist Application

This application enables the cardiologist a live monitoring and control of trainings. The collected data are recorded and can be analyzed afterwards in the training history. Moreover the application offers functionalities for managing patients, users and smartphones.

All running trainings and those waiting for a supervisor are offered in a list view. Selecting one of them opens the view for live supervision. An overview lists all sensor values of the participating patients (see Fig. 11). For a quick and easy medical assessment of the patient's constitution the sensor values are colored (green, yellow, red). The thresholds for each sensor were specified by the cardiologist in the forefront. A traffic light highlights the most important sensor value for a medical assessment, in this case the pulse value. Furthermore, the sensor data history of the last 30 s is displayed. It is visualized by a horizontal colored bar under the sensor. At every time, the physician can influence the training by pressing the "slower" or "emergency" button which sends a notification to the group leader. An additionally entered text specifies the problem in detail. Details of the patients are displayed after selecting one of them out of the list view. Those details are personal information and medical information like medication and diagnosis. Processes of sensor values are visualized over a specified period (Fig. 12), colors ease the classification in thresholds.

All sensor data, events and notifications are stored in a data base persistently and are accessible for the cardiologist. Next to common information about the training—the distance covered, the duration and weather conditions—all sensor data, the events and notifications are visualized graphically. By selecting a time period, it is possible to zoom into the diagram. Single sensors can be hidden or

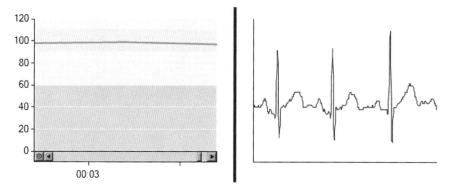

Fig. 13 Cardiologist application: excerpt of a training session in the history (pulse and ECG)

enabled to get a lucid illustration. All data are colored regarding the thresholds (see Fig. 13).

The management part offers the functionality to create, modify and delete users which can access the Cardiologist Application. Additional to common data, you can specify a role based access control. Next to users, you can also manage the patients, e.g. assigning sensors to them and specify thresholds for each sensor. Finally, every smartphone, used in the GlobalSensing system is indexed and labeled for an easy assignment to patients.

6 Evaluation

This evaluation was aimed to verify the technical reliability and the acceptance of the application in terms of design, usability and ergonomics of the surface for a defined target age group. The study was conducted using the "cognitive walkthrough" [25]. This method is a usability inspection method, which is allocated to the analytical evaluation processes. At this juncture (semi-) finished systems were tested by users under real conditions. Users need to think through tasks and activities. Subsequent questionnaires and interviews can discover existing problems and improvement opportunities [26].

The target group of this project are people with cardiovascular diseases. These patients, who will use the device, can be divided into two groups. The first group would be former patients of cardiac rehabilitation, who got information while the rehabilitation about the application and were recommended to use it. Familiarities with this medium can not be expected from these former patients. The second group consists of people, who are interested in the system itself. In this group is an assumption of familiarity with this instrument. The distribution of the age of the study participants corresponds to current rehabilitation statistics. According to the report of cardiac rehabilitation 2012 increases the rehabilitation services in men from the age of 36 and in women from the age of 61 continuously [27].

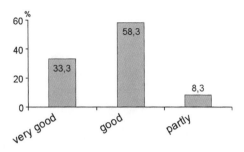

Fig. 14 Results of question number 13: "Do you like this kind of training?"

Overall 12 participants, including 5 women (41.7 %) and 7 men (58.3 %) were tested in the age range of 42 years up to 65 years (average 56 ± 2 years). All subjects knew smartphones, four (33.3 %) of them own a smartphone and ten (83.3 %) participants reported that they know how to use a touch screen. One (8.3 %) subject announced that he or she wears a belt to measure the heart rate while exercising.

The evaluation starts with a presentation of the GlobalSensing system comprising a live demonstration. Thereafter, the participants tested the Patient Application itself. The Cardiologist Application and the Group Leader Applications were served by professional staff. Finally, all subjects filled out a usability questionnaire with 17 questions about the acceptance and user friendliness of the system and the graphical user interface. The results were pretty good.

The implementation of the different applications worked properly. Both could be used in all functions described without failures. In addition to the technically satisfactory operation of an application, it is important that the graphical user interface is clearly and attractively. The results of the questionnaire indicated helpful suggestions for the further development of the system.

The majority of the subjects (58 %) rated the interface as easy and clearly to read and to use. The other participants had only slight problems in understanding the instructions, which can be remedied in an update version. One subject criticizes the size of the script, which would be too small to read out. All participants were able to track the start and end of the training session by the group leader. Finally, the participants were asked whether they like this kind of training. The results are shown in Fig. 14. Without exception, all subjects would recommend this (cycling) hiking application to patients with cardiovascular diseases.

7 Conclusion

The user study clearly demonstrates that the main objectives of the GlobalSensing system were achived, to motivate patients with cardiovascular disease to a regular physical activity and, to give them safety at the same time.

One prerequisite for safety is a stable internet connection. Fading connection qualities and complete link failures are important environmental characteristics. Although the GlobalSensing system ensures that all data will be transferred as soon as possible and the most current data are prioritized, the routes must be selected carefully with respect to the availability of the mobile internet. Otherwise, lacking connectivity may hinder the live monitoring. An interesting solution for that problem would be a satellite backpack, carried by the group leader, that provides a wireless access point for smartphones.

The likewise important medical requirements regarding a meaningful monitoring of individual training sessions and an age-appropriate user interface are met. Helpful suggestions to improve the system were obtained by the study and will be evaluated on participants of ambulatory heart groups on a next stage in the practical field test.

References

1. Statistisches Bundesamt: 2010: Herz-/Kreislauferkrankungen verursachen rund 41% aller Todesfälle. Wiesbaden (2010). URL https://www.destatis.de/DE/PresseService/Presse/Pressemitteilungen/2011/09/PD11_354_232.html
2. Statistisches Bundesamt: Herz-/Kreislauferkrankungen verursachen die höchsten Krankheitskosten. Wiesbaden (2008). URL https://www.destatis.de/DE/ZahlenFakten/GesellschaftStaat/Gesundheit/Krankheitskosten/Aktuell.html
3. Yusuf, S., Hawken, S., Ounpuu, S., Dans, T., Avezum, A., Lanas, F., McQueen, M., Budaj, A., Pais, P., Varigos, J., Lisheng, L.: Effect of potentially modifiable risk factors associated with myocardial infarction in 52 countries (the INTERHEART study): case-control study. Lancet **364**(9438), 937–952 (2004)
4. Sattelmair, J., Pertman, J., Ding, E., Kohl 3rd, H., Haskell, W., Lee, I.: Dose response between physical activity and risk of coronary heart disease: a meta-analysis. Circulation **124**(7), 789–795 (2011)
5. Lee, C., Folsom, A., Blair, S.: Physical activity and stroke risk: a meta-analysis. Stroke **34**(10), 2475–2481 (2003)
6. Nocon, M., Hiemann, T., Müller-Riemenschneider, F., Thalau, F., Roll, S., Willich, S.N.: Association of physical activity with all-cause and cardiovascular mortality: a systematic review and meta-analysis. Eur. J. Cardiovasc. Prev. Rehabil. **15**, 2239–2246 (2008)
7. Balady, G., Williams, M., Ades, P., Bittner, V., Comoss, P., Foody, J., Franklin, B., Sanderson, B., Southard, D.: Core components of cardiac rehabilitation/secondary prevention programs. Circulation **115**(20), 2675–2682 (2007)
8. Predel, H.G., Tokarski, W.: Einfluss koerperlicher Aktivitaet auf die menschliche Gesundheit. Bundesgesundheitsblatt **48**, 833–839 (2005)
9. Kotseva, K., Wood, D., Backer, G.D., Bacquer, D.D., Pyoeraelae, K., Keil, U.: Cardiovascular prevention guidelines in daily practice: a comparison of Euroaspire i, ii, and iii surveys in eight European countries. Lancet **373**(9667), 929–940 (2009). doi:10.1016/S0140-6736(09)60330-5
10. Bjarnason-Wehrens, B., Held, K., Karoff, M.: Herzgruppen in Deutschland—Status quo und Perspektiven. Herz **31**, 559–565 (2006)
11. Busch, C., Litvina, A., Willemsen, D.: Kardiale telerehabilitation auf basis einer flexiblen Plattform für verteilte systeme—das OSAmI-D projekt. e-Health pp. 218–221 (2009)
12. Sports Tracker: Website. URL http://www.sports-tracker.com

13. Smartrunner: Website. URL http://www.smartrunner.com
14. Runtastic: Website. URL http://www.runtastic.com
15. Buttussi, F., Chittaro, L., Nadalutti, D.: Bringing mobile guides and fitness activities together: a solution based on an embodied virtual trainer. In: Proceedings of the 8th Conference on Human-Computer Interaction with Mobile Devices and Services, MobileHCI '06, pp. 29–36. ACM, New York (2006). DOI http://doi.acm.org/10.1145/1152215.1152222
16. Pollard, J.K., Fry, M.E., Rohman, S., Santarelli, C., Theodorou, A., Mohoboob, N.: Wireless and web-based medical monitoring in the home. Med. Inform. Internet Med. **27**, 219–227 (2002)
17. Anliker, U., Ward, J.A., Lukowicz, P., Tröster, G., Dolveck, F., Baer, M., Keita, F., Schenker, E.B., Catarsi, F., Coluccini, L., Belardinelli, A., Shklarski, D., Alon, M., Hirt, E., Schmid, R., Vuskovic, M.: Amon: a wearable multiparameter medical monitoring and alert system. Inf. Technol. Biomed. IEEE Trans. **8**, 415–427 (2004)
18. Ferraiolo, D., Cugini, J., Kuhn, D.R.: Role-based access control (RBAC): features and motivations. In: Proceedings of 11th Annual Computer Security Application Conference, pp. 241–248 (1995)
19. Titscher, G.: Psychologische Aspekte der kardiologischen Telemedizin. Austr. J. Cardiol. **13** (2006)
20. Pautasso, C., Zimmermann, O., Leymann, F.: Restful web services vs. "big" web services: making the right architectural decision. In: Proceeding of the 17th international conference on World Wide Web, WWW '08, pp. 805–814. ACM, New York (2008)
21. Clark, J., Murata, M.: Relax NG specification. http://www.oasis-open.org/committees/relax-ng/spec-20011203.html (2001)
22. OSGi Alliance: OSGi service platform core specification & service compendium—release 4, Version 4.2 (2009)
23. Lefort, L., Henson, C., Taylor, K., Barnaghi, P., Compton, M., Corcho, O., Castro, R.G., Graybeal, J., Herzog, A., Janowicz, K., Neuhaus, H., Nikolov, A., Page, K.: Semantic Sensor Network XG Final Report. Tech. rep., W3C Semantic Sensor Network Incubator Group (SSN-XG) (2011)
24. Borg, G.: Perceived exertion as an indicator of somatic stress. Scand. J. Rehabil. Med. **2**, 92–98 (1970)
25. Dix, A., Finlay, J., Abowd, G., Beale, R.: Human Computer Interaction, 3rd edn. Prentice Hall, New Jersey (2004). URL http://www.hcibook.com/e3/
26. Wharton, C., Rieman, J., Lewis, C., Polson, P.: The cognitive walkthrough method: a practitioner's guide. Usability Inspection Methods, pp. 105–140. Wiley, New York (1994). URL http://dl.acm.org/citation.cfm?id=189200.189214
27. Bund, D.R.: Reha-bericht 2012: Die medizinische und berufliche Rehabilitation der Rentenversicherung im Licht der Statistik (2012)

Part IV
AAL Platforms

Representation of Integration Profiles Using an Ontology

Ralph Welge, Bjoern-Helge Busch, Klaus Kabitzsch,
Janina Laurila-Epe, Stefan Heusinger, Myriam Lipprandt,
Marco Eichelberg, Elke Eichenberg, Heike Engelien,
Murat Goek, Guido Moritz and Andreas Hein

Abstract The Integration and commissioning of AAL systems are time consuming and complicated. The lack of interoperability of available components for Ambient Assisted Living has to be considered as an obstacle for innovative SMEs. In order to ease integration and commissioning of systems knowledge based methods should be taken into account to enable innovative characteristics of AAL systems such as design automation, self-configuration and self-management. Hence, semantic technologies are suitable instruments which offer the capability for mastering the problems of interoperability of heterogeneous and distributed systems. As an important prerequisite for the emergence of knowledge-based assistance functions a standard for unambiguous representation of AAL-relevant knowledge has to be developed. In this paper, the development of an AAL-ontology is proposed as a formal basis for knowledge-based system functions. A prototype of an AAL specific ontology engineering process is presented through the modeling example of a formal representation of a sensor block which is part of an AAL-Integration Profile proposed by the RAALI consortium.

1 Interoperability in the Context of Ambient Assisted Living

In order to enable senior citizens to grow old gracefully in independence from other people and institutions, it is mandatory to reconstruct their familiar environment in respect to their specific restrictions and demands. Technical solutions

---

R. Welge (✉) · B.-H. Busch · K. Kabitzsch · J. Laurila-Epe · S. Heusinger · M. Lipprandt ·
M. Eichelberg · E. Eichenberg · H. Engelien · M. Goek · G. Moritz · A. Hein
ENS—Freies Institut fuer Technische Informatik, Steckelberg 4, 21400 Reinstorf, Germany
e-mail: rw@embedded-network-solutions.de

B.-H. Busch
e-mail: bhb@embedded-network-solutions.de

and services adopting the domain of AAL take the key position for the success to overcome the effects and implications of the demographic change. Due to their often overwhelming system architectures and their similarity to ambient intelligence solutions in general, the integration, installation and putting into operation of AAL-assistance systems is complex and elaborate. Due to the fact that most of the specific AAL-components are currently still remaining in the development process, standardized elements, parts and multi-sensor/multi-actuator networks from the building automation, telemedicine and ICT are applied for the orchestration of the hardware substructure of human centered assistance systems. Thereby, experts for the system deployment and installers are faced with still open interoperability issues. Manufacturers of AAL-system have either to rely on proprietary solutions or to care about a large number of partial disjoint standards and norms.

Nowadays, the obvious lack of interoperability regarding AAL-systems and components is commonly stated as one significant obstacle for innovation and development, especially for SME's. One promising approach to find a sufficient solution regarding interoperability criteria is granted by methods and techniques from the semantic web. Semantic technologies offer a couple of ideas and strategies to handle large, complex, heterogeneous and decentralized systems. Thereby, only the introduction of methods of knowledge processing is appropriate for the attainment of worthwhile targets as design automation, self-configuration of autonomous, distributed systems and the fully automatic self-management of AAL-systems. But initially, a machine-recognizable, formal representation for the unambiguous description of system-related knowledge has to be created. After that, necessary processes of deployment, launching and management of AAL-systems can be supported by knowledge based services. This chapter proposes the development of an AAL-ontology as a formal representation for knowledge-based system components. As an example, the preferred ontology engineering process is outlined through the modeling of a function block covering a RAALI-integration profile for sensor components. Hereafter, the prototypal executed ontology engineering process is the springboard for the integration of domain experts within the standardization process of AAL-ontologies.

Paper structure: In Sects. 1.1–1.3, the reasons for the use of ontology-based approaches for the representation of AAL-integration profiles including general the development effective properties of AAL-systems are explained. Section 2 deals with the state of the art regarding standards and norms and in addition, methods and techniques for the system design from the domain of telemedicine and building automation. Thereby, semantic based technologies which are suitable for the representation of integration profiles are objects of special attention. In Sect. 3.1, the first results from the funded BMBF-project *Roadmap AAL-Interoperabilitaet—RAALI*, the integration profiles, are part of the discussion. The ontology engineering process itself is subject of Sect. 3.2. This includes an expanded modeling example of a sensor node by the function block based approach and the aid of descriptive methods, depicting the current state of research.

1.1 Relevant Properties of AAL-Systems

Current AAL-systems have distinct specific attributes in comparison to classic technology developments. This fact regards the development process of components as well as the stepwise integration of such systems. Therefore, following aspects have to be considered:

(a) AAL-systems do not have a product life cycle in the terms of classical systems. In respect to the paradigm of an aging environment/home, it is necessary that the components are selected due to the altering needs of their residents. This implies that the assistance is mutable. From this point of view, the interplay of system components whose market entries are temporally wide apart from each other is an important property of aging environments. Furthermore, the different stages of an AAL-product as development, launching and the usual runtime can be associated by cyclic interacting processes which need to be synchronized.
(b) Common AAL-infrastructures are heterogeneous and characterized by the orchestration and integration of unusual components which typically do not belong to the domain of AAL. A complete setup of services and components from one single manufacturer or company is currently not available. Therefore, actual AAL-approaches subsume many devices of different distributors which lead directly to a significant workload for system integrators. These experts must handle about components from the telemedicine, building automation, ICT and probably, proprietary sensor solutions like UWB-sensors for vital sign acquisition.
(c) Mainly, AAL-systems depend on the integration of existing components. AAL-system deployment is therefore characterized by the use of descriptive, integration oriented methods because usually the system designer isn't an expert in the specific domain of the target component. (e.g. not an expert for building automation) and hasn't got any knowledge about the implementation.

1.2 Application Based Integration Profiles

One promising approach for the realization of interoperable systems is the definition of application based integration profiles. In the last ten years, these profiles have been approved in the area of medical IT. As a reference, the surveillance of vital signs in home care domains utilizing integration profiles from the *Integrating the Healthcare Enterprise-Initiative* (IHE) and the *Continua Health Alliance* (CHA) is a vivid example of practice [1–3]. These profiles describe the components including their interfaces by the aid of common standards in order to achieve the required Plug'nPlay functionality. The project *RAALI* adopts the principles behind these profiles to the domain of AAL. In addition, the development of a library of function block, covering manifold devices and software services of AAL, is part of the work in order to simplify the outline of complex AAL-systems.

1.3 Knowledge Management as a Way to Solve Interoperability Issues

In order to overcome current innovation obstacles, it is promising to pursue three main objectives:

- Simplification of the AAL-system deployment process through design automation.
- The initial start-up by the aid of self-configuration approaches.
- Maintenance of running AAL-systems by techniques of self management.

In accordance to current interoperability issues, the semantic technologies take a decisive role because the included knowledge bases services are appropriate to solve problems evoked by system runtime, heterogeneous structures and the diversity of system components. The interaction of the different stages in the AAL-product lifecycle as deployment, integration, initial start-up, maintenance and operation is the main reason for lifelong knowledge management. The introduction of knowledge based systems for design automation and the reuse of gained knowledge about the engineering process is crucial for the consecutive procedures of self-management and self-configuration.

1.4 Ontology Based Approaches for Life-Long Knowledge Management

The essential part of every knowledge based system is an ontology. Ontologies are known as a promising concept to describe a thing (resp. AAL-integration profile) or an object in a universal manner so everybody including machines gets a better understanding about the feature of interest and its properties. Logic based languages as *OWL 2*, established reasoning infrastructures for the implementation of inference processes and in addition, a couple of valid tools are a solid basis to start from. The introduction of an AAL-ontology, however, requires special considerations; the complexity of single AAL-components and complete AAL-systems is a challenge for ontology design. Each available technical component can be an element of an AAL-system if this system is designed for the specific functions of the high level application context. The apparent dynamic of AAL-systems prevents the design and finalization of AAL-ontologies. In addition, the large number of existing and pronounced products complicates the definition of a universal system of concepts. On the other hand, it is essential to evaluate and maintenance. AAL-ontologies continuously if they are determined to take the key role within a knowledge management infrastructure for the warranty of stable system functions during the development process or during runtime. To fulfill these demands in a sufficient manner, it is essential to integrate manufacturers, designer, system architects and installers within an unending standardization process.

2 State of the Art

2.1 Description Methods

2.1.1 Description Methods for Medical IT-Solutions

As mentioned in Sect. 1.2, the definition of use case based integration profiles for the realization of interoperable systems has been approved in the area of medical IT. These complementary integration profiles belong to a higher level of abstraction compared to established communications standards. Well known application scenarios from the health care system are the integration profiles from the initiatives *Integrating the Healthcare Enterprise* (IHE) and the *Continua Health Alliance*. In order to achieve a maximum of interoperability by the systematically use of standards, the IHE was founded by user and companies in 1998. For this purpose, the typical work flows in health care institutions were modeled and adopted to integration profiles which cover the transactions between the involved IT-systems in accordance to internationally accepted standards of biomedical technics.

2.1.2 Description Methods for Building Automation

Typical aspects of building automation components and topologies are described by the standard *IEC 61499* [4]. This standard comprehends the definition of a system itself, device, application and function block. The automation system can be described via a network of distributed sensor-/actuator components which are dedicated to a distinct process. The representation regards hierarchical aspects which is essential for the decomposition of the system due to a arbitrary number of levels. The aspect of decomposition can be applied for processes and components as well as for function blocks which encapsulate applications on a lower level (e.g. driver). Based on the input–output orientation, it is easy to combine and link different objects. The processing is executed within the basis-function blocks; the execution control charts of these elements are connected. Thereby, a automaton based approach including event chains resp. state sequences can be implemented. The *IEC 61499* standard provides a textual representation as well as a non-standard graphical representation for the system design itself. In the field of building control the *VDI guidelines 3813–3814* regulate with 48 different function block types for sensors, actuators, HCI, etc. algorithms the system design. The *VDI guideline 3813* enables the description of so-called automation schemes, which can be used as a rough draft in the planning phase. This means that system designers can focus on the functional relationships of the design and must not care about the detailed implementation knowledge. Thus, the *VDI 3813* grants a technology-independent specification of systems and their interaction.

2.1.3 Description Methods for ICT-Systems

A basic task in the modeling of information processing systems is the mapping of distributed processes and their communication. There are manifold possibilities for the modeling of often transforming, embedded technologies and systems. Basically, one can distinguish between constructive and declarative/descriptive methods. In the ICT sector, constructive methods are preferred. Constructive methods provide language elements which are important for the deployment of a first abstract system model. In the following step, the model can be defined, configured and implemented by semi-automatic or fully-automatic procedures. In the area of ICT one well known language is *Specification and Description Language* (SDL). SDL [5] leads directly to the development of a system structure which is called hierarchical decomposition (refer to Fig. 1) whose integral parts finite consist of *CEFSMs* (Communicating Extended Finite State Machine). Furthermore, in order to improve the modeling process, the target is an ideal machine with infinite resources like memory, processor time and program threads. This leads together with the message-orientated communication between the processes (analogous to UML State Charts) to a loose coupling of the CEFSMs. The result is a hierarchically structured, automaton-based model with a message based communication. A recent, more powerful language, but with a much broader focus is *the Unified Modeling Language* (UML). UML builds also on an automatic model based on.

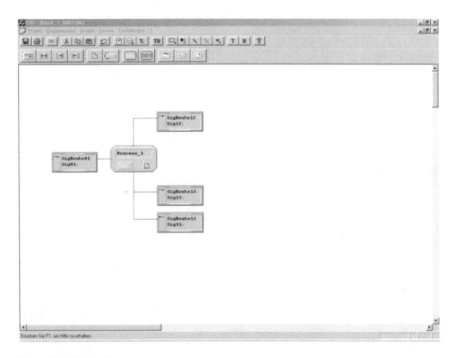

Fig. 1 SDL-block diagram

2.2 Semantic Technologies

In the context of modern information systems as an ontology is the explicit specification of conceptualization of an application area considered [6]. In accordance to the general understanding of information sciences, ontologies are technical artifacts [7–9, 16] which are composed of a vocabulary and the coherent explicit assumptions regarding the meaning of the vocabulary. For the description of the vocabulary, logic-based languages can be used with their most prominent representative, the *Web Ontology Language* (OWL).

2.2.1 The Ontology Language OWL 2

The logic-based ontology language OWL [10] is a W3C recommendation from 2004 with the successor OWL from 2009. One central target of OWL is the description of complex ontologies; for practical use there has to be a balance between the inference and the expressiveness. The widespread standard *OWL-DL*, a predecessor of *OWL-2*, is based on the expressive description logic *SHOIN (D)*. OWL-DL includes the semantics of the class description logic *Attributive Language with complement (ALC) plus transitive roles (r+)*, whereby the class $ALCr+$ is abbreviated with S. Other language elements H (sub role relationships), O (closed classes), I (inverse roles), N (number of restrictions) and D (data types). The standard OWL2 of the W3C has a lot of modifications compared to OWL-DL which became necessary due to practice with OWL-DL. Based on the logic SHOIN considering a number of restrictions, the logic SHOIQ was designed. On this basis, including expansions and language features which influence the handling but not the expressiveness of the dedicated language, the *OWL-2 SROIQ* underlying logic has been developed.

2.2.2 Basic Ontologies

In [11] several types of ontologies were introduced. There is a distinction between top-level ontologies, domain ontologies, task ontologies and application ontologies. Thereby, it has been proven that it is much more effective to separate common knowledge from domain specific knowledge and to store it within top-level ontologies. Basic ontologies, also known under the terms upper ontology, top-level ontology, foundation ontology and hyper ontology describe common knowledge that should be used on all domains and applications. Examples for basic ontologies are *DOLCE*, *OpenCyc* and *Sumo*. *DOLCE* is a result of the so-called *WONDERWEB* project which has been accomplished by Nicola Guarino and his team from 2002 to 2004. This in OWL formulized and in several versions available ontology is still the central pattern for the design for basic ontology approaches. A widespread used basic ontology is *DUL (DOLCE + DnS Ultralite)*.

DUL is a differentiated axiomatized framework which offers all the basic concepts and roles for the modeling of systems such as physical artifacts, abstract etc. The generic approach of DUL allows use in any application area.

2.2.3 Ontology Design Pattern

Ontology Design Pattern (ODP) are modeling patterns which can be suggested as implementation independent solutions for commonly recurring problem classes [12]. ODPs gained prominence in the context of the Description and Situations (DnS) Fontology. DnS is context-sensitive description of types and relationships while expanding the descriptive characteristics of DOLCE. Content ontology design pattern are primarily discussed in the context of DnS.

2.2.4 Sensor Ontologies

From the abundance of the currently available sensor ontologies, the exemplary representation of the W3C Semantic Sensor Network Ontology (SSN) is best suited for the creation of AAL ontologies because this ontology is a direct result from the analysis of most of the relevant existing sensor ontolgies. From the work of the OGC Sensor Web Enablement under the name [14], a service-oriented architecture and and a couple of standards have emerged. Inter alia there are four languages, which deal with capabilities of sensors, the measurement variables and other characteristics. Besides the classification of sensors and a process-oriented view of measurement systems, interoperability and data exchange are discussed. However, the semantic interoperability for the construction of self-organizing sensor networks is still not supported. The goal of the SSN—W3C Semantic Sensor Network Incubator Group was to create an abstract view of sensor networks. The main objectives of the work were the self-organization in terms of installation, management and retrieval as well as the understanding of a sensor network and its data by services of higher order. In the first stage of the development, an ontology for the representation of single sensors and complete networks for the usage within web applications was designed. The classification of the sensor components themselves as well as the coherent conclusions regarding their properties, related measurement values, the origin of sensor data and the orchestration of sensor nodes is supported by the development macro tools. It is aimed to adapt the existing standards of the Open Geospatial Consortium (OGC) . The SSN ontology is based on domain ontology as a basic ontology. It is an alignment of the SSN ontology and the DOLCE UltraLite Upper Ontology to normalize the ontology structures. In addition, it is aimed to integrate other ontologies and linked data resources. Sensor networks are completely different from IT systems (e.g. SOAs). Therefore, for the completion of the SSN ontology it is essential to regard the event driven characteristics as well as the spatialtemporal context of the data to consider. In addition, the ontology does not cover specific application domains, measurement units, time and space, and mobile aspects.

3 Methodology

Actually in respect to the state of research, there are no common techniques for the description for AAL technologies which cover the entire spectrum of the necessary technologies for all possible AAL applications. For this reason, in the following an approach is proposed, which can serve as a graphical representation of integration profiles.

3.1 Block Diagrams for Integration Profiles

After a survey and deep analysis of existing AAL-systems and components, a function block based description method for heterogeneous distributed systems was proposed by the RAALI-project (refer to Fig. 2) which implements graphical representations for services, actuators or sensors. Thereby, the general function block (FB) is only defined by input–output relations similar to simple mathematical functions with parameters and return values, and serves as a blueprint for much more complex structures e.g. block types as sensors or HCIs. Expanding the general FB by a sensory component, graphically characterized by an additional input, the function block for a sensor is fully described. An actuator is described in the same manner—only the direction of the input–output indicating arrow is opposite and the index switches from S (sensor) to A (actuator). In the case of a user interface (HCI—Human Computer Interface), there is a bidirectional arrow that symbolizes both user input and system output. If the feedback path is removed, the HCI is reduced to an actuatory or a sensory component. Besides the pure graphical representation each component is described in detail a separate report. Through the connection of the inputs and outputs of the individual components, it is possible to arrange them into a logic and functional sequence. This requires conformity of the respective affiliated inputs and outputs. To reduce the complexity of real scenarios by a higher level of abstraction, the encapsulation of functional related components through super blocks is beneficial. By this way, it is possible to aggregate individual functions to complex structures only by the adjustment of the input–output relationships. For additional details about the use of function blocks refer to [15]. The selected form of description is available by a informal representation. For the introduction of automatic assistance functions as design automation or the self-configuration of AAL-systems, it is necessary to gain a formal representation of AAL-integration profiles.

3.2 Engineering Process of an AAL-Ontology

In addition to the block diagrams for the AAL-integration profiles presented in Sect. 3.1, as an output of the project Standardisierung eines semantischen

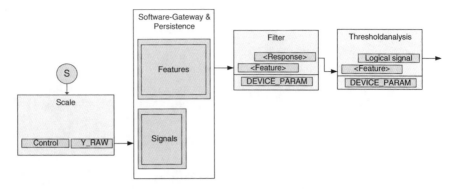

Fig. 2 Block diagrams for AAL-integration profiles

Laufzeitsystems zur Foerderung der Interoperabilität von AAL-Komponenten a formal representation has been proposed. Thereby, the expertise of the included ontology engineers is as important as the involvement of domain experts who hold the knowledge about the area to be modeled. In similarity to the software engineering area, diverse engineering approaches focused on interdisciplinary development processes have been established. An OTK methodology for the design of ontology based knowledge management infrastructures based on CommonKADS was proposed by [16, 17]. In accordance to classical project management methods the ontology engineering process consists of the consecutive steps feasibility study; kick off, refinement, evaluation and application and evolution. Usually, this process is iterative and executed by multiple times. In the following the development of the prototype is described. For this reason, the region under examination is restricted to a non-representative group of users, stakeholders and experts.

3.2.1 Feasibility Study

During the Feasibility Study possible applications and solutions were identified and analyzed due to their applicability and relevance for practice by the users and stakeholders. The following applications have been selected as the basis for much more complex scenarios:

- Device Discovery and Selection.
- Data Discovery and Binding.

In order to preserve the proximity to existing components at the beginning a sensor was modeled instead of a general function block. The core element of the feasibility study is the collection of data sources. Own project experiences and conducted expert interviews allow a reduction of the set of relevant sensors for AAL. The DIN standard 1319–1 .. 4 [19] provides valuable information regarding the general characteristics of sensors, measurement methods and data analysis, allowing a structural point of view to such systems. In particular for sensor ontologies, there exists a lot of information from the OGC and the W3C Sensor Network Incubator Group. As the same in the area of AAL, it is aimed to abstract complex sensor network infrastructures and their binding to IT-systems also as the encapsulation of services. While the actual prototypical engineering process is based more on methodological sources, future approaches primarily will integrate expert knowledge, manufacturer knowledge and product knowledge.

3.2.2 Kickoff-Phase

In accordance to [20] an ontology requirements specification document (ORSD) was provided at the beginning of ontology development.

Objectives and the Role of the Ontology: It became apparent that two different views at one and the same function block are required. The system integrator has a purely conceptual view at a function block. At this point it is important to note that the system integrator is thinking about a specific measure and description for the entity heart instead of the biological unit itself including parameters which describe the coherent sequential values of heart beat. This means, it is only important to distinguish between different measurement and description concepts and to refer to associated data bases. Following concepts were selected for the reference of sensory databases. The concepts (see Table 1) are incorporated into a so-called description pattern—one ontology design pattern which is suitable to represent descriptions consisting of manifold concepts.However, automatic support functions require a pure technical view which describes the physical device completely—this fact implies a description of device properties, parameters and values, measurement processes and the communication interfaces. Contrary to the conceptual view which describes the intention of a function block, the technological view is dedicated for the provision of detailed information for the automatic knowledge processing (e.g. for the identification of

Table 1 Concepts from the integrator's perspective

Concept	Example
Domain	Building automation
Procedure	Conditions, plan
Aim of measurement	Detection of residents
Target value	Status variables (door closed)
Location	Room
Measurement frame	5 h campaign
Platform	EnOcean
Operation model concept	Measurement principle
Interface concept	wired/wireless communication, 802.15.1 BT, 802.15.4/Zigbee
Product line-up	wired/Identiy of the manufacturer, product family

interoperable sensors which deliver data to a processing function block. Both views are mutual complements.

Methodology: The main structural idea deals with the description of the various function blocks by ontology modules. Thereby, a function block is a graphical representation of an ontology module. Thinking about an orchestration of different functions blocks covering a scenario or in particular an AAL-application, this array of blocks is equivalent to the ontology with additional meta-information. For the modeling, the usage of OWL 2 and DUL − DOLCE + DnS Ultralite (for the basic ontology) was selected. Based on the work of Gangemi [12] each function block is projected by a so-called ontology design pattern (ODP). In order to improve the interoperability of parallel approaches all ODP-concepts and ODP-roles are derived from the basic ontology DUL (DUL-Alignment). The last step consists of the differentiation of the function blocks. The semantic part of work is done by a device specific implementation of sensor-ODP.

Competency Questions: In addition to non-functional demands the specification of competency questions (CQ) is decisive for the success and impact of the whole engineering process (refer to Table 2). The competency questions give a first impression about the necessary vocabulary, its classes and roles. Due to the questions it becomes clear which answer shall be derived from the upcoming ontology. This is exactly the point where the SSNs had been expanded [13].

Concept Retrieval: Finally, the ontology relevant objects are derived from the responses to the competency questions and the questions itself. Due to the fact that the W3C SSN-ontology has been adapted to this procedure, the vocabulary exists of integration oriented concepts and roles.

Refinement-Phase: One important step in ontology engineering is the refinement phase, essentially for the transform of semiformal ontology into a machine-recognizable representation. In order to formalize the ontology, two fundamental methods are recommended by [16, 17]: The Top-Down method as well as the Bottom-Up-method. The Bottom-Up-method is based on procedures and tools for the automatic text analysis. If documents contain all the system relevant information, the semi- or fully automatic generation of taxonomies leads to the complete summary of concepts of a domain. Due to the fact that all possible concepts will be regarded, a consecutive process much more focusing on the aspects of central issues will be executed which induces many efforts due to system complexity. For this reason the bottom-up method is not used. On the other hand, the

Table 2 Competency questions

Use case class	Competency questions
Data discovery and linking	Which observations are sufficient to criteria as domain, task, measurement object, location, time window, platform, operation model, network interface and identity
Device discovery and selection	Which devices are sufficient to criteria as domain, task, measurement object, location, time window, platform, operation model, network interface and identity

preferred top–down approach leads from an abstract view on the emerging knowledge model to an increasing specialization of concepts and roles.

In this context, the reuse of established basic ontologies like DUL as well as the W3C domain ontology SSN is beneficial for practical usage. This condition allows the construction of self-consistent, evaluated conceptual framework which can be expanded systematically by the consideration of the competency questions. For our approach, we selected a couple of diverse ODPs including the DnS pattern as the most important one. The DnS-pattern (refer to Fig. 3) consists of a distinct pattern describing situations (S-ODP) and a description pattern itself (D-ODP). The D-ODP is used to associate concepts (DUL: Concept) by the aid of the role (DUL: uses concept) with a context, represented by (DUL: description). It can be interpreted as an abstract, conceptual description of a context e.g. a sensory data source. The S-ODP is used for the technological view. In the S-ODPs, the mapped situation is expressed by a sum of entities (DUL: Entity) under the usage of the role (DUL: is setting for). The DnS-ODP is a composition of both pattern and relates a description (DUL: description) to a situation. Thereby, it is possible to relate single entities of the S-ODP to concepts of the D-ODP and associate them with a specific context. The competency questions typical for this pattern are:

- Which sensor situation complies with the sensor description?
- Which sensor descriptions can be accomplished by a sensor situation?

The classes and roles from the W3C SSN ontology were applied due to compatibility aspects (refer to Fig. 4). In order to improve the conceptual view (from

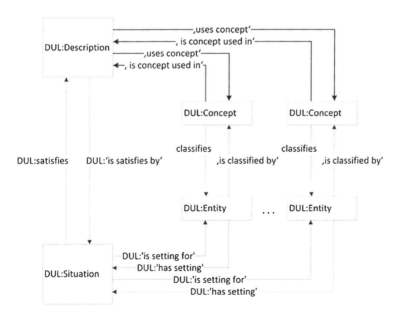

Fig. 3 DnS pattern (www.ontologydesignpattern.org)

AAL-system integrators), additional classes have been introduced which were derived from the class DUL: Concept (refer to Fig. 4—dotted area). The introduced classes are not physical objects and therefore not derived from DUL: Physical object. The classes care about the process of information exchange and cover universal properties of sensor components and their data for the implementation within the context itself. The concept Domain Concept relates the entity feature of Interest to a specific context.Through the D-ODP, it is possible to integrate the instance of a usual, network-compatible scale within the context of the application dry-weight monitoring—the instance gets a context related role. Thereby, entities can be embedded within various situations and get a context independent meaning. If the entity telephone is an integral element of an emergency indication system, there exists also an entity for other possible contexts (account of telephone charges). But however, by the classification of the universal component telephone through the concept emergency indication devices, the telephone gets its role within the context emergency indication system. In the following section the new concepts from the prototypical engineering process are introduced. In particular, it is expected that the knowledge will be extended by repetitive runs of the process considering domain experts. The concepts are therefore only a basis of discussion. Contextualization of measurement objects:

- *Domain Concept* specifies a domain within a finite set of AAL-domains (BA, telemedicine). Domain Concept contextualizes the SSN class Feature of Interest, which super class can be either DUL: Event or DU: Object. By this way, Domain Concept associates an arbitrary object or event with a domain.
- *Task Concept* contextualizes the SSN class Feature of Interest with the focus on the planning of the measurement campaign. Best practice suggestions from manufacturers are expected (specific application regarding exact rules for the handling).

Contextualization of object properties:

- *Objective Of Measurement Concept* relates properties of a measurement object to the target of application (The detection of residents in a room through the evaluation of BA sensors like door contacts).
- Objective Of Measurement Concept describes the result of the measurement or the feature extraction process (possibly complex processes executed over different data processing layers).
- *Physical Quality Concept* associates the properties of a measurement objects with the characteristics of the measurement process itself.

Contextualization of the sensor:

- *Location Concept* is responsible for the spatial definition of the operation area. Possible instances are indoor or outdoor. A fine granular differentiation is planned within the taxonomy itself.

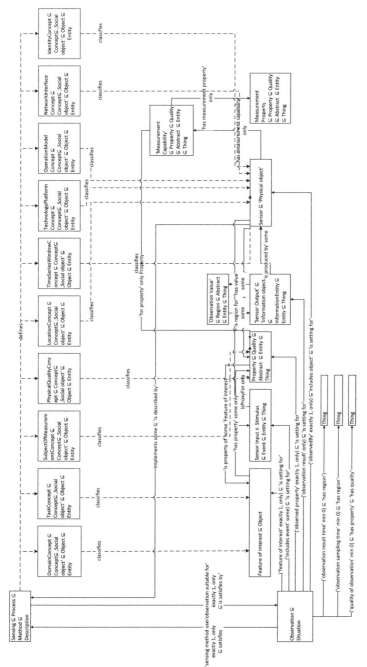

Fig. 4 AAL-sensor ontology (SSN [13] and AAL D-ODP

- Time Series Window gives information about the data properties regarding time. This is important for the interpretation of dynamic processes and various contexts.
- *Technology Platform Concept* is a concept for the limitation of the inference based search processes. The examination area is restricted to a subset of available AAL-components which can be associated to a distinct technology platform. The concept contextualizes DUL: Sensor, a physical object executing real measurement processes. Furthermore, the concept is appropriate for the description of home automation concept from a systemic perspective. This means that a building automation system of a manufacturer X which implements LON and an IP-based control level and provides a BACNet-IP interface is mapped to a sensor in accordance to the BACNet-standard. This implies that larger infrastructures can be subsumed by data points, named by the term sensor. For real scenarios, it is obvious that specific implementation characteristic should be considered in practice.
- *Operation Model Concept* describes concepts for sensor functions, result classes and working principles Network Interface Concept contain interface concepts for e.g. the wireless communication. In particular, the restriction to standards enables a limitation of the examination area.
- *Identity Concept* enables the system integrator to search for components of single manufacturers or second source products during the design of non-interoperable AAL-systems. Thus, the Identity Concept is another approach for specialization outside the taxonomy.

Formalization: For following reasons it was possible to skip the usual first steps in ontology development (creation of the taxonomy):

- A basic ontology following DUL was used.
- A complete domain ontology was available (SSN).

Instead, the taxonomy has been extended to the classes described above. All necessary parts were taken from the DUL-vocabulary. Considering the upcoming meetings with domain expert, any prognoses regarding new concepts cannot be made. But however, new concepts will be specializations of super concepts (e.g. contained in DUL). The refinement phase is closely associated with the evaluation phase. Detected errors from the different evaluation processes (user-, technological- or ontological view) enter a cyclic repetition of the refinement phase for successive improvement.

4 Discussion of Results

After the completion of the first prototypical refinement phase, as an integral part of an AAL-ontology a formalization of a sensor function block was derived. According to technological aspects the ontology module was analyzed and will be evaluated through subsequent expert workshops. Referring to [16] evaluation approaches regarding the technical view, the user position resp. the application view and the ontological view (deployment view) will be executed. Analysis and evaluation from the technological perspective: In addition to the analysis of linguistic conformity several modeling problems regarding the satisfiability of class definition have been fixed by the aid of different inference machines. The analysis of the runtime properties of the coherent inference processes, of consumed memory and the scalability due to the growth of A- and T-box is the next step after finishing the expert workshops and the implicated model modifications and instantiations of real products. The start of the evaluation of the prototype regarding the user perspective is planned for the first workshop. There is the question whether the defined ontology meets the requirements gathered in the previous steps. At this point it is beneficial to refer to the central ORSD and the related competency questions. The result of the workshop is still uncertain. Analysis and evaluation from the modeling perspective: An acclaimed approach from 2000 for checking the consistency of ontological taxonomies is the Onto-Clean method. OntoClean works with expressions from the classical philosophy (e.g. essence, rigidity, identity, unity) and proposes the establishment of a meta-notation for the ontology classes. A meta-notation expands every conceptual part of an ontology through the appendage of coherent properties due to consistence criteria. In particular for concepts which are interrelated by super-/subclass relations. For further details refer to [21]. The analysis of taxonomies and their evaluation will be executed after the integration of the expert workshops.

5 Conclusion and Outlook

The research results of the BMBF-project RAALI provide a graphic based description approach for the deployment of AAL-systems. But however, for the technical support in accordance to the principles of design automation, it is indispensable to formalize the RAALI integration profiles. Therefore, this chapter focuses on the prototype of an ontology engineering process which is determined to collect and aggregate AAL-specific knowledge through expert interviews and workshops, ending with the formalization of the gathered information. After a successful test case, the proposed ontology engineering process will be repeated with experts from the area of building automation, telemedicine and telecommunication. By this way, a representative impression of the expert knowledge adopting the involved disciplines of AAL can be achieved and formalized to an ontological representation. The processes will be repeated several times to increase the quality of knowledge significantly. It is assumed that these expert workshops end in June 2013 in form of a proposal for the standardization of AAL-ontologies.

References

1. Integrating the Healthcare Enterprise (IHE): IT infrastructure technical framework, Revision 8.0. http://www.ihe.net/Technical_Framework/index.cfm#IT (2011)
2. Integrating the Healthcare Enterprise (IHE): Patient care coordination technical framework revision 7.0 (2011)
3. Continua Health Alliance: Continua Design Guidelines, Version 2012
4. DIN EN 61499-1, Funktionsbausteine fuer industrielle Leitsysteme—Teil 1: Architektur (IEC 61499-1:2005); Deutsche Fassung EN 61499-1:2005, Beuth Verlag
5. ITU-T Rec. Z.100 (11/99) Specification and description language. www.itu.int/ITU-T/studygroups/com10/languages/Z.100_1199.pdf
6. Gruber, T.: Towards principles for the design of ontologies used for knowledge sharing. Int. J. Hum. Comput. Stud. **43**, 907–928 (1993)
7. Noy, N., Hafner, C.: The state of the art in ontology design—a survey and comparative review. AI Magazine. **36**(3)
8. Maedche, A., Staab, S.: Ontology learning for the semantic web
9. Fensel, D.: Ontologies: Silver Bullet for Knowledge Man-Agement and Electronic Commerce. Springer, Berlin (2001)
10. Boris Motik, Peter, F., Patel-Schneider, Bijan Parsia, (eds.) W3C Recommendation: OWL 2 Web Ontology Language: Structural Specification and Functional-Style Syntax. http://www.w3.org/TR/2009/REC-owl2-syntax-20091027/, (2009)
11. Guarino, N.: Formal ontology and information systems. In: Proceedings of FOIS'98 (Formal Ontology in Information Systems). IOS Press, Trento, Italy (1998)
12. Gangemi, A., Presutti, V.: Ontology Design Patterns. In: Handbook on Ontologies, 2nd edn. Springer (2009)
13. W3C Incubator group report 28 June 2011: Semantic sensor network XG final report. http://www.w3.org/2005/Incubator/ssn/XGR-ssn-20110628/ (2012)
14. OGC White Paper—OGC sensor web enablement: Over-view and high Level architecture. http://www.opengeospatial.org/pressroom/papers (2012)
15. Lipprandt.M , et al: Beschreibungsmethodik für AAL-Integrationsprofile. Proceedings GMDS (2012)
16. Sure, Y., Studer, R.: On-To-knowledge methodology final version. EU-IST-Project IST-1999-10132 On-To-Knowledge (2002)
17. Sure, Y., Staab, S., Studer, R.: Ontology engineering methodology. In: Handbook on ontologies, pp. 135–152, 2nd edn. Springer, ISBN 978-3-540-70999-2 (2009)
18. Schreiber, G., Akkermans, H., Anjewierden, A., de Hoog, R., Shadbolt, N., Van de Velde, W., Wielinga, B.: Knowledge Engineering and Management—The CommonKADS Methodology. Massachusetts Institute of Technology
19. DIN 1319-1..4, Beuth
20. Suárez-Figueroa.MC., Gómez-Pérez,A., and Boris Villazón-Terrazas: How to write and use the Ontology Requirements Specification Document. In: Proceeding OTM '09 Proceedings of the Confederated International Conferences, CoopIS, DOA, IS, and ODBASE 2009 on the Move to Meaningful Internet Systems: Part II, pp 966—982, Springer, Berlin, Heidelberg (2009)
21. Guarino, N., Welty, C.: An Overview of Ontoclean. In: Staab, S., Studer, R. (eds.) Handbook on Ontologies, pp 201–221, 2nd edn. International Handbooks on Information Systems. Springer (2009)

Methods and Tools for Ontology-Based Configuration Processes of AAL Environments

Tom Zentek, Alexander Marinc and Asarnusch Rashid

Abstract In the last few years, technologies with a semantic middleware were established under the AAL platforms. These technologies simplify the reaction to various and rapidly changing needs of assisted elderly. Building on this established semantic basis, the set-up and configuration of individual use cases can be simplified. Up to now it is hardly possible to set up an AAL environment without technical knowledge. This paper presents how the process of set-up and configuration of AAL environments based on ontologies could proceed. The support starts with the developers of the use cases, continues with the integration in the middleware and ends up in the maintenance during operation. At the end, different configuration support tools based on the semantic middleware universAAL will be described.

1 Introduction

In the last few years, the care for the elderly as well as the support of caring relatives and the nursing personnel has advanced significantly with the help of the socio-technological approaches of Ambient Assisted Living (AAL). Use cases have and are still being developed for different "need" scenarios [1]. Thus, it is possible to cover the rapidly changing requirements of this group much better.

T. Zentek (✉) · A. Rashid
FZI Reserach Center for Information Technologies, Haid-und-Neu-Str. 10-14,
76131 Karlsruhe, Germany
e-mail: zentek@fzi.de

A. Rashid
e-mail: rashid@fzi.de

A. Marinc
Fraunhofer Institute for Computer Graphics Research IGD, Fraunhoferstr. 5,
64283 Darmstadt, Germany
e-mail: alexander.marinc@igd.fraunhofer.de

However, it is shown, that there will never exist just one solution for all need scenarios.

In current research approaches, the necessary adaption of the AAL environment to the person in need of care is only possible with expert knowledge [2].

For their support, the use cases utilize interconnected sensors and actors that pass on their information to a middleware. For instance, a door lock mechanism might be connected to electronic devices and remind the house resident of the stove still being switched on when leaving the apartment.

Even in a simple use case there is a variety of possibilities to adjust the use case to the resident and his environment.

Which devices are to be monitored? Are there several alternatives to leave the apartment? How is the reminding message supposed to be delivered to the resident? What should be done in case the resident does not react to this message? Presently, these are all questions that still require a complex configuration of an expert and thus interfere with the expansion of AAL systems.

In the following chapters, an ontology-based approach to this problem as well as the implementation and the first evaluation will be presented.

Chapter 2 begins with the description of the configuration requirements in the individual aspects of an AAL environment as well as the correlating problem-solving approaches from other domains.

The configuration process that was developed based on these considerations will be demonstrated in Chap. 3. It begins with the developer of use cases and ends with maintenance operations that influence the configuration of the AAL environment. Out of this, ontology-based approaches will be presented that reduce complexity by means of a simple configuration and thus significantly simplify the construction of an AAL environment.

Subsequently (see Chap. 4) tools based on the solution and their implementation will be presented and finally will be evaluated in Chap. 5.

2 State of the Art

In order to gather the configuration requirements of AAL environments correctly, the entire process from planning to daily use needs to be considered. Furthermore, the applied framework affects the possibilities in configuration.

Due to the consolidation and fermentation of a variety of AAL frameworks that had been developed in research projects over the last years [3–7], one can suggest that today semantic middlewares have established themselves. These use formal languages such as OWL, OW-S and BPEL to gather declarative knowledge, service requests and skills as well as procedural knowledge in a machine readable form and to realize use cases. So far, this formal description has been a very suitable but idle basis for configuration aspects.

In literature, one can find approaches for ontology-based configurations in other domains.

In [11], Ardito et al. describe how orthogonal data sources can be connected for configuration purposes by means of ontologies and mapping processes. Other papers (see [12]) attempt to simplify configurations with the help of a reasoning based on a semantic description. However, none of these approaches in its present form is directly applicable to AAL and the requirements involved.

Further interesting preliminary works can be found in the field of smart homes. For instance, there are publications [13, 14] about the configuration and personalization of intelligent houses adjusted to the residents' requirements. Unfortunately, semantic approaches and a medical context in terms of AAL are not an issue at this so far.

3 Configuration Requirement and Conception of Configuration Processes for AAL Environments

A reference model for AAL home environments [8] was applied to investigate the configuration requirement of AAL environments and semantic middlewares. For the analysis, it had to be partially extended and instantiated.

The model describes four phases for the integration of use cases in an AAL environment:

1. Planning
2. Installation
3. Configuration
4. Maintenance.

Each of these phases is fragmented into individual process stages and associated components that support the configuration process. Furthermore, the reference model served as a basis for expert interviews.

Hereby, the most important roles that are involved in the integration of use cases could be identified: end user, developer, case manager, technician (for further information see [9]).

Depending on the process stage of the reference model, different requirements can be shaped for each of those roles. Altogether, four general and seven technical requirements were identified pointing towards the configuration aspects (for further information see [10]).

Subsequently, the individual roles of the concerned parties and the requirements together with the AAL environment and the middleware were allocated to the individual configuration components.

The precise analysis of the individual configuration components is particularized in [2]. Therefore, in terms of an overview only the most important aspects concerning the integration of a new use case are described in Fig. 1.

The developer programs the software (AAL application) of the AAL use case (e.g. warning of the stove being switched on when leaving the apartment).

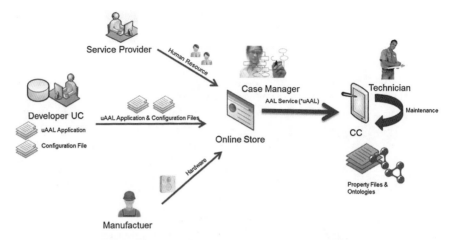

Fig. 1 Overview of the configuration requirements in AAL environments

Here, one has to take into consideration which items need to be adjusted to the resident and the environment later. These parameters should be imported into a suitable configuration file at an early stage.

Furthermore, a use case might need additional sensors (e.g. measuring electricity consumption) and/or actors (e.g. giving a warning). Thus, the manufacturer should enclose detailed meta information to facilitate the configuration.

Human resources can form the third part of the AAL use case, when their service is part of a use case (e.g. contacting the nursing personnel when warning is ignored). According to the given example, the configuration information for contacting other persons would be helpful in in case of an emergency.

All three components could be brought together and sold in an online store by a case manager, who has got the possibility to bring the individual use case components together in a suitable number and scaling and who can form the interface between end user and technician. In order to accomplish this task, he will add further configuration information to his adequately compounded use case if necessary.

Subsequently, the technician integrates the use case into the AAL environment with the help of a configuration program (control center, CC). When doing so, he uses the information enclosed to configure the use case to the special user requirements. It is decisive not to forget that in many cases one person assumes several roles in the process and operates with a variety of configuration components.

3.1 Configuration Files

At many interfaces, the exchange and the transfer of configuration information is performed by the help of computer files. Figure 1 terms the most important interfaces and files.

A use case is represented by a *.uAAL file. This is an archive that contains all the required information. The case manager assembles the use case and adds further information. The mandatory part of the *.uAAL is compounded of:

- Application
- Configuration file of the application
- Meta information of the application
- End user license agreement.

A large variety of further configuration information concerning human resources or integrated hardware can be added optionally.

The composition of new higher-value use cases by combining already existent use cases is supported by an additional XML file.

The developer generates a configuration file apposite to his AAL application. When doing so, he should be provided with the possibility to generate it automatically from the source code which is performed with the help of annotations and leads to an XML configuration file.

The completed *.uAAL file can be exchanged between case manager and technician and be integrated into the system. For this purpose the middleware has to provide a configuration program (CC) that processes the *.uAAL. This program assists with the instantiation of the parameters given in the configuration file and creates a property file. The newly installed use case accesses this file to operate adequately.

Furthermore, the configuration program supports the middleware in maintenance tasks. By focussing information, tasks such as deinstallation or exchange of defect or outdated hardware can be easily accomplished.

3.2 Configuration Ontologies

Besides configuration files for the exchange of information, in an onotology-based system all information is stored in a semantic representation. There is one ontology describing AAL environments and one illustrating user profiles (extract see Fig. 2). Furthermore, each use case can extend the semantic representation by its own information in form of ontologies.

Especially the formal, machine readable description helps facilitating the configuration process decisively. Information that is already available in the system does not have to be entered again and new knowledge can be generated.

Many of the semantic information in the system are reflected in other configuration components. Thus, the configuration program of the middleware does not have to replace or exchange a stove sensor that is already integrated in other use cases for a new memory case but can reuse it.

Semantic services provide information on the abstract possibility to scan the status of the stove.

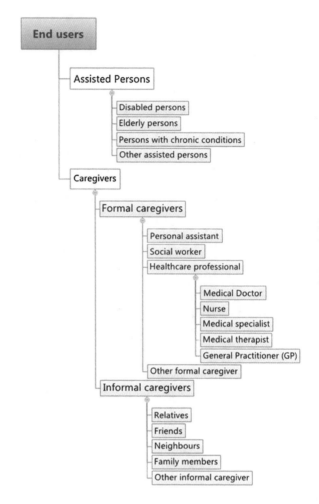

Fig. 2 Ontology of the user profile [2]

Already when assembling the appropriate use case, the case manager is provided with information concerning the existent persons (user profile ontology; Fig. 2) of an AAL environment. The users' preference or demand of assistance can have some influence more easily and an already existent care-giver can perform new tasks in other use cases.

The description of the AAL environment (AAL space profile) also supports the case manager; when selecting a suitable use case for a resident the costs can already be estimated, as it is known which hardware is already implemented in the AAL environment and which one still has to be integrated due to the local circumstances (e.g. monitoring several exits).

Besides the reduction of complexity, an ontology-based configuration also helps to save expenses and to facilitate the processes of integration.

4 Implementation of the Configuration Processes Using the Example of UniversAAL Living Labs

Having identified all the important configuration components, some tools that have been developed are presented in the following chapter. Each tool supports a single configuration component and the associated role. All tools together form a tool suite for configuration. Figure 3 presents the tools that are described in the following subchapters.

The configuration extractor (CE) supports the developer in generating a configuration file for his use case. Having been successfully uploaded to the online store, the use case can be integrated into an AAL environment. This process is supported by the control center (CC) which generates the required property files and extends and instantiates the ontologies.

The implementation of the tools was made on the semantic middleware universAAL [7]. This is a knowledge-based service infrastructure that is currently being developed in an EU research project and is operating in eight associated Living Labs.

4.1 Support During Development

Having demonstrated how developers participate in the configuration, it is now to be presented how they can be assisted by suitable tools.

The CE tool was developed for this purpose. It offers the possibility to transfer source code parameters that are highlighted by comments to a configuration file. Several tags are utilized to perform this task. The most important ones are <listpanel>, <panel> and <element> which structure individual records. After reading the comments the CE provides the possibility to carry out some adaptions or extensions. Later, this will help the CC to display the configuration parameters in an appropriate way. Figure 4 shows the CE's GUI for the adaption of the parameters and the meta information (e.g. help, text, label). Furthermore, the CE provides an expert mode. In this mode the configuration file that is to be developed

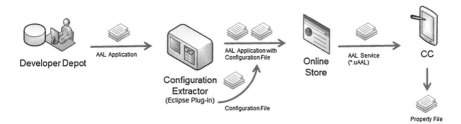

Fig. 3 Overview over the developed tools

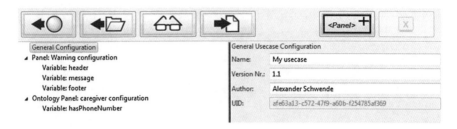

Fig. 4 GUI of the CE

Fig. 5 XML scheme of the configuration file of the CE

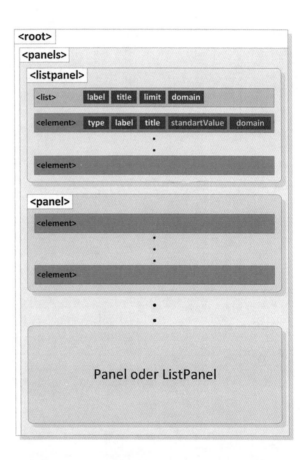

can be edited directly in an XML format. As soon as all input has been completed, a configuration file is generated following the scheme in Fig. 5 and taking up the tags that have already been described.

4.2 Support During Integration

After the use case has been distributed via online shop (see Fig. 3), it can be integrated into the middleware with the help of the CC.

First step of the integration is the installation. For this purpose the use case file (*.uAAL) is opened and it is ensured that it contains all the required files. Subsequently, the end user license of the uAAL-file is shown to the user. After receipt of the license, the configuration file generated by the CE is opened. According to its defaults a GUI is generated asking the user questions concerning the configuration of the use case. (see Fig. 6). With all specifications being complete the CC generates a property file, if necessary, and extends and instantiates the ontologies for the use case. Subsequently, the use case's software is installed and started as appropriate. The CC records all the installation steps for possible maintenance tasks.

4.3 Support During Maintenance

The components for maintenance (see Fig. 6) are at the end of the integration of use cases and are always available during the operation of the AAL environment. Many maintenance tasks use information from previous configuration steps or intervene actively in the configuration of the environment.

The most important implemented component is the use case manager (overview). It gives an overview over currently installed use cases and keeps many information

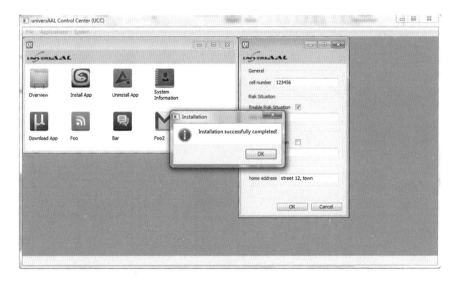

Fig. 6 GUI of the configuration tool (CC)

on hand (e.g. involved software components, version number, involved configuration files). Furthermore, it can be seen which use cases are presently active in the AAL environment.

While the use case manager only accesses information, they are actively modified by the uninstaller. This component removes a use case with all its data and the modifications of the system configuration (e.g. ontologies). Here, the removal of the use case from the use case manager happens, too.

5 Results of the Evaluation

During the first evaluation the feasibility of the developed configuration processes was to be verified with the help of experts in a real environment. Secondarily, the usability of the applied tools was also investigated. To achieve these goals, the method of the Living Labs was chosen. This method offers the possibility to test the implemented processes and to advance the conception [15, 16]. For the more special field of mass deployments in general, there is a paper [17] which confirms the suitability of Living Labs for this purpose. Thus, it is perfectly possible to investigate configuration processes with the help of Living Labs.

In our experimental assembly, the middleware and the CC (see Fig. 6) from the project universAAL were installed in eight Living Labs. The task had been to upload a new use case from the online store and to integrate it into the middleware with the help of the CC (see Fig. 3). The task was successfully completed when the use case was subsequently working in the AAL environment. Each Living Lab received an instruction how to use the configuration tools alongside the process.

The feedback was collected in the form of a questionnaire. Up to this point, there is feedback from four out of eight Living Labs. All use case integrations were performed by developers. Additionally, all four were experienced with the middleware. In 75 % of the cases, the task was successfully completed. In one case, the integration was canceled for unknown reasons. Apart from this, there were no problems in any of the integrations and only once an authorized developer was consulted for a request.

In the free text fields, suggestions and proposals for an improvement of the tools and the configuration process were made.

6 Prospect

The next step will be the progression of the CC and the configuration process based on the results of the evaluation.

Apart from the parts of the process that have already been tested, there are tasks to approach such as an extensive planning of the configuration especially for much wider scenarios with a large amount of use cases. For this purpose, a simulation of the environment can help to meet the users' requirements much more precisely and to prevent an unnecessary expenditure of time and costs.

Furthermore, it is necessary to think about the initial equipment of the AAL environment. The middleware, too, has to be installed before use cases can be integrated and used. Especially this basic functionality involves a significant expenditure of configuration.

However, the most interesting step for the moment is the more intensive utilization of the middleware's semantic interiority for the configuration of use cases. Only by this, a massive simplification (as it is shown in Sect. 2.1) can be achieved. Furthermore, it is considered, to generate configuration files already in the CE instead of XML-files. In order not to burden the developer with the complete modeling of the very same, an ontology repository is thinkable, from which a suitable configuration technology can be selected and extended if necessary. The same should be considered for hardware and human resources.

These ontologies can be parsed all the same in the CC to generate a configuration GUI. After the input of the configuration parameters, the instances can be entered into the semantic middleware more easily. By means of an appropriate matching and reasoning the information content in the middleware about the AAL environment rises quickly. This increase in information can then be utilized in all places with configuration requirement.

References

1. VDE-Positionspapier: Intelligente Assistenz-Systeme-im Dienst für eine reife Gesellschaft, VDE Berlin (2009)
2. Zentek, T., Wolf, P.: Ontologiebasierte Konfiguration in AAL Umgebungen. GI (2011)
3. Wolf, P., Schmidt, A., Otte, J.P., Klein, M., Rollwage, S., König-Ries, B., Dettborn, T., Gabdulkhakova, A.: openAAL—the open source middleware for ambient-assisted living (AAL), in 'AALIANCE conference, Malaga, Spain (2010)
4. Mikalsen, M., Hanke, S., Fuxreiter, T., Walderhaug, S., Wienhofen, L. W.: Interoperability services in the MPOWER ambient assisted living platform, vol. 150, pp. 366–370, IOS Press (2009)
5. Schmidt, A. et al.: SOPRANO ambient middleware: Eine offene, flexible und marktorientierte semantische Diensteplattform für Ambient Assisted Living. In: 2. Deutscher Kongress Ambient Assisted Living, VDE Berlin (2009)
6. Fides-Valero, A., Freddi M., Furfari, F., Tazari, M. R.: The PERSONA framework for supporting context-awareness in open distributed systems. In: Proceedings of the European Conference on Ambient Intelligence, AmI (2008)
7. Hanke, S. et al.: An open and consolidated AAL platform. VDE, Berlin (2011)
8. Zentek, T., Rashid, A., Wolf, P., Kunze, C.: Mit Plug&Play zur intelligenten Wohnumgebung: Ein Referenzmodell zum Einrichten und Verwalten einer Ambient Assisted Living Umgebung (2009)
9. universAAL deliverable 3.5-A Service management and personalisation tool. (2011)
10. universAAL deliverable 2.2-A universAAL Generic Platform Services, AAL platform services and ontology artefacts (2011)
11. Ardito, C., Barricelli, B.R., Buono, P., Costabile, M.F., Lanzilotti, R., Piccinno, A., Valtolina, S.: An ontology-based approach to product customization. In: Proceedings of the Third

international conference on End-user development (IS-EUD'11), pp. 92–106, Maria Francesca Costabile, Yvonne Dittrich, Gerhard Fischer, and Antonio Piccinno (eds.). Springer-Verlag, Berlin, Heidelberg (2011)
12. Dong, M., Yang, D., Su, L.: Ontology-based service product configuration system modeling and development. Expert Syst. Appl. **38**(9), 11770–11786 (2011)
13. Cetina, C., Trinidad, P., Pelechano, V., Cortes, A.R.: Mass customisation along lifecycle of autonomic homes (2009)
14. Kawsar, F., Nakajima, T., Fujinami, K.: Deploy spontaneously: supporting end-users in building and enhancing a smart home. In: Proceedings of the 10th international conference on Ubiquitous computing (UbiComp '08), pp. 282–291. ACM, New York, NY, USA (2008)
15. Bergvall-Kåreborn, B., Ståhlbröst, A.:. Concept design with a living lab approach. In: Proceedings of the 42nd Hawaii International Conference on System Sciences (2009)
16. Niitamo, V.P., Kulkki, S., Eriksson, M., Hribernik, K. A.: State-of-the-Art and Good Practice in the Field of Living Labs (2006)
17. Eriksson, M., Niitamo, V.P., Kulkki, S.: State-of-the-art in utilizing Living Labs approach to user-centric ICT innovation—a European approach (2005)

The Robot ALIAS as a Database for Health Monitoring for Elderly People

Tobias Rehrl, Jürgen Geiger, Maja Golcar, Stefan Gentsch,
Jan Knobloch, Gerhard Rigoll, Katharina Scheibl,
Wolfram Schneider, Susanne Ihsen and Frank Wallhoff

Abstract Health plays an important role with increasing age, therefore keeping track of health data is a very essential task for elderly people. Many different Information and Communication Technology systems were developed to measure health data. We integrated a health monitoring system on a mobile robotic platform, which is designed as communication platform for elderly people. To realize

---

T. Rehrl (✉) · J. Geiger · M. Golcar · S. Gentsch · J. Knobloch · G. Rigoll
Lehrstuhl für Mensch-Maschine-Kommunikation, Technische Universität München,
Munich, Germany
e-mail: rehrl@mmk.ei.tum.de

J. Geiger
e-mail: geiger@mmk.ei.tum.de

M. Golcar
e-mail: golcar@mmk.ei.tum.de

S. Gentsch
e-mail: gentsch@mmk.ei.tum.de

J. Knobloch
e-mail: knobloch@mmk.ei.tum.de

G. Rigoll
e-mail: rigoll@mmk.ei.tum.de

K. Scheibl · W. Schneider · S. Ihsen
Gender Studies in Ingenieurwissenschaften, Technische Universität München, Munich,
Germany
e-mail: katharina.scheibl@tum.de

W. Schneider
e-mail: wolfram.schneider@tum.de

S. Ihsen
e-mail: ihsen@tum.de

F. Wallhoff
Jade University of Applied Sciences, Oldenburg, Germany
e-mail: frank.wallhoff@jade-hs.de

this health monitoring system on a robotic platform for elderly people, it is important to address their needs and wishes, therefore, a user survey was conducted to figure out important issues relevant for elderly people. Taking the outcome of the user survey into account and following a generic design principle, a health monitoring system was realized, which features sensor-based and manual input for single or multiple users. The system interaction is browser-based and allows remote access to the health data stored on the robotic platform for authorized persons.

1 Introduction

The intended purpose of the AAL-Joint Programme project *Adaptable Ambient LIving ASsistant* (ALIAS) is the development of a mobile robotic platform (see Fig. 1), which should be able to interact with elderly people and thus offer assistance in their everyday life situations [1, 2]. Consequently, the robotic platform aims to support people living alone at home or even in residential homes for the elderly and prevent them from becoming lonely. ALIAS shall interact with elderly users, monitor them and provide social inclusion and assistance in the daily routine.

One aspect of the project is health monitoring, thus, a database concept for health monitoring of vital functions was developed for the ALIAS robot. The ALIAS project has a user-centered design approach, therefore a user survey was conducted to determine important features of the database. The recorded vital functions were stored on the robotic platform, however, a remote access for the data is also possible, therefore a web-based approach for recording and visualizing the data was chosen.

Fig. 1 The ALIAS robot

Table 1 Most widespread diseases among elderly people (given in percentage) according to [3]

Diseases	Percentage
Musculoskeletal disease	45.2
Other circulatory disease	38.8
Eye disease	34.5
Hypertension	20.8
Heart attack	20.7
Disease of the joints/arthritis	14.2
Bronchitis/emphysema/Asthma/other respiratory disease	13.7
Diabetes mellitus	13.4
Urogenital disease	12.9
Infections	11.5
Hip fracture	10.3
Ear disease	9.7
Rheumatic disease	8.7
Dermatosis	9.0
Cerebrovascular disease	7.7
Metabolic disease	6.4
Dementia	5.0

In general, for elderly people, living and staying healthy is very important. Elderly people are more vulnerable to diseases and their quality of life is strongly affected by their health. Elderly people suffer from many diseases. Table 1 shows some of the most widespread diseases among elderly people according to a study performed by Hellström [3]. For this study, 448 persons (294 female, 152 male) who were dependent of help by others were questioned.

In this table, many different diseases are listed which affect elderly people. With the scope of the project, the ALIAS robot can assist elderly people whose quality of life is restricted by diseases by providing a wide spectrum of functionalities. Some of the diseases in Table 1 are not in the scope of the ALIAS project, for example everything that has to do with physical assistance. The main purpose of the ALIAS robot is to communicate and it has no physical manipulators. Thus, diseases like musculoskeletal diseases or hip fracture cannot be addressed by the project. However, some of the diseases are dealt with in the ALIAS project. By providing a multimodal dialog system, the ALIAS robot can help people suffering from ear or eye diseases, for example. People suffering dementia can be assisted by providing reminder functions and cognitive assistance.

In this paper the development and realization of a health monitoring system for the ALIAS robot is described. Together with additional systems for recording health data, the ALIAS robot can impersonate a helpful assistant to record, monitor and store health data.

Important health data include vital signs like body temperature, pulse rate (or heart rate), blood pressure, and respiratory rate. By providing a system for monitoring of these data, the ALIAS robot can assist elderly people suffering from diseases like hypertension, Diabetes Mellitus, Cerebrovascular disease (which can be caused by hypertension) or diseases of the circulatory system. The recorded biosignals can be inspected by authorized persons from remote through the ALIAS system, e.g. a person's doctor or relatives. Additionally, the signals might be saved for later on-site inspection.

The remainder of this paper is organized as follows: Sect. 2 describes related work for health monitoring. In Sect. 3 relevant physiological parameters are introduced. The results of the user survey for creating a database concept for health monitoring are presented in Sect. 4. The system design is delineated in Sect. 5. In Sect. 6 the developed database concept is explained. The system interaction is topic of Sect. 7. Use cases for the health monitoring are presented in Sect. 8. The paper ends with a conclusion.

2 Related Work

According to the progressive prevalence of diseases with increasing age, there are many activities to apply Information and Communication Technologies (ICTs) for compensating age-related diseases or providing health care. These ICT-systems need to take special requirements like reduced cognitive capacities, sight loss, hearing loss and minor experience with interactive systems of elderly people into account.

There exist many solutions for health monitoring. An overview over different health monitoring systems is given in [4, 5]. This includes systems for remote monitoring, for example the system presented in [6] called MOMEDA, a personalized medical information system. In [7], a wireless PDA-based system is presented. It is designed for use during intrahospital transports of patients. Physiological data are recorded and visualized on a screen. In [8] ubiquitous wireless computing is considered, where different sensors measure various values of all kind of sorts. These sensors are combined to an *Infrastructure For Elderly Assistance* as the task for the future where the sensors serve as health indicators or full-time attendants. This approach is supported by [9] with the *Mobile Robot Peekee II* representing such a mobile infrastructure, here a mobile robot incorporates the sensors, thus no changes in the environment have to be conducted. *MobiSense* is another mobile health monitoring system for ambulatory patients introduced in [10]. This system is able to detect postures of people like lying, sitting and standing while it is also able to manage its' own resources like reconfiguration of parts. Another health or activity monitoring system is shown in [11], this physical activity monitoring system has a built-in vital sign measurement and fall detection. Again, this system includes a wearable device collecting physical and activity data with various sensors, e.g. a 3-axial accelerometer. A similar approach has already been introduced in the year 2000 in [12] with presenting a *Home Telemonitoring Framework*, which is divided in two categories: the daily activity monitoring category for the elderly and the vital sign monitoring category for patients recovering at home. Thus, the home care needs were separated in different levels. The system itself was divided in the home monitoring unit mainly responsible for the data storage, a hospital monitoring center and a communication network connecting these components.

It can be seen, that the topic of health monitoring has a lot of different aspects, and, therefore many different technical systems providing different solutions were developed. One important factor is the type of health data, which is used for the health analysis.

3 Physiological Parameters Relevant for Elderly People

For health monitoring systems, there are many potential biosignals to record. The most important physiological parameters are the four standard vital signs: body temperature, blood pressure, heart rate (or pulse) and respiratory rate.

Several other biosignals have been proposed as fifth or sixth vital sign, but none of them has been officially universally adopted. Examples include the oxygen saturation, pupil size, equality and reactivity to light, perception of pain, blood glucose level, Body Mass Index (BMI) or galvanic skin response (GSR).

3.1 Vital Signs

The four standard vital signs are body temperature, blood pressure, pulse rate (or heart rate) and respiratory rate.

Body temperature [13] is measured with a thermometer and can be an indicator for several abnormalities. Average core body temperature is 37.0 °C. Elevated body temperature can be a sign for a systemic infection or fever but can also be caused by hyperthermia. Temperature depression (hypothermia) can be caused by alcohol consumption or dehydration, for example. When measuring and interpreting body temperature, it is always important to review the trend of the patient's temperature. This makes the ability of a system like the ALIAS health monitoring system to record and store biosignals a very important feature.

Blood pressure is recorded as two values: a high systolic pressure and the lower diastolic pressure. The difference between these two values is called the pulse pressure. The blood pressure can be measured using an aneroid or electronic sphygmomanometer. Normal blood pressure values are 120 mmHg (millimeters of mercury) for the systolic and 80 mmHg for the diastolic. The systolic being constantly over 140–160 mmHg is defined as elevated blood pressure (hypertension), whereas low blood pressure (hypotension) is generally considered as systolic blood pressure less than 90 mmHg. Using a device like the ALIAS health monitoring system, thresholds can be set to supervise blood pressure. The pulse is the physical expansion of the artery. Its rate is recorded as beats per minute and can be measured at the radial artery at the wrist or at other places of the human body.

Another way to measure it, is by directly listening to the heartbeat using a stethoscope. The normal reference range for an adult is 50-80 beats per minute. Respiratory rate [14] is the number of breaths per minute. A normal adult's rate is between 12 and 20 breaths per minute.

3.2 Physiological Data Processing With ALIAS

The health data can be transmitted to the robotic platform directly by a recording system or by a user who recorded the data on his own and stores it via the multimodal ALIAS user interface into the database. Examples are blood pressure and body weight. The ALIAS health monitoring system can then perform several tasks with these data. The system can store data in a short-term or long-term way, provide statistical analysis of the data, and visualize these data.

4 User Survey

For innovation processes it is essential to integrate the perspectives of the potential users into the development of the mobile platform [15]. Therefore, the relevant and heterogeneous target groups have to be considered. Only then it is possible to implement their needs and preferences to the robotic platform [16, 17]. The acceptance of health monitoring modules in general and the discussed database in particular have been investigated in a quantitative user survey with 79 senior students (35 women/44 men) in 2011. The mean age of the polled seniors was 70 years.

The seniors estimated the possibility of health monitoring as positive (50 %), some are not sure (27 %) and the rest (23 %) are skeptical about modern health monitoring technologies. Among the skeptical persons, the reasons for this rejection were that they consider themselves as "too young" to use such a help and they want to test the database first to value the personal benefit. Unfortunately this was not possible in this setting.

Also the survey evaluates what kind of health data should be recorded and saved on the mobile robot platform (see Table 2). The results show that seniors are more open to the measurement of the pulse (61 %) and the blood pressure (63 %) than to the measurement of the vital data such as breathing (47 %), weight (46 %) and body temperature (46 %). It was striking that more than every fourth senior does not want the weight to be measured. Altogether the desired data can be summarized with the following statement of a senior "if it is important for a certain kind of illness". Some participants were so enthusiastic that they wish to store further data e.g. MRT/CT-picture files, important blood values etc.

The survey showed that seniors prefer a comfortable and safe way of using the health database. Almost half of the participants wants that the medical devices are

Table 2 Health data that should be recorded on the mobile robot platform

	In any case (%)	Indifferent (%)	In no case (%)
Pulse	61.2	28.4	10.4
Blood pressure	63.4	25.4	11.3
Blood sugar	54.7	29.7	15.6
Weight	46.3	26.9	26.9
Body temperature	45.5	37.9	16.7
Breathing	47.0	36.4	16.7

directly connected to the robot so that the data can be saved automatically (44 %). 23 % are uncertain about this technology and the rest refuses the automatic data saving and transfer generally (23 %). Nevertheless, the seniors are skeptical about a password-secured electronic health record. Only every third participant wants to save the health data on an external server. Reversely, more than 70 % of the seniors would save the data on the hard drive of ALIAS at home. It remains to be seen if the seniors—as soon as they recognize the potential and the personal benefit of the electronic health record—give higher approval ratings.

For the interaction with the system, it is very important to the seniors that they can examine their data entry before it is saved (for 78 % seniors this should be included in any case). Because of that when the robot is developed it should be paid attention to summarize the data in synopsis before it is actually saved, to announce the storage and to let this be confirmed by the user. Another point for the interaction is presentation of the stored data, 71 % of the seniors wish a presentation in graph form and 74 % a presentation in table form, respectively.

Altogether, it could be seen that seniors are open for health monitoring or concepts for physical monitoring of vital functions, but first they have to get familiar with a technical assistant system to save their health data. Some seniors pointed out that for this purpose ALIAS has to be "reliable". That is for the target groups a crucial point of view.

5 System Design

In order to properly design the health monitoring within ALIAS, it is necessary to identify what the system should realize from the users' point of view, therefore the user survey was conducted, see Sect. 4. The derived design decisions for the system are introduced in the following. The decisions concern the recorded health data (via sensors or manual entry) and how the user interacts with the system in a general way, therefore, different activity diagrams will be presented to explain the proceeding.

Table 3 Minimum/maximum values for the recorded health data according to [18]

Value type	Maximum value	Minimum value
Systolic	140	105
Diastolic	90	60
Blood sugar	110	70
Heart rate	100	60
Respiration	20	12
Temperature	37.5	35.8
Weight (BMI)	24.9	18.5

5.1 Recorded Physiological Data

The database for the ALIAS health monitoring comprises the following entries: (1) blood pressure with the systolic and diastolic value, (2) blood sugar, (3) heart rate, (4) respiration, (5) temperature, (6) weight.

For all these values minimum and maximum values for alarm triggering were defined. For the weight the body mass index (BMI) was chosen to provide the corresponding minimum and maximum values. In the following Table 3 the minimum and maximum values for the health data are shown.

The ALIAS health monitoring has two different sensor systems to obtain health data. The first system is the *g.tec g.MOBIlab* [19], the second system is the *Zephyr HxM BT* [20]. The first system is a professional medical device providing a lot of different features, whereas the second one is a consumer device, which is easier to handle. Both sensors are shown below in Fig. 2.

The g.tec g.MOBIlab system records the following biosignals and physiological features: electrocardiograph (ECG) (with several corresponding values like heartrate (HR) heartratevariability (HRV) etc.), galvanic skin response (GSR) and respiration.

The Zephyr HxM BT system records the following biosignals and features: heart rate, speed and distance.

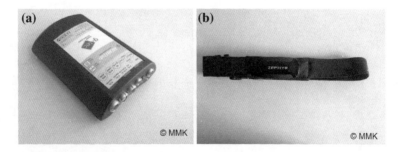

Fig. 2 Integrated sensors, **a** g.tec g.MOBIlab, **b** Zephyr HxM BT

In addition to the sensors, the user can also enter health data manually into the database. The user is able to manipulate the following entries manually: blood pressure, blood sugar, heart rate, respiration, temperature and weight.

According to the results of the user survey (see Sect. 4) the recorded data is stored on the robotic platform and not on a remote server. However, a remote access for the data is also possible, since a web-based approach for recording and visualizing the data is chosen, thus authorized persons (doctors, relatives) are able to access the data if they have the permission.

5.2 Activity Diagrams

Activity diagrams are used to visualize the corresponding processes the system needs to perform in order to realize the interaction between system and user. In the following, the diagrams for entering health data will be shown and briefly explained. There are two ways of entering data into the database: sensor-based and manually.

For the sensor-based entering, the user initiates live data recording, therefore between the two sensor devices can be chosen: g.tec g.MOBIlab or Zephyr HxM BT, respectively. When a device received the user input, the data recording is started, the data is queued, until enough data are stored to start the monitoring and the corresponding visualization. After a dataset value has been displayed, the system stores it into the underlying database. When the user chooses to stop the recording and forwards his choice to the system by pushing a button on the screen, the logging, monitoring and storing of the data is stopped abruptly and thus the live visualization is finished. The corresponding activity diagram is shown in Fig. 3.

For the manual data entry, there is one special case, since for blood pressure three values (systolic level, diastolic level, heart rate) have to be entered. For all other health data recorded with the proposed database, only one value has to be entered. According to the user survey (see Sect. 4), the user wants to check the data before they are finally stored in the database and additionally a warning message should be shown if critical values (see Table 3) are exceeded. In the following only the activity diagram for the blood pressure data entry is shown, since the proceeding for the single value health data follows that scheme.

When the blood pressure entry is initiated by the user, the system waits for the systolic and diastolic values as well as the heart rate to be entered. After the user has entered these values and pushes the confirm button, the system displays the entered values again to the user and offers him/her two options: The first option is to edit the data in case the entered values turn out to be wrong. In this case, the values are aborted and again the system waits for new values to be inserted. During this process, the underlying database is never touched, so the system makes sure that the database does never contain any false values. The second option for the user is to confirm the values he/she entered. Only if the values are confirmed by the user, the system stores the data into the database. At the same time, the system

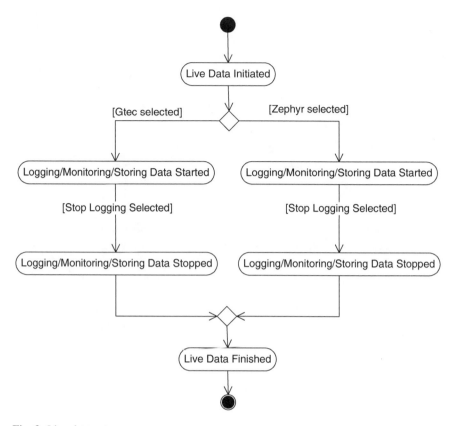

Fig. 3 Live data entry

checks if the entered values are alarming, i.e. if one (or more) of the values are either lower than the minimum value stored in the database or higher than the corresponding maximum value (see Table 3). The corresponding activity diagram for the blood pressure entry is shown in Fig. 4.

6 Database Concept

6.1 Basics

There are different types of databases available and they provide different kinds of features, more extensive information about databases can be found in [21, 22]. In general a database is a collection of data, which has a certain kind of interrelation and can be accessed via a specific set of functions or programs. For databases a specific architecture was defined comprising three different layers (see Fig. 5):

The Robot ALIAS as a Database for Health Monitoring for Elderly People 235

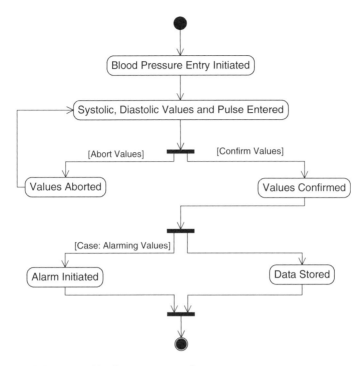

Fig. 4 Manual data entry: blood pressure example

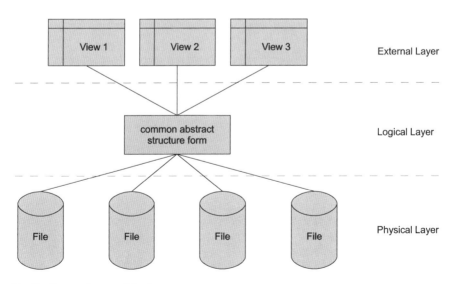

Fig. 5 Abstract layers of databases

physical layer (containing the data file on physical disk drives), logical layer (first abstraction level: data is represented in a common abstract structure form) and the external layer (second abstraction level: the form of the data accessed by the users). Different database models (flat-files, Hierarchical Model, Network Model, Relational Model, Object-Relational Model, etc.) have evolved over time, the ALIAS health monitoring database follows an object-relational approach.

6.2 Database Concept for the Health Monitoring

The database concept for the health monitoring system envisioned for the ALIAS project has to fulfill several properties, which are relevant for handling health data of elderly people. First, the data should be stored over longer time periods enabling to perceive changes in the course of the data (e.g. blood pressure). Second, the database should be able to handle different users and their related health data (this is relevant for the usage of the ALIAS system in care facilities). Third, the health data should be accessed either directly on the ALIAS system or via remote access, thus enabling doctors and the authorized health care providers to access the health data.

In order to achieve a generic, resource-efficient database, an elaborate database concept has to be developed. Figure 6 depicts the generated database concept consisting of the following entities: user, dataset, dataset value and dataset value type. These four entities will be described in the following.

User: In order to enable various users to share the same ALIAS robot, the table *user* was created. Here the user id, corresponding to the login name, is used as the

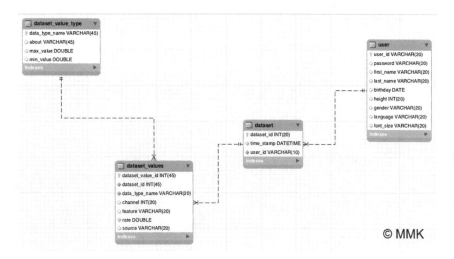

Fig. 6 Database concept for the health monitoring

primary key of the user. This ensures that each user can be uniquely identified by his chosen user id, at the cost that no two users can choose the same id. Furthermore, the user has to choose a password. The user id and the corresponding password are used to grant registered users access to the database and into the system. In addition the following information about the user is provided by the user-entity: **First Name, Last Name, Birthday, Height, Gender, Language, Font size, Login and Password** (hashed). Finally, a user can be associated with various datasets. This is realized by a one-to-many relationship between a user and datasets.

Dataset: In order to uniquely identify a dataset, it contains an automatically generated *dataset id*. The timestamp enables to reconstruct, at which date and time a dataset was recorded. Furthermore, a dataset has a foreign key which assigns it to a specific user. A single dataset can contain multiple dataset values, which is realized by a one-to-many relationship between dataset and dataset values. The dataset relation is necessary, because a sensor may record multiple biosignals and extract many health features thereof at one time.

Dataset Value: The automatically generated field *dataset value id* allows to reference each dataset value in a unique way. In order to differentiate between multiple connected sensors, from which dataset values are transferred to the system, the field *channel* is invented. The field *feature* allows capturing further information of the sensor, if necessary. The dataset value itself is stored by the field *rate* and the sensor from which the dataset value is originated is captured by the field *source*. The dataset values table contains two foreign keys. The first assigns the current dataset-value to its corresponding data set. The second assigns a specific dataset-value-type to it.

Dataset Value Type: The *dataset value type*-table enables the resource-effective and generic framework to be realized: This table holds the various dataset value types that can be recorded, for example blood pressure, respiration or heart rate. These different types are recorded only once and are then assigned to the dataset values. It contains the field *data type name*, which also acts as a unique identifier for the dataset value type. Further information about the dataset value type can be stored in the field *about*, if necessary. The fields *minValue* and *maxValue* define the thresholds of the lower and upper alarm limits if any is set.

7 System Interaction

The ALIAS health monitoring module is provided as a web application, which means that it runs in a web browser. Thus, it is possible to allow remote access to the data, so that it can be inspected by authorized persons (i.e. a person's doctor, relatives etc.). The integration into the ALIAS dialog system is easily realizable, because a web browser is integrated into the graphical user interface (GUI) of the ALIAS system. For the manual data entry it is possible to use the touchscreen of the ALIAS robot or apply speech input.

In the following several system interaction steps will be described. The first step registration, is not intended to be done by the user, but in cooperation with either the nursing staff (in case the system is used in care homes), or together with a person of trust (relative, doctor) (in case the ALIAS system is used at home).

Registration: The registration is necessary to create an account for a user in the database, so that the data can be stored. The following data are required for the registration procedure: first name, last name, birthday, height, gender, language and font size. Additionally, the user has to choose a login, which serves as a unique identifier (i.e. user id) and a password, which is stored in hashed form. These two values serve in the future for logging into the system. When the user pushes the *Create-Button*, the system checks, if the login is already assigned (no two users can choose the same login). Therefore, if the login is already assigned, the user has to choose another one. Otherwise, the account is created and the user data are stored into the underlying database. The registration mask is shown in Fig. 7.

Login: Before any manipulation on the system can be performed by a user, he/she needs to log him/herself into the system by using his/her chosen login and password. The login procedure is only possible, if an account was already created by the user. When the user pushes the *Login-Button* after he/she inserted his/her id and corresponding password, the system checks, if the values are correct. If they are not, an error message is sent to the user. The user can repeat the login procedure or press the button *Forgot your password?* to recover the login. The login mask is shown in Fig. 8.

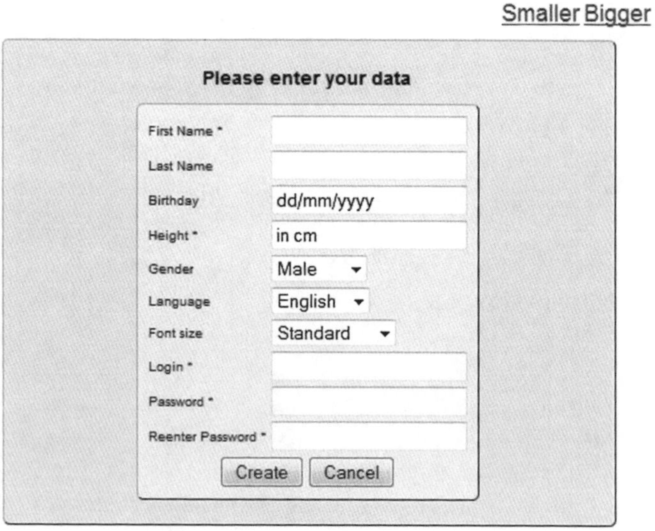

Fig. 7 User registration

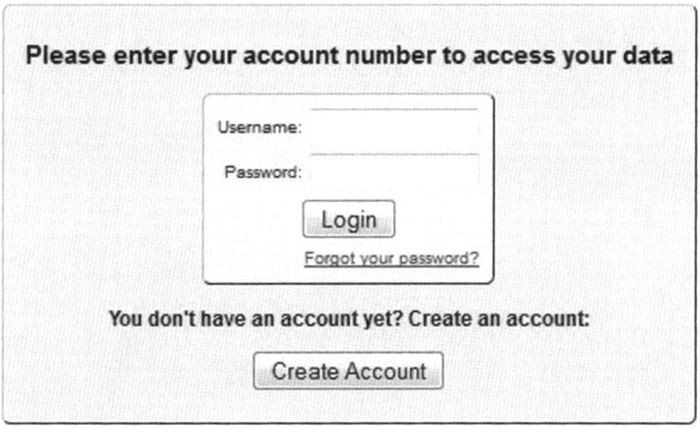

Fig. 8 User login

Fig. 9 Home display (larger on the actual screen)

After a successful login, the home display is shown in the web browser (see Fig. 9). This home display consists of a menu bar, where health data entries are shown, within the field *Live Data*, the connected sensors can be started.

Live Data Entry: For the live data recording the user has to selected one of the sensors first. After one of the sensors was selected, the live data recording starts. All recorded channels are shown, therefore the charting library *Highcharts* [23] is used. An example for visualization of a recording of health data from the g.tec g.MOBIlab sensor is shown in Fig. 10.

In general, Highcharts offers a good possibility to add interactive charts in a web application. Furthermore, Highcharts is compatible with all modern browsers like Chrome, Firefox, Opera etc. and it also supports the iPhone/iPad, besides it offers numerous chart types like line, spline, area, areaspline, etc. and allows multiple modifications.

The recording of the live data follows the *What You See Is What You Get* (*WYSIWYG*)-principle, since a recorded session is displayed in the same way as it was shown while the recording, because there were no additional points recorded or skipped during a live recording.

Manual Data Entry: The manual data entry consists of three steps: In the first step, the data values are entered by the user. Afterwards the user has the possibility to check his/her entered values. The final step consists of the data storage in the database. In the following these three steps are exemplary shown for blood pressure data entry. In the first step the user has to select blood pressure in the home display. Afterwards the overview for the blood pressure is presented (see Fig. 11).

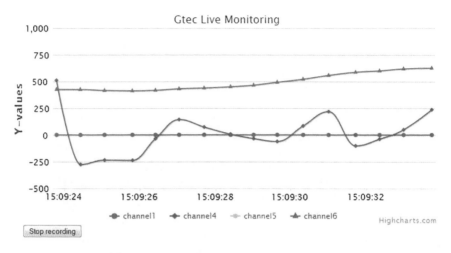

Fig. 10 Live data recording: g.tec g.MOBIlab

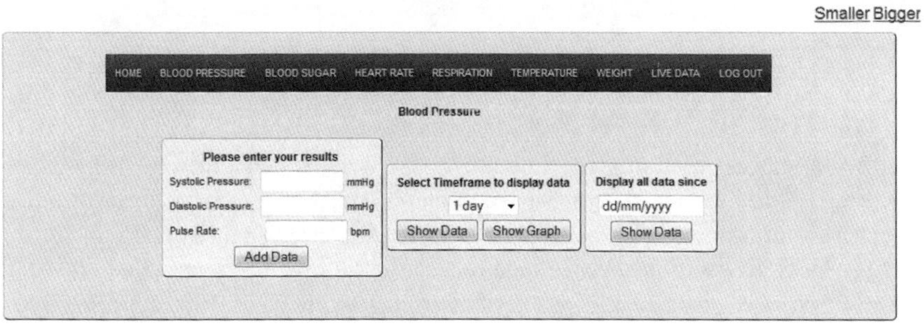

Fig. 11 Manual data entry: entering blood pressure values

This overview consists of three entries: adding data, displaying data for a specific time frame (week, month), display data since a specific date.

For the blood pressure three values have to be entered: systolic pressure, diastolic pressure and heart rate. After all data entries were provided the user has to press the *Add Data*-button to proceed to the next step. In this step, all the entered data is displayed (see Fig. 12) and the user has now two possibilities: Either he/she can edit the data once again, in case he/she made a mistake while the data entry, or he/she confirm the entered values.

After the user confirmed the entered values, he/she gets a confirmation on the screen, see Fig. 13, that the values are now stored in the database.

Presentation of Stored Data: The user has two possibilities to view previous data, he/she can choose to display the data by selecting a timeframe or by inserting a specific date. In addition to time selection, there are two different presentation styles: presentation in table form (see Fig. 15) and presentation in graph form (see Fig. 14). This procedure for the visualization was wished by the users (see Sect. 4).

The Robot ALIAS as a Database for Health Monitoring for Elderly People 241

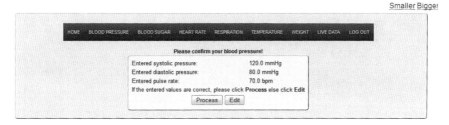

Fig. 12 Manual data entry: confirmation of blood pressure values

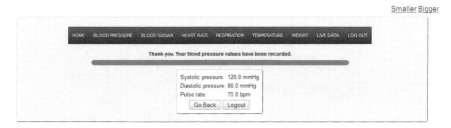

Fig. 13 Manual data entry: blood pressure values storage

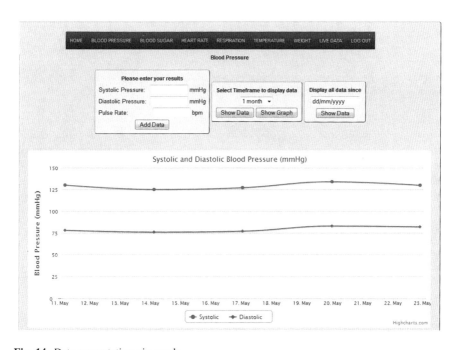

Fig. 14 Data presentation via graph

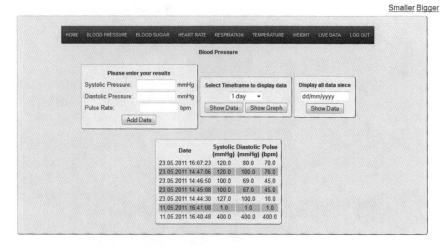

Fig. 15 Data presentation via table

The realization of the system interaction sticked to the user survey from Sect. 4. The system design and interaction followed the wishes and needs of the elderly users. In general, the health monitoring on the ALIAS robot is designed in a generic way, since the system is designed for usage by multiple users in nursing homes or for a single user.

8 Use Cases

Different kinds of use cases are thinkable depending on the task and the wishes of the user. In general, there are two kinds of categories of use cases. The first kind of case provides a direct feedback of the obtained data towards the user, whereas the second kind is more related to monitoring operations. These monitoring operations can be subdivided into a long-term monitoring, acquiring health data for a longer time period, to obtain data, where small irregularities can be more detectable, the second category is short-term monitoring, where the health data is recorded for a couple of minutes.

8.1 On-line Cases

Exercising with ALIAS The health data are directly provided as feedback for the user, while performing a physical exercise. The online feedback can help the user to rate his/her personal performance and might adjust his/her training according to the health data.

Gaming with ALIAS The health data can be shown while the user is playing with ALIAS, so either the level of physical effort can be estimated for instance by playing with the Nintendo Wii, or it is possible to extract the stress level as well as the state of the mental effort during gaming. If necessary and desired the data even can feed back to the games, for a closed loop control of the parameters defining the difficulty of the game levels, with the aim to optimize (maximize) the performance level of the user.

8.2 Monitoring Cases

Diary Monitoring of Vital Functions The diary monitoring of vital functions provides the possibility to create a profile of specific vital functions over time. This profile can be used to estimate the change of the vital functions of the user with external events. For instance, a cardio-training is set up for the user to enhance his/her fitness, thus a change of the resting pulse rate can be detected. Another possible example is to track the influence of different drugs on the vital functions.

Long-Term Monitoring The objective of long-term monitoring is to detect small irregularities in the health data of the user for a longer monitoring period (e.g. 24 h).

9 Conclusion

In this paper a health monitoring system is presented, which is part of the robotic platform ALIAS. The design of the health monitoring system followed a user-centered approach, thus a survey was conducted to gain information about wishes and needs of elderly people with regard to: what kind of health data shall be recorded, where shall the data be stored, what are important steps for the system interaction. The outcome of the survey was considered in the realization and implementation of the database, additionally, the database was designed in a way that it can be applied to multiple user in nursing homes or a single user living alone.

Acknowledgments The ALIAS project is funded by the Ambient Assisted Living (AAL) Joint Programme, by the German BMBF, the French ANR and the Austrian BMVIT.

References

1. Rehrl, T., Blume, J., Geiger, J., Bannat, A., Wallhoff, F., Ihsen, S., Jeanrenaud, Y., Merten, M., Schönebeck, B., Glende, S., Nedopil, C.: ALIAS: Der anpassungsfähige Ambient Living Assistent. Proc. Deutscher AAL Kongress (2011)
2. Scheibl, K., Geiger, J., Schneider, W., Rehrl, T., Ihsen, S., Rigoll, G., Wallhoff, F.: Die Einbindung von Nutzerinnen und Nutzern in den Entwicklungsprozess eines mobilen Assistenzsystems zur Steigerung der Akzeptanz und Bedarfsadäquatheit. Proc. Deutscher AAL Kongress (2012)
3. Hellström, Y., Hallberg, I.R.: Perspectives of elderly people receiving home help on health, care and quality of life. Health Soc. Care Community **9**(2), 61–71 (2001)
4. Pattichis, C.S., Kyriacou, E., Voskarides, S., Pattichis, M.S., Istepanian, R., Schizas, C.N.: Wireless telemedicine systems: an overview. IEEE Antennas Propag. Mag. **44**(2), 143–153 (2002)
5. Pantelopoulos, A., Bourbakis, N.G.: A survey on wearable sensor-based systems for health monitoring and prognosis. IEEE Trans. Syst. Man Cybern. **40**(1), 1–12 (2010)
6. Pavlopoulos, S., Prentza, A., Kyriacou, E., Marinos, S., Stasis, A., Kalivas, D., Koutsouris, D.: Mobile medical data (MOMEDA)-a personalized medical information system. *User Acceptance of Health Telematics Applications: Education and Training in Health Telematics* 125 (2000)
7. Lin, Y.H., Jan, I.C., Ko, P.C.I., Chen, Y.Y., Wong, J.M., Jan, G.J.: A wireless PDA-based physiological monitoring system for patient transport. IEEE Trans. Inf Technol. Biomed. **8**(4), 439–447 (2004)
8. Mufti, M., Agouridis, D., ud Din, S., Mukhtar, A.: Ubiquitous wireless infrastructure for elderly care. Proceedings of the 2nd International Conference on Pervasive Technologies Related to Assistive Environments, PETRA'09, pp. 22:1–22:5. ACM, New York, US, (2009)
9. Brell, M., Meyer, J., Frenken, T., Hein, A.: A mobile robot for self-selected gait velocity assessments in assistive environments: a robotic driven approach to bring assistive technologies into established homes. Proceedings of the 3rd International Conference on Pervasive Technologies Related to Assistive Environments, PETRA '10, pp. 15:1–15:8. ACM, New York, NY, USA, (2010)
10. Waluyo, A. B., Yeoh, W.-S., Pek, I., Yong, Y., Chen, X.: Mobisense: Mobile body sensor network for ambulatory monitoring. ACM Trans. Embed. Comput. Syst. **10**(1), 13:1–13:30 (2010)
11. Dinh, A., Teng, D., Chen, L., Shi, Y., McCrosky, C., Basran, J., Del Bello-Hass, V.: Implementation of a physical activity monitoring system for the elderly people with built-in vital sign and fall detection. Sixth International Conference on Information Technology: New Generations 2009, ITNG '09, pp. 1226–1231 (2009)
12. Bai, J., Zhang, Y., Cui, Z., Zhang, J.: Home telemonitoring framework based on integrated functional modules. Proceedings of the 22nd Annual International Conference of the IEEE Engineering in Medicine and Biology Society, vol. 1, pp. 778–781 (2000)
13. Kelly, G.: Body temperature variability (Part 1): a review of the history of body temperature and its variability due to site selection, biological rhythms, fitness, and aging. Altern. Med. Rev. **11**(4), 278–293 (2006)
14. Tortora, G.J., Anagnostakos, N.P.: Principles of anatomy and physiology. (1987)
15. Chesbrough, H.W.: Open Innovation: The New Imperative for Creating and Profiting from Technology. Harvard Business Press (2006)
16. Gassmann, O., Enkel, E.: Open Innovation/Die Öffnung des Innovationsprozesses erhöht das Innovationspotenzial. Zeitschrift Führung und Organisation (zfo) **75**(3), 132–138 (2006)
17. Reichwald, R., Piller, F., Ihl, C., Seifert, S.: Interaktive Wertschöpfung: Open Innovation, Individualisierung und neue Formen der Arbeitsteilung, Gabler Verlag (2009)
18. Asmussen-Clausen, M., Menche, N.: Pflege heute: Lehrbuch für Pflegeberufe, Elsevier, Urban & Fischer (2008)

19. gtec medical engineering, online
20. Zephyr, online
21. Oppel, A.: Databases a Beginner's Guide, 1st edn. McGraw-Hill, Inc., New York (2009)
22. Silberschatz, A., Korth, H., Sudarshan, S.: Database Systems Concepts, 5th edn. McGraw-Hill, Inc., New York (2006)
23. highcharts, online

Part V
Interaction

Visual and Haptic Perception of Surface Materials for Direct Skin Contact in Human–Machine Interaction

C. Brandl, A. Mertens, J. Sannemann, A. Kant, M. Ph. Mayer and C. M. Schlick

Abstract Due to the high demand for technical support systems to preserve the personal mobility of elderly people, this study evaluates the visual and haptic perception of ten materials for direct skin contact of the forearm in human–machine interaction. The expectations and requirements of 48 subjects are analyzed for different age groups and correlations between the different sensory stimuli are investigated. Results are recommendations for so far less common materials that have achieved a high level of acceptance as well as evidence for difficulties in the merely visual presentation of new products and materials.

Keywords Perception · Surface material · Skin contact · Acceptance · Human–Machine Interaction · Elderly people · Mobility aid

1 Integration of Technology in the Care Processes of Elderly People and People in Need of Care

Due to the demographic change the care of elderly and people in need of care becomes particularly challenging. In the future the amount of elderly people unable to look after themselves is expected to rise [1]. As this development is accompanied by a decreasing working population, the likelihood of a shortage in the nursing personnel and home care staff is increasing [2].

C. Brandl · A. Mertens (✉) · J. Sannemann · A. Kant · M.Ph. Mayer · C. M. Schlick
Chair and Institute of Industrial Engineering and Ergonomics of RWTH Aachen University, Bergdriesch 27, 52062 Aachen, Germany
e-mail: a.mertens@iaw.rwth-aachen.de

C. Brandl
e-mail: c.brandl@iaw.rwth-aachen.de

The approach that is pursued by Ambient Assisted Living (AAL) is to enable restricted people to live longer independently at home with help of assistive technologies. Just like communication—and monitoring devices that are designed for special needs [3], technology is meant to support and maintain the mobility of a person in his domestic environment. Electric wheelchairs and walking frames are established solutions to enable autonomous locomotion, but there is no universal solution to support all relevant steps for personal mobility in everyday life (see Fig. 1) [4]. Especially due to age-related loss of muscle [5] and the biomechanical demands on the musculoskeletal system [6] the transfer from a sitting to a standing position and vice versa is a motion that a lot of people cannot realize themselves without help although they can walk autonomously or with support [7].

As these two steps are part of the transfer from bed to a wheel chair also elderly people and people with disabilities of the musculoskeletal system, who cannot walk by themselves, would profit from a technical device that supports sitting up and down. Therefore the concept of such a "universal" mobility aid for an active support of personal mobility was developed in a user-oriented process. The functional principle is based upon resting on the forearms close to the body to ensure an ergonomic flux of force throughout the whole motion. The person does not have to change his position in order to move directly after sitting up. Based upon the knowledge that existing technical devices are only used seldom due to a lack of acceptance [8] an empiric study (N = 48) was performed to study the acceptance of different materials when the subjects lean on their forearms. As part of the developing process the goal was to analyze the Human–Machine Interaction in respect to the direct contact on the forearms. Therefore the following problems were evaluated:

- What are the expectations and requirements of potential users regarding the contact surface of a mobility aid?
- How are the different chosen materials assessed by the users?
- In how far does the purely visual perception comply with the bimodal perception (visual and tactile)?

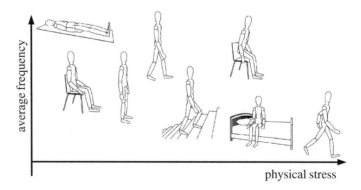

Fig. 1 Systematization of the steps of personal mobility in relation to the frequency and physical stress

2 Related Papers and Basics

In preparation for the study many other studies and papers have been found that discuss the haptic perception of different surfaces. However these studies have been made for different fields of products, e.g. automobile interior [9] or clothing [10] or basic studies that discuss different influences on haptic perception [11]. The focus of these studies is the perception on the finger tips or palms which cannot be transferred to the forearm.

Another related field that is discussed in many other papers is the integration of technology in health care. These works deal mainly with the currently used systems in the care for the elderly or the transport of patients [12, 13]. Some works also deal with the "being touched" by a robot but in these cases the reaction of the participants was only tested with one kind of material [14] or the technical systems had totally different functions [15]. Therefore no valid conclusions can be drawn over the perception of different materials on the forearms of elderly people. The following paragraph describes the main function and factors of haptic perception on the forearm and which of these are relevant for the design of the study.

2.1 Physiology of Haptics

The somatosensory system is in charge of recording the information from the skin surface, the locomotive system and the intestines of humans [16]. Whereas the perception of the intestines is neglected in the following, the other two aspects will be described in detail.

The sense of touch is build up of two subsystems. The sense of skin (see Fig. 2) records information on vibration, pressure, heat, cold and tissue damage (pain). The proprioception (the sense of the locomotive system) is capable of determining the location and movement of extremities and the impact of external forces.

2.1.1 Function of the Skin Senses

The Merkel–Ranvier cell complex registers as most important stimulus pressure. It is a slow adapting system with a small, sharply limited, receptive field which means a high spatial resolution for mechanic impulses. Therefore experts draw the conclusion that especially these cells are responsible for the perception of the hardness of an object when it is touched with the fingers. Ruffini corpuscles detect pressure as well but they are even more sensible for shear forces. They adapt slowly and have a large receptive field which causes a low spatial resolution. During the resting on a material shear forces should occur in correlation to the coefficient of friction which could be registered by the Ruffini corpuscles. Meissner- and Pacinian corpuscles are relevant for the registration of vibration. When a finger is moved over a textured

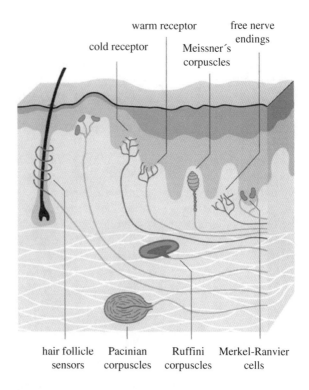

Fig. 2 Touch receptors of the skin [16]

surface vibration occurs, which is registered by the receptors. The assumption is that these cells record the roughness of a surface. Cold receptors detect a negative discrepancy to the indifference temperature (~ 30 °C). The used materials are at room temperature (~ 21 °C) so a reaction of the cold receptors can at least in the beginning of the resting contact be assumed. A large-area contact supports the impression of a cold surface. The free nerve endings react to great pressure in combination with a deformation of the skin in the contact area thus the impression of grip and a certain stickiness can be explained [16].

2.1.2 Functioning of Motion Sensors

The proprioceptors record the position and the movement of extremities as well as the acting forces. The joint receptors detect a stretch of the joint capsule whereas the Golgi tendon organ detects muscle tension. It is assumed that during the resting and touching a force that is correlated to the coefficient of friction is produced between the surface and the upper arm and fingers. Depending on how great the reaction forces are in respect to the pressing forces the according receptors should be capable of detecting the resulting resistance and thus indirectly the coefficient of friction [16].

2.1.3 For the Study Irrelevant Somatosensory Senses

Some sensory cells could be declared irrelevant for the study as they do not influence the perception of a surface. Warm receptors for example detect a positive discrepancy with an indifference temperature of 30–45 °C. The analyzed materials were not preheated but placed in the study with room temperature. As the room temperature of about 21 °C is far below the indifference temperature the warm receptors could be neglected in this analysis. Hair follicle sensors record a movement of the body hair. It cannot be excluded that these have a part in the first contact with the surface but trough a standardized experimental design the first contact was equalized and therefore has no influence on the variance of the case study. The hair follicle sensors can also be neglected in the touching experiment because the finger tips are hairless [17].

3 Methods

In order to answer the research questions, with regard to the perception of direct skin contact with different surface materials through technical support systems, a five-stage study design was developed:

1. Gathering of general expectations and requirement of different surface material attributes. (independent from materials)
2. Determining the optical perception and assessment of various materials
3. Evaluation of a 20 s passive contact of the underarms with various materials concerning the temperature cognition (blindfolded) forced by the instructor
4. Evaluation of contact perception by active resting on the forearms (blindfolded)
5. Bimodal exploring of surface material (visual and tactile perception)

To compensate learning effects and non-quantifiable interactions between the various materials, the order of the material appearance to the study objects was permuted according to Latin Square.

3.1 Study Subjects

To analyze relevant parameters for the product design the divergent experience, personal circumstances and the range of capacities within the potential target group must be taken into account already at the study planning process. To determine valid and applicable recommendations for the selection of a specific material for the contact surface of a mobility support, the study profiles were adequate distributed concerning socioeconomic status, workplace, living situation, and required special assistance. This minders the homogeneity of the samples, but on the other hand the applicability of the study increases through generic findings. All subjects participated on a voluntary basis without any compensation.

In total 48 person (♀ = 26, ♂ = 22) between the age of 52 and 93 years participated in the study (M = 82.3; SD = 8.1). The captured items related to the medical history revealed four participants with a diseased central nervous system (8 %), one participant with skin disease (2 %), and 17 participants with a limited visual acuity (35 %), whereas only four participants (8 %) showed limitations despite using visual aids. Concerning limitations in the personal mobility it appeared that 67 % of the participants perceived it 'difficult' or 'nearly impossible' to get up from an armchair or tub. While only 22 % reported getting up (or sit down) from a chair and toilet as difficult compared to the previous assessed circumstances. 68 % of the study participants reported 'no problem' or 'slight problems' with getting up from a bed and 47 % had difficulties to climb stairs. The descriptive analysis shows that different starting positions—without human or technical support—cause diverse barriers for elderly people.

3.2 Study Performance

Various survey tools were developed in order to quantify the expectations and perception of the participants prospectively and during the study. Negative formulating was avoided to prevent stigmatizing by the participants during the study (i.e. 'health restrictions' instead of 'disease' or 'appearing challenges' instead of 'difficulties') [18].

3.2.1 Participants' Profile and Expectations

Not only demographic factors were captured, but also information in terms of relevant diseases, possible restrictions in the ability to see and personal mobility of the participant were assessed via a questionnaire. In the following block of questions the expectations of the participants regarding the surfaces for direct skin contact with a mobility aid were gathered with a six-stage Likert scale for the ten subsequent attributes: 'good grip', 'inconspicuous', 'soft', 'easy to clean', 'comfortable', 'clean appearance', 'robust', 'pleasant warm', 'not scratchy' and 'precious'.

3.2.2 Perception of the Surfaces

Before initiating physical contact with the surfaces the participants were asked to evaluate the visual appearance of the materials by means of semantic differentials with a six-stage Likert scale. The items used are 'cheap/high quality', 'slippery/maximum grip', 'hard/soft', and 'warm/cold'. Therefore materials were placed with one meter distance in a slanted position in front of the participant. For this study the front side of the equipment was placed at the tables' edge centered before the participant. The evaluation of the participants was reported in a written form during the experiment by the instructor.

Through passive contact the participants were asked to rate the perceived temperature for each material. Therefore the instructor forced physical contact on the underarms of the participants with a 20 s constant pressure.

Following perceived temperature was assessed through physical contact by bracing the arms on the surface. Thus additional aspects were determined which were measured with a six-stage Likert scale. The participants therefore got the instruction to lean their upper part of the body over the equipment and to stress their underarms with the highest possible portion of their body weight. The handhold should further be held.

In the previous describes study steps, the participants wore blindfolds so only somatosensory stimulants influenced the evaluation of the different materials. The last step assessed the cognition of the participants during the bimodal exploring with the same four semantic potentials as in the visual perception. For this the blindfold was removed after announcement. The participants were asked to examine the material in front of them for approximately 20 s visually, but also by the use of their hands. In the end the participants were requested to give an opinion on the ability of the material to be used for a mobility aid.

3.3 Experimental Setup

For the experimental setup it was necessary to enable passive contact with the material surface through the instructor and to ensure adequate stability of the equipment during the active leaning on the underarms. Therefore a rotary arm construction pursuant to the relevant anthropometric measures (DIN 33402-2) was constructed (see Fig. 3). The carrier plate for the material samples can be fixed at 15° so that the participant is always enabled to lean over the material sample with the same angle.

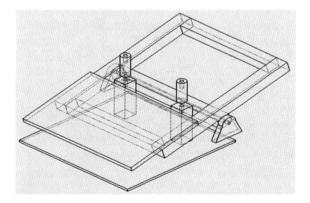

Fig. 3 Experimental setup as CAD draft

3.3.1 Surface Material

Before deciding on specific material samples for the contact surface firstly an analysis was initiated in order to find out which materials are used in health care by different manufacturer for their equipment. It appeared that only a limited number of materials are used for direct skin contact (i.e. arm rests of wheelchairs). Solid materials out of plastic and respectively cushions covered with fabric or (artificial) leather have the widest dissemination. Next to these established comparatives more materials were identified, that were already used in different fields. The total of ten chosen materials can be classified in two groups (see Table 1): the first group contains only solid materials, whereas the second group consists of different material covers, which are fixed on identical polyurethane plastic foam (volumetric weight of 14 kg/m^3) carriers (see Fig. 4). Due to the fact that the material samples are partly inhomogeneous material combinations and the material thickness of the solids are not uniform, it was not possible to use valid evidence based parameters from corresponding datasheets in terms of thermal characteristics and static friction.

3.3.2 Measurement of Material Characteristics

As these factors will probably have an influence on the perception of the surfaces through the participants, these characteristics were objectively recorded with the help of measurements and then related with each other. The conducted measurements only illustrate a heuristic approach, which should point out the relative relation between the materials in order to match them with the subjective perception of the participants.

Thermal Characteristics

To assess the perceived temperature in the study after 20 s skin contact adequately, the cooling of a reference object in compliance with DIN EN ISO 10456 in a defined period was recorded and represents an objective benchmark. The measurement instrument was an infrared thermometer (TFA-Dostmann Pro Scan Laservisier 31.1118) fixed on a tripod. A 79.8 g polished aluminum plate with 1 cm height and 6 cm diameter was used as reference object. The reference object was heated up to 55 °C and placed on each material sample. The temperature measurement was carried out at point t_0 directly after the heating. The measurement was repeated at t_{20} and t_{30} after 20 respectively 30 s and the differences (Δ_{20} and Δ_{30}) to t_0 were determined (see Table 2).

All measurements were carried out two times. Measurement repeats show that differences of temperatures are below 1 %.

Table 1 Tested surface materials and their characteristics

	Color	Material characteristics	Photograph of the surface
Solid materials			
Aluminum	–	AlMg$_3$ polished	
Rubber	Black	Rubbergranulate (Polyurethan) Shore Hardness A ~75°	
Wood	Oak bright stained	Particle-board with melamine resin facing veneer with wood structure	
Plastic	Grey	PVC Shore Hardness D ~80°	
Sponge rubber	Black	Polyethylen-Foam Closed cell Shore Hardness A ~35°	
Natural cork	Natural-colored	Compressive strength: 0.5 MN/m^2 (at 15 % compression)	

(continued)

Table 1 (continued)

	Color	Material characteristics	Photograph of the surface
Materials for covering			
Artificial leather	Black	100 % Polyester Coating: 100 % PVC Leatherlike look	
Lambskin	Natural-colored	Backside: tanned	
Neoprene	Black	95 % Polyester, 5 % Polyurethan Crosswise elastic	
Fabric	Brown	Cottonfabric	

Fig. 4 Material samples with covering

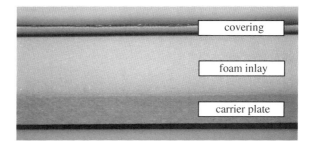

Table 2 Overview of the measured temperature loss and the static friction coefficient

Material	Δ_{20}	Δ_{30}	μ mean
Sponge rubber	0.35	0.55	0.33
Lambskin	0.55	0.9	0.33
Fabric	0.9	1.35	0.32
Neoprene	0.95	1.2	0.27
Natural cork	1.06	1.6	0.23
Artificial leather	1.25	1.65	0.79
Rubber	2.5	3.25	0.58
Wood	2.5	3.4	0.22
Plastic	3.15	4.15	0.2
Aluminum	16.4	21.2	0.45

Static Friction

To facilitate the correlation between the semantic differential 'slippery/maximum grip', evaluated by the user, and the objective characteristics of the different surface materials, the static friction was assessed with a heuristic measurement. Therefore an Inclined Plane Test was accomplishes, which procedure is in accordance with the method described by DIN EN ISO 12957-2.

The already under Sect. 3.3.2.1 described polished aluminum plate was used as reference object. The process was conducted the following way: first, the reference object was placed on the particular material sample. Afterwards the sample was inclined until the reference object started to slide. The measured angle α was assessed with an angle meter (DMW 40L by Bosch) with a measuring tolerance of $\pm 0.1°$. Each measurement was repeated five times and the arithmetic mean was determined (see Table 2).

3.4 Subjective Perception of Material Parameters of the User Study

As to be seen in Table 3 the cold receptors for temperature stimulus lies between 15 and 30 °C. The subjective perception of cold is assessed during the passive contact of underarm and material as well as during the actively resting of the

Table 3 Allocation of subjective assessed stimuli and resulting material parameter

Body part	Stimulus	Occurrence in study	Derivable parameter
Cold receptors	Temperature (15–30 °C)	Passive contact Resting on forearms	Thermal conductivity Heat transfer Heat capacity
Merkel–Ranvier cells Ruffini corpuscles Free nerve endings	Pressure	Bimodal exploring Resting on forearms	Compression hardness
Ruffini corpuscles	Shear forces	Resting on forearms	Static friction coefficient
Golgi tendon organs Joint receptors	Forces on the locomotive system	Bimodal exploring	Sliding friction coefficient

underarms on the material. The identified material parameters were heat exchange, heat conductivity, and heat capacity. These parameters might influence the refrigeration of the skin. Merkel–Ranvier cells and Ruffini corpuscles as well as free nerve endings are able to capture pressure stimulus.

The identified material parameter in this case is the hardness. Ruffini corpuscles in compliance with Golgi tendon organs and joint receptors are able to gather reaction force on the skin and the human locomotive system.

During the resting on the forearm and bimodal exploring the participants were asked to report in this regard. The approximate material parameter, which influence the subjective perception are, static- and dynamic-friction coefficients.

3.5 Experimental Variables

Independent variables are the different material surfaces as intra-subject factor. Age and personal mobility of the participants serve as control variables. Dependent variables are the four semantic differentials 'cheap/high quality', 'slippery/maximum grip', 'hard/soft', and 'warm/cold' both assessed during visual perception and during bimodal exploring.

3.6 Hypotheses

With regard to the dependent variables following hypotheses were proposed and evaluated during the user study:

H_1: The perception and assessment of material characteristics in the form of semantic differentials during bimodal exploring differs for all materials.
H_2: The purely visual assessment of the material characteristics differs from the assessment after the bimodal exploring in semantic differentials.

4 Results

4.1 General Requirements for Contact Surfaces and Direct Skin Contact

Figure 5 describes requirements and priorities elderly persons have concerning surface materials for technical support systems. They were requested during the first project phase independently from material conditions. It emphasized that all participants prioritize 'good grip' as the most important factor. This characteristic shows the highest mean value ($M = 5.72$) and the smallest standard deviation ($SD = 0.71$) of all items, which points out a consensus in terms of the relevance of a secure and controllable interaction, independent from age or mobility.

The second highest priority, an inconspicuous presentation of the surface ($M = 4.9$, $SD = 0.83$), is a significant indicator for the fear to be recognized as frail and dependent [19]. Characteristics in the field of comfort such as 'comfortably warm' or 'not scratchy' ($M = 2.13$, $SD = 1.12$; $M = 2.04$, $SD = 0.95$) were reported as less important. Lowest priority has been given to the characteristic 'precious' ($M = 1.17$, $SD = 1.60$). It is clearly recognizable that the average user does not use assistive technologies as a status symbol.

Looking at the control variable 'age' and corresponding requirements in terms of material characteristics, tendencies showed that with increasing age safety characteristics ('good grip', 'robust') and convenient handling ('easy to clean') are

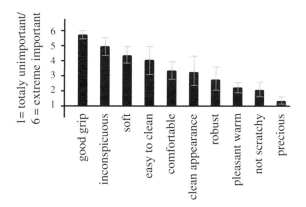

Fig. 5 Prioritizing of material surface characteristics with average values (independent from material)

prioritized in contrast to younger participants. A reverse effect is proven for characteristics related to comfort ('soft', 'comfortable', 'not scratchy'), since the participants tend to prioritize these characteristics the lower their age is.

Tendencies concerning the influence of the personal mobility were identified. People with limited personal mobility (more than three steps of the personal mobility rated as 'difficult' or 'hardly possible') graded characteristics which focus on perception of third parties ('inconspicuous' and 'precious') higher. This may be due to the fact that corresponding support systems are needed on long-term and stigmatizing can be reduced through inconspicuous and precious design [20].

4.2 Perceived Surface Characteristics

In Fig. 6 the results of visual perception and the perception during bimodal exploring of the ten different surface materials are represented. A correlation analysis for each material shows only occasionally strong correlations ($r > 0.50$) between perception after visual assessment and after bimodal exploring with regard to the four examined semantic differentials. Only for common materials (aluminum, wood, plastic, fabric) correlate visual evaluation and evaluation after skin contact significantly related to semantic differentials. Therefore participants often perceived materials differently through touching then through visual assessment, unless they did have experience with comparable materials. The results let assume that the evaluation of elderly people often not reflects the real characteristic in terms of innovative material characteristics they only are able to assess visually. Hence the hypothesis H_2 can be verified.

Through this effect the acceptance of new systems can be negatively influenced, as in any case a discrepancy exists between reality and subjective perception. Thus products may be refused by the user due to wrong or negative expectations even before use or as a consequence of unfulfilled expectations and disappointment.

A single factor variance analysis with repeating measures (ANOVA) of the evaluation of the assessment through bimodal exploring was performed with regard to the perception of semantic differentials by examined materials. For long-term acceptance by the user it is crucial to have interaction between visual and tactile perception because this reflects the influences during the use of a potential mobility assistance best [21]. At this point in time a grading concerning the suitability of a certain material for a surface of a mobility assistance tool is not applicable and only different perceptions are described objectively. Grading in terms of suitability from perspective of the user is provided in Sect. 4.3 with regard to the prioritized characteristics in Sect. 4.1.

It turned out that significant differences exist for the semantic differential 'slippery/maximum grip' within the different materials ($F = 7.43$; $df = 9$; $p < 0.01$). As a result natural cork, rubber, sponge rubber and fabric were graded with maximum grip, whereas plastic, neoprene and aluminum were perceived as very slippery.

Fig. 6 Visual and bimodal perception of ten different surface materials

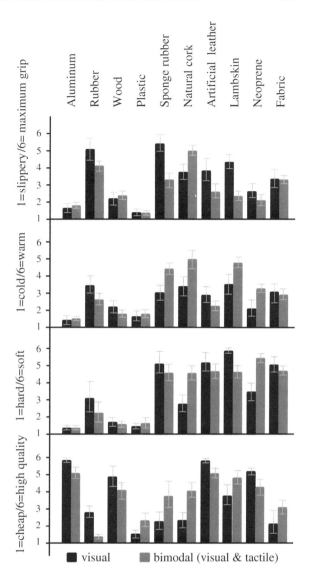

Also huge significant differences (F = 6.50; df = 9; $p < 0.01$) exist for the semantic differential 'cold/warm' between the ten tested materials. Natural cork was perceived as warmest followed by lambskin and sponge rubber. Aluminum, wood and plastic was perceived as cold as expected.

For the semantic potential 'hard/soft' likewise significant differences were identified (F = 5.44; df = 9; $p = 0.03$). However, Fig. 6 shows a strong cluster of the materials to very hard (aluminum, plastic, wood, rubber) and very soft (lambskin, artificial leather, sponge rubber, natural cork and fabric). The conducted paired sample t Test with Bonferroni-correction within these clusters does not result with a significant difference for all combinations.

For the semantic potential 'cheap/high quality' significant differences were verified (F = 7.21; df = 9; $p < 0.01$). Rubber and plastic were experienced as very cheap, whereat aluminum, artificial leather, lambskin and neoprene were graded high-quality.

Consequently it was possible to identify a significant effect of the independent variables on the dependent variables—the perceived material characteristics, for all semantic differentials. Due to the fact that not all factor levels for 'hard/soft' differ significantly, the Hypothesis H_1 must be rejected.

4.3 Assessment of Surface Suitability

For the final assessment of the suitability of a material for a mobility aid from the participants perspective, the four semantic differentials introduced in Sect. 4.1 were weighted as well as the suitability from the participants perspective (see Table 4). The semantic differentials 'slippery/maximum grip' was allocated to the requirement 'good grip', 'hard/soft' is in accordance with the requirements in Table 4 (materials arranged downwards according to the meeting of requirements). 'Soft' surface materials 'cold/warm' corresponds with the characteristic 'pleasant warm'. 'Cheap/precious' was weighted in accordance with the prioritizing of the requirement 'precious'. The results show that next to the established materials such as sponge rubber, also natural cork and lambskin reach a high acceptance among the users. The target group specific requirements were even better met as with the standard materials. Cork for example is already used in different situations with high mechanical load for example as floor covering or as handlebar grips (see www.ergon-bike.com). Especially rather unknown materials show the previous described pattern, that materials which are only visually assessed are not perceived adequately.

4.4 Evaluation Alignment of Objective Material Characteristics

The collected relative material parameter static friction and thermal characteristics partly differ strongly from the subjectively perceived material characteristics.

Although natural cork has the third lowest static friction coefficient it was evaluated to have the best possible grip. Artificial leather was measured with highest static friction coefficient and was evaluated through the user only low above the average. From these results the conclusion could be drawn that the standard technical methods to assess material characteristics are not sufficient in order to determine the user preferences concerning the contact surface for the underarms adequately. The usage of an aluminum reference unit was not able to

Table 4 Evaluation of the materials as contact surfaces for mobility support tools (appearing in decreasing order of meeting the requirements)

Material parameter	Weight (%)	Natural cork	Sponge rubber	Lambskin	Neoprene	Artificial leather	Fabric	Wood	Aluminum	Rubber	Plastic
Good grip	25	5.0	3.3	2.3	2.1	2.8	3.2	2.6	1.8	4.1	0.4
Softness	20	4.9	4.8	4.9	3.3	2.2	3.0	1.8	1.6	2.7	1.8
Heat	15	4.8	4.6	4.8	5.4	4.7	4.8	1.5	1.3	2.1	1.6
Precious	10	4.1	4.0	4.8	4.3	5.1	3.0	4.2	5.1	0.3	2.2
Suitability	30	5.7	4.4	4.9	5.5	5.5	3.4	3.8	2.8	1.4	2.2
Σ (weighted)		5.1	4.2	4.2	4.1	4.0	3.4	2.8	2.3	2.3	1.6

represent the perception during strong pressure contact. It was further not possible to simulate conclusions from the perceived friction with human skin. Especially sweat as liquid, which may lead to adhesion effects, and the different age dependent characteristics of the skin were hard to picture [22]. The evaluation of the thermal characteristics in terms of the measured cool down of the reference unit, is an indicator for the specific thermal transition during direct skin contact with one of the surface materials. The model further is only partly suitable to make a statement concerning the preferences of potential users. The sponge rubber for example was measured with the lowest cooling, but it was not perceived as most comfortable in terms of warmth. This phenomenon may be explained with the fact that the reference unit was too light to produce full contact with the surface and therefore an isolating air cushion hindered the thermal activity.

5 Summary and Conclusion

The conducted study gives specific references towards the material selection for the surface of a mobility aid with direct skin contact. In this context relevant connections and problems between the purely visual and the haptic perception of a surface could be determined.

Safety and the inconspicuousness are clearly the most relevant factors for the target group which answers the research question which requirements the target group has on the contact surface. Characteristics such as comfort or especially the value play a minor role in the requirements.

Materials that are often used in this domain were often rated extremely negative and a high discrepancy between the perception and the scientifically measured parameters could be detected.

A significant finding for the construction of new assistive systems is the fact that the visual perception does not correlate with the haptic perception of materials. This effect was detected more strongly for materials that the subjects did not have contact with before. This is due to the lack of experience the test person has with these surfaces.

Summarizing there can be said that especially the use of natural products has a very high acceptance in the senior population. The participatory approach towards finding new materials and design parameters can help to increase the acceptance of technical support systems and thus the potential of the integration of technology into the treatment process of elderly people and people in need of care.

Acknowledgments We want to thank all subjects who participated in this case study. This research is part of the project "Tech4P" which was supported by the Federal Ministry of Education and Research (01FG1004). The Project Management Agency is DLR - part of the German Aerospace Center.

References

1. Peters, E., Pritzkuleit, R., Beske, F., Katalinic, A.: Demographischer Wandel und Krankheitshäufigkeiten—Eine Projektion bis 2050. In: Saarländisches Ärzteblatt 9/2010
2. Statistisches Bundesamt: Bevölkerungs- und Haushaltsentwicklung im Bund und in den Ländern. Onlineveröffentlichung. (2011)
3. Mertens, A., Jochems, N., Mayer, M., Schlick, C.: Ergonomische Analyse und Gestaltung der Mensch-Rechner-Interaktion für die Benutzung telemedizinischer Dienstleistungen. Zeitschrift für Arbeitswissenschaft **66**, 150–168 (2012)
4. Schlag, B. (ed.): Leistungsfähigkeit und Mobilität im Alter. TÜV Media GmbH, Köln (2008)
5. Skelton, D.A., Greig, C.A., Davies, J.M., Young, A.: Strength, power and related functional ability of healthy people aged 65–89 years. Age Ageing **23**(5), 371–377 (1994)
6. Riley, P.O., Schenkman, M.L., Mann, R.W., Hodge, W.A.: Mechanics of a constrained chair-rise. J. Biomech. **24**(1), 77–85 (1991)
7. Kivinen, P., Sulkava, R., Halonen, P., Nissinen, A.: Self-reported and performance-based functional status and associated factors among elderly men: the finnish cohorts of the seven countries study. J. Clin. Epidemiol. **51**(12), 1243–1252 (1998)
8. Läubli, T., Gassert, R., Nakaseko, M.: Human-centered design in the care of immobile patients. In: Kurosu, M. (ed.) Human Centered Design, HCI 2011, LNCS 667, pp. 321–326. Springer, Berlin (2011)
9. Bahn, S., Lee, C., Nam, C.S., Yun, M.H.: Incorporation affective customer needs for luxuriousness into product design attributes. Hum. Factors Ergon. Manuf. **19**(2), 105–127 (2009)
10. Broega, A.C., Noguiera, C, Cabeco-Silva, M.E., Lima, M.: Sensory Comfort Evaluation of Wool Fabrics by Objective Assessment of Surface Mechanical Properties. In: AUTEX 2010 World Textile Conference
11. Lederman, S.J., Thorne, G.: Perception of texture by vision and touch: multidimensionality an intersensory integration. J. Exp. Psychol. Hum. Percept. Perform. **12**(2), 169–180 (1986)
12. Mukai, T., Shinya, H., Nakashima, H., Kato, Y., Sakaida, Y, Guo, S., Hosoe, S.: Development of a Nursing-Care Assistant Robot RIBA That Can Lift a Human in Its Arms. In: The 2010 IEEE/RSJ International Conference on Intelligent Robots and Systems (2010)
13. Kasagami, F., Wang, H., Araya, M.: Development of a Robot to Assist Patient Transfer. In: 2004 IEEE International Conference on Systems, Man and Cybernetics (2004)
14. Chen, T.L., King, C.-H., Thomaz, A.L., Kemp, C.C.: Touched by a Robot: An Investigation of Subjective Responses to Robot-initiated Touch. In: HRI '11, Proceedings of the 6th International Conference on Human-Robot Interaction, pp. 457–454 (2011)
15. Stiehl, W.D., Liebermn, J, Breazeal, C. Basel, L., Lalla, L. Wolf, M.: Design of a Therapeutic Robotic Companion for Relational, Affective Touch. In: 2005 IEEE International Workshop on Robots and Human Interactive Communication, pp. 408–415 (2005)
16. Lang, F., Lang, P.: Basiswissen Physiologie, 2nd edn. Springer, Berlin (2007)
17. Priewe, J, Tümmers, D. (Hrsg): Somatoviszerale Sensorik. In: Das Erste: Kompendium Vorklinik—GK1, pp. 1198–1214. Springer, Berlin (2007)
18. Porst, R.: Fragebogen, Ein Arbeitsbuch. Verlag für Sozialwissenschaften, Wiesbaden (2008)
19. Ziefle, M., Röcker, C., Holzinger A.: Medical Technology in Smart Homes: Exploring the User's Perspective on Privacy, Intimacy and Trust. In: The 3rd IEEE International Workshop on Security Aspects of Process and Services Engineering 2011, pp. 410–415
20. Wilkowska, W., Ziefle, M.: Perception of privacy and security for acceptance of E-health technologies: exploratory analysis for diverse user groups. User-centred-design of pervasive health applications (UCD-PH'11) (2011)
21. Jakobs, E.-M., Lehnen, K., Ziefle, M.: Alter und Technik. Eine Studie zur altersbezogenen Wahrnehmung und Gestaltung von Technik. Aachen, Apprimus (2008)
22. Escoffier, C., de Rigal, J., Rochefort, A., Vasselet, R., Lévêque, J.-L., Agache, P.G.: Age-related mechanical properties of human skin: an in vivo study. J. Invest. Dermatol. **93**, 353–357 (1989)

Human-Robot Interaction: Testing Distances that Humans will Accept Between Themselves and a Robot Approaching at Different Speeds

Alexander Mertens, Christopher Brandl, Iris Blotenberg, Mathias Lüdtke, Theo Jacobs, Christina Bröhl, Marcel Ph. Mayer and Christopher M. Schlick

Abstract Service robotics has great potential for helping people to live independent lives in their own homes. However, if this potential is to be fully exploited in the near future, research and development cannot limit itself to solving the technological challenges involved. The only way to develop service robots that people will accept is to get potential users involved in the process as early as possible. With that in mind, this study investigates human-robot interaction from the perspective of a service robot approaching the user at varying speeds. We developed an empirical study to measure the distance that humans will accept between themselves and a robot when approached by that robot. The results show that the robot's speed and the test subject's body position significantly affect the accepted distance. We also found that the physical appearance of humanoid service robots has no substantial bearing on the accepted distance.

Keywords Human-robot interaction · Service robot · Acceptance · Speed

1 Introduction

While the use of robots is accepted practice in many areas of industry today, service robotics still has many problems to overcome. To date, the spectrum of tasks performed by service robots has been limited to providing simple services. Current research (e.g. ARMAR, Care-O-bot® 3, DESIRE, Justin and RIBA),

---

A. Mertens (✉) · C. Brandl · I. Blotenberg · C. Bröhl · M. Ph. Mayer · C. M. Schlick
Chair and Institute of Industrial Engineering and Ergonomics of RWTH,
Aachen University, Bergdriesch 27, 52062 Aachen, Germany
e-mail: a.mertens@iaw.rwth-aachen.de

M. Lüdtke · T. Jacobs
Fraunhofer Institute for Manufacturing Engineering and Automation,
Nobelstraße 12, 70569 Stuttgart, Germany

however, is developing and testing ways of having robots perform complex tasks. These projects are increasingly focusing on health and nursing care. In domestic settings, mobile service robots have great potential for helping people live independent lives in their own homes for longer. This might include transferring physically strenuous activities to the robot, or using the robot to compensate for existing impairments in the user's mobility.

Unlike industrial robotics, service robotics is oriented towards humans. It is therefore important to know as much as possible about what potential users want from the technology, and to pay sufficient attention to these needs when developing the systems. Developers must take account of user requirements concerning, for example, the kind of human-robot interaction that necessarily occurs when using a robot. In an industrial setting, humans and robots often work in separate areas. But this is often impossible for the kinds of tasks a service robot carries out. To perform their tasks, service robots move around freely within their user's living space. As a result, the design of human-robot interaction plays a decisive role in whether or not users accept the technology.

One of the first steps in human interaction with server robots is the user seeing the machine and identifying it as a robot. Next, the user will apply their own mental models to produce expectations about the robot's function and behaviour. The way that these first steps unfold depends largely on the physical appearance of the robot.

Robot designs are generally classified according to how much they resemble a human. The two ends of the spectrum are "machine-like" and "human-like". The "uncanny valley", a qualitative model often used in the design of service robots, describes familiarity as a function of the degree of human-likeness in relation to movement or lack of movement [1]. Because a causal relationship exists between familiarity and acceptance, the literature often speaks specifically about how a robot's appearance or behaviour affects whether or not a user accepts it [2]. A recognised goal in robot development is to positively influence acceptance by producing an anthropomorphic appearance [3]. However, anthropomorphism does not always benefit acceptance—it can actually confuse users [4]. A robot's appearance can therefore affect acceptance or familiarity to differing degrees.

Alongside appearance, behaviour is also important when it comes to developing effective robots [5]. According to the "uncanny valley" model, these are the two main factors influencing familiarity. Depending on how human-like it is, a mobile robot can produce a much higher or lower level of familiarity than a stationary robot [1]. With regard to acceptance, robot behaviour is considered more important than key physical aspects (e.g. its size and design) and more important than its ability to produce facial expressions and gestures [4].

If a person likes a robot's overall appearance but rejects its behaviour, this can result in a certain level of disappointment [6]. Unlike robot appearance, the field has yet to succeed in producing authentic anthropomorphic robot behaviour (e.g. gestures, facial expressions and motor functions). A robot that closely resembles a human will thus disappoint users when they interact with it, since its behaviour does not back up the expectations that its appearance creates. To avoid this kind of disappointment in the following empirical study, we deliberately chose a robot that did not look human.

For service robots to be able to perform their tasks satisfactorily, they must interact with humans in the same space.

With this in mind, Walters et al. investigated the distance that people considered comfortable to maintain when approaching a stationary robot to interact with it [7].

Irrespective of that distance, service robots must be able to come within reach of humans when performing fetch-and-carry tasks. These are especially important for people with limited mobility. As a result, the way in which a robot approaches a person is a crucial factor to consider when designing human-robot interaction. The following section will present the current state of scientific knowledge on the effects of various robot-human approaches, and will use this information to develop research questions and relevant hypotheses.

Our aim was to develop and carry out an empirical study investigating the factors that influence the distance that people will accept between themselves and a service robot approaching them.

2 Current State of Scientific Knowledge

A further study by Walters et al. shows that about 40 % of subjects tolerated shorter distances between themselves and an approaching robot than they did between themselves and another person (human–human interaction) [8]. We can therefore assume that these subjects did not perceive the robot as a social being [8]. This finding shows that it is important to consider a robot's appearance when setting the distance at which humans and robots will interact. Another study, this time by MacDorman, showed that the degree of human-likeness affects how eerie or familiar a person considers a robot to be [9].

In the study's conclusion, however, MacDorman says that the perceived human-likeness is not the only factor affecting how eerie or familiar a person finds the robot [9]. Therefore, in what follows, we will present other factors that might influence the accepted distance.

The study by Walters et al. measured the spatial distance at which 28 test subjects of varying ages became uneasy in the presence of an approaching, non-humanoid robot [8]. However, the distance sensor stopped the robot automatically when it was 0.5 m away from the subject. This means that the study was unable to investigate the human response to the robot at close range. If we compare the results of this study with those produced by humans approaching a stationary robot, we find that there are differences in the distances measured. Mizoguchi et al. investigated different speeds of approach and found that the faster the robot approached the person, the larger the acceptable distance was [2]. They also found that when the robot approached at the same average speed but with different speed profiles, it affected the level of familiarity that subjects felt towards the robot [2].

The appearance and speed of a robot affect user acceptance. It is therefore important that even at the stage of developing the technology for service robots, sufficient consideration is given to the design of their appearance and behaviour. The standard DIN EN ISO 9421-210 describes a process for designing usable interactive systems. It recommends taking the intended use as the starting point for developing these kinds of systems, and to pay attention to it throughout the entire process. In ambient assisted living situations, a variety of scenarios exist in which service robots could conceivably help humans. In these scenarios, the user might be standing, sitting or lying down. Different body positions alter people's perceptions. For example, if the angle of the head changes, it can cause a person's subjective judgement of the vertical to deviate from the objective situation [10]. The position of the body also affects the time needed to get out of the way of an approaching robot. If a person is lying down, it will take more time and effort to move out of the robot's path than it would if they were standing up. This influence on the distance that a person will accept between themselves and an approaching robot can be described as perceived safety. Developing safety mechanisms to protect humans from harm is an important issue in the field of human-robot interaction as a whole [11–13]. It has also been shown that robot behaviour affects the way in which users perceive safety [14].

Furthermore, it makes sense to consider that age might affect a user's ability to interact with technological systems. Numerous effects of this nature have been found to apply to computers. The effects can be the result of age-related changes in physical and mental abilities [15]. Well-known factors include a person's attitude to technology [16], their knowledge and experience of technological systems [17], and their technology-related self-efficacy [18]. The interaction of these elements is underpinned by the individual's overall affinity for handling electronic devices [19]. A final hypothesis will therefore address how age affects the accepted distance between human and robot.

To summarise, research in the field of human-robot interaction relating to robot approaching human should focus on the following:

- Influence of different, non-anthropomorphic physical appearances of robots
- Influence of different speeds and speed profiles
- Influence of the intended use on interaction with the robot

Past studies were carried out as Wizard of Oz experiments, which could have had a non-quantifiable impact on the results. The investigator stopped the robot when the subject expressed that wish. The distances measured in this way do not reflect the actual accepted distances, which is why we optimised the methodology for our study. Further, existing studies have mostly been unable to investigate human-robot distances of less than 0.5 m, even though the robot must enter this space if it is to come within reach of the person.

3 Acceptance

In this context, acceptance means the active willingness of the person in question to make use of a state of affairs that they perceive to be new. This state of affairs, or innovation, includes complete services, ideas, products and processes—as well as specific characteristics (e.g. design, quality, interfaces and behaviours). The decision on whether or not to accept a given innovation can vary depending on who is judging the situation. The crucial factor here is how the person processes information [20], as this helps them compare their expectations with the way specific characteristics actually appear. Decisions on whether or not to accept individual characteristics culminate in an overall decision on whether or not to accept the innovation. This means that changing a single characteristic can affect overall acceptance. And because the expectations that a person has of the innovation can change over time, these too can affect overall acceptance [21]. Most of the changes can be represented by the different phases of acceptance [22]. These phases are divided into: motivation, awareness, first contact, and use. If, for example, a person in the awareness phase sees a very human-like robot, they will have significantly higher expectations of witnessing anthropomorphic behaviour in the subsequent phases (cf. Sect. 1).

For developers to design the characteristics of an innovation in such a way as to ensure optimal acceptance, they must be aware of the factors that determine acceptance. Depending on the innovation in focus, the literature discusses a variety of influences that can affect acceptance [23–25]. Overall, the terms (perceived) usefulness and (perceived) ease of use are posited as the primary factors that determine acceptance.

The test subjects in the following study assumed the role of a user of a service robot. We selected subjects who had no experience with the service robot in question, and little experience with robots in general. This meant that the empirical study covered the phases of motivation, awareness and first contact.

4 Methodology

Our empirical study investigated, for a number of variables, the distance that subjects would accept between themselves and an approaching robot. We used the Care-O-bot® 3 (Fig. 1), a service robot that performs tasks which require it to interact with humans. It can fetch and carry household items and drinks, lay the table, and open drawers and doors. The Care-O-bot® 3 is not humanoid—its design deliberately avoids human attributes. The robot's functions are split between the front and back. The "working side" faces away from the user and houses all the technical components (e.g. its manipulator) that cannot be covered. The "serving side", which has no visible technical components, is where the human–machine interaction happens. This side has a fold-away tray, which

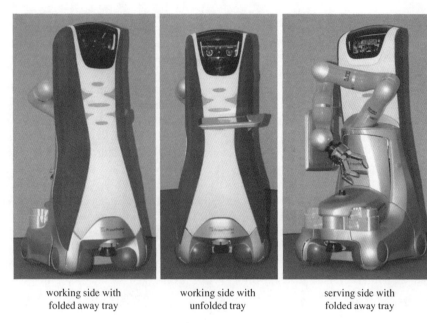

working side with
folded away tray

working side with
unfolded tray

serving side with
folded away tray

Fig. 1 Three views of the Care-O-bot® 3

functions as the main interface between the user and the robot. In addition to transporting objects, the tray also has an integrated touchscreen which is used for inputting and outputting information. The Care-O-bot® 3 is approximately 1.45 m tall and occupies a 0.6 m diameter floor space [26].

4.1 Task and Test Subjects

To carry out the task, subjects were given a button which they could use to stop the robot. The subjects held the button in both hands as the robot approached them. They were asked to press the button as soon as they felt that the distance between themselves and the robot was unacceptable.

Thirty test subjects aged between 20 and 75 ($\bar{x}$: 43.33a; SD:19.01a) took part in the empirical study. The participants, 17 women and 13 men, were divided into three age groups. AG_1 had 15 subjects between 20 and 39, AG_2 had eight subjects between 40 and 59, and AG_3 had seven subjects between 60 and 75. Asked to state their highest educational qualification, 13 subjects said that they held university degrees, while 11 said it was their *Abitur* (university entrance examination taken at German secondary schools). Two held doctorates, and two had completed an apprenticeship. Two other subjects said that their highest qualification was the *Mittlere Reife* (roughly equivalent to GCSEs in the UK or a high-school diploma in the US) and, respectively, a qualification from the *Handelsschule* (vocational business high school). Eleven subjects said that they work in the field of technology or science. Ten subjects said that they work or used to work in the field of social studies or the humanities. Nine said that they work or used to work in business or administration.

4.2 Independent and Dependent Variables

The following independent variables were considered as within-subject factors: the robot's appearance (e), the robot's speed (v), the subject's body position (k) and the robot's task (a). The age group, also an independent variable, was considered as a between-subject factor.

The distance that the subjects accepted between themselves and the approaching robot was measured as a dependent variable.

4.2.1 Robot Appearance

The front and back of the robot look different and serve different purposes. To see how each side influences the user, the investigators had the robot approach the subjects with each side showing in turn. As Fig. 2 demonstrates, neither side is humanoid in appearance. One side is more aesthetic, while the other looks more technical.

4.2.2 Robot Speed

Section 2, which dealt with the current state of scientific knowledge, discussed how the speed and speed profile affect the distance that subjects are willing to accept between themselves and the robot. Therefore, it seemed sensible and

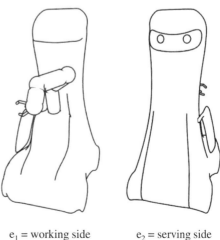

Fig. 2 Factor levels in the robot's appearance

e_1 = working side e_2 = serving side

necessary to consider both of these variables. With this in mind, we carried out the empirical study with three constant speeds and two speed profiles.

The robot's top speed is 0.75 m/s. DIN EN ISO 10218-1 considers 0.25 m/s to be a safe speed if the robot is making hazardous movements. At the moment, this standard only applies to industrial robots. But DIN EN ISO 13482, which is currently being drafted, will expand it to apply to non-industrial, non-medical household and assistive robots. We therefore used 0.25 m/s as the minimum speed for our service robot. The third constant speed (0.5 m/s) is the mean of the maximum and minimum speeds. The robot's maximum acceleration is 0.8 m/s^2. This means that it covers:

- 0.04 m to accelerate to 0.25 m/s
- 0.16 m to accelerate to 0.5 m/s
- 0.35 m to accelerate to 0.75 m/s

At all three of these speeds, which are shown in Fig. 3, the robot travels in a uniform motion.

We used the two speed profiles to investigate whether subjects would accept less distance between themselves and the robot when it had a non-uniform motion than they would when it moved in a uniform way, where the average speed was the same in both cases. To make it theoretically possible to reduce the accepted distance, we developed the speed profiles so that the robot travelled at a higher speed when it was further away from the subject and at a lower speed when it got closer. We achieved this by reducing the robot's speed as it approached the subjects. The difference between the two speed profiles was based on the following considerations. One profile was smooth so that it would protect the technical components in the drivetrain (e.g. electric motors and fan belts) and thereby extend the robot's lifetime or make it possible to produce relatively cheap drivetrain parts for service robots in the future. The other profile was graduated so that the subjects could clearly see that the robot was slowing down. Both profiles had an average speed of 0.5 m/s. This meant we could compare the results with those for the constant speed of 0.5 m/s.

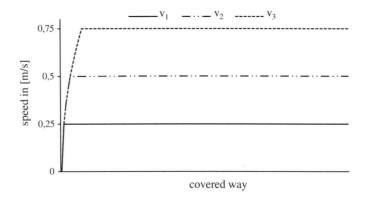

Fig. 3 v-s graph of the three constant speeds

Human-Robot Interaction

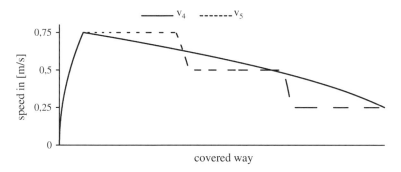

Fig. 4 v-s graph of the two speed profiles

For both profiles the robot's maximum acceleration was 0.8 m/s². At the start, the robot accelerated to the maximum 0.75 m/s. It covers about 0.35 m before it reaches that speed. We calculated the speed profiles for the remaining distance to the subject. In the smooth profile, the robot decelerated evenly from 0.75 to 0.25 m/s. The graduated profile involved three equal stages of motion at speeds of 0.25, 0.5 and 0.75 m/s. Between these, there were two downward "ramps" where the robot decelerated at 0.8 m/s². Figure 4 shows the two speed profiles.

The five factor levels of the independent variable speed are described as follows:

- Speed $v_1 = 0.25$ m/s
- Speed $v_2 = 0.5$ m/s
- Speed $v_3 = 0.75$ m/s
- Speed profile $v_4 =$ smooth
- Speed profile $v_5 =$ graduated

4.2.3 Body Position of the Test Subjects

With regard to an ambient assisted living environment we considered the positions standing, sitting and lying (Fig. 5) to be best-suited to testing the most common scenarios in which a service robot would support its user.

4.3 Procedure

We conducted the empirical study in four stages:

- Preliminary interview
- Introduction
- Distance measurement
- Concluding interview

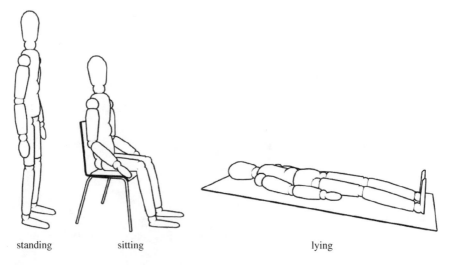

Fig. 5 Factor levels of the body positions adopted by test subjects

The preliminary interview, which we conducted before introducing the robot, established the subjects' demographic details, their experience with robots and their affinity for technology [21].

During the introduction stage, the subjects were allowed to inspect the robot. In a structured interview, the subjects discussed how they felt about the robot and were encouraged to think about how it works. At the end of the introduction stage, the test subjects had all received the same information on how the robot works and on the tasks it performs.

During the distance measurement stage, we recorded the extent to which the independent variables affected the distance that the subjects would accept between themselves and the approaching robot. The subjects were positioned—either standing, sitting or lying—so that each of them began the test at the same distance from the robot's starting position. As shown in Fig. 6, a desk for two investigators was set up to the left of the test subject. One investigator was tasked with starting the robot and recording the distances. The other investigator was responsible for safety during the test.

The safety plan consisted of three separate levels. First, the speeds and speed profiles were programmed in such a way that the robot would automatically stop as soon as it got within 0.1 m of the test subject. Second, a distance sensor would stop the robot if it got closer than 0.1 m to the subject. Third, the investigator responsible for safety could use an emergency stop button to abort the test manually. The robot began approaching each of the subjects from five metres away. The subject pressed a button to stop the robot when they felt it got too close. The distances recorded reflect the shortest distance that the subjects accepted

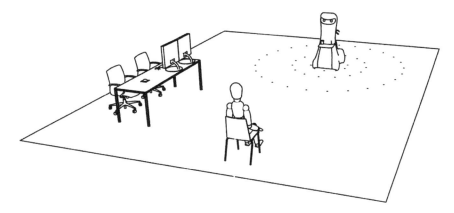

Fig. 6 Test set-up for distance measurement

between themselves and the robot. The distances were measured via the robot's sensors, which have an overall accuracy of ±0.02 m. The independent variables were altered one by one during the course of the experiment. To avoid knock-on effects, the changes were made according to a Latin square design. Thus, the two factor levels for the robot's appearance, five for its speed and three for the subject's position gave a total of 30 different approaches. The robot carried these out with the tray folded away.

In the concluding interview, subjects answered questions about their impressions of the robot and, again, about their affinity for technology [21]. They also evaluated the speeds and speed profiles.

4.4 Hypotheses

Based on research into the current state of scientific knowledge, we posit the following hypotheses:

H_1: The robot's appearance significantly affects the distance that the subject will accept between themselves and the robot.

H_2: The robot's speed significantly affects the distance that the subject will accept between themselves and the robot.

H_3: The subject's body position significantly affects the distance that the subject will accept between themselves and the robot.

H_4: The subject's age group significantly affects the distance that the subject will accept between themselves and the robot.

Table 1 Influence of reaction time ($t_{reaction}$) on the distance travelled by the robot during this time ($s_{reaction}$) as a function of the subject's age (A) and gender (G), and of the robot's speed

A	G	v (m/s)	$t_{reaction}$ (s)	$s_{reaction}$ (m)
20	Female	0.25	0.203	0.051
20	Female	0.75	0.203	0.152
20	Male	0.25	0.170	0.043
20	Male	0.75	0.170	0.128
75	Female	0.25	0.305	0.076
75	Female	0.75	0.305	0.229
75	Male	0.25	0.285	0.071
75	Male	0.75	0.285	0.214

4.5 Accounting for Age-Related Differences in the Measured Distances

The sensors on the Care-O-bot® 3 allowed the investigators to record the distances at the point in time when the subjects pressed the button. But this does not precisely reflect the actual distance that they accept, as it takes a little time for them to react and press the button. Reaction time is the time it takes a person to react to a given stimulus. A stimulus can come in a variety of forms and can target different receptors or senses. In the case of our study, a visual stimulus (distance) triggered a motor response (pressing the button). The reference value for the average reaction time for a motor response triggered by a visual stimulus is 220 ms [27]. However, because reaction times change throughout a person's life and because this study included people of different ages, we considered the reaction times differently depending on age. Following Haas [28], we recorded the reaction time for each person based on their age and gender. The distance that the robot covered within that time ($s_{reaction}$) was calculated for each subject at each speed and then subtracted from the distance measured for a given approach. Table 1 shows some examples of the distance travelled in the reaction time.

Therefore, the effects of age on human performance, which primarily concern reaction time in this empirical study, do not affect the dependent variable because this was accounted for in the distances measured.

5 Results

The results from the distance measurement stage are presented separately to those from the preliminary and concluding interviews.

5.1 Accepted Distance Between Human and Robot

The three null hypotheses (H_{01}, H_{02} und H_{03}) were checked using a multifactor analysis of variance with repeated measurements, as the subjects were tested multiple times in the various factor levels. The null hypothesis was rejected at a significance level of $\alpha = 0.05$. We investigated the influence of the three factors appearance (e), speed (v) and body position (k).

All the requirements for a multifactor analysis of variance were fulfilled, with the exception of sphericity [29]. Mauchly's sphericity test found that only the factor concerning body position fulfilled the sphericity requirement. For the other factors, we used the Greenhouse-Geisser correction to degrees of freedom. Table 2 shows the results of the multifactor analysis of variance. There were no significant interactions between appearance (e), speed (v) and body position (k).

The analysis shows that the appearance factor (i.e. whether the robot approached the subject with its serving side or working side showing) had no significant influence on the distance ($F(1.29) = 0.197$, $p = 0.660$, $\eta_p^2 = 0.007$). This means that we cannot reject null hypothesis H_{01}.

The results for the factor speed ($F(2.077; 60.235) = 237.175$, $p < 0.001$, $\eta_p^2 = 0.891$) show that the speed significantly affected the distance that the subjects would accept between themselves and the robot. Thus, we can reject null hypothesis H_{02} and adopt the alternative hypothesis H_2. The post hoc tests with a Bonferroni correction show that all the speeds, except v_1 and v_4, differed significantly ($p < 0.001$). Speed v_1 ($\bar{x} = 0.821$) and speed profile v_4 ($\bar{x} = 0.830$) produced near-identical mean distances. The mean distance of speed profile v_5 ($\bar{x} = 0.71$) is actually significantly ($p < 0.001$) shorter than the mean distance of

Table 2 Results of the three-factor analysis of variance with repeated measurements for the distance that subjects will accept between themselves and the approaching service robot

Source	Sum sq.	df	F	p	η_p^2
e	0.18	1	0.197	0.660	0.007
Error:	2.599	29			
v	31.287	2.077	237.17	<0.001	0.891
Error:	3.826	60.235			
k	9.845	2	13.665	<0.001	0.320
Error:	20.894	50.431			
e * v	0.023	2.759	0.399	0.737	0.014
Error:	1.695	80.025			
e * k	0.016	1.467	0.145	0.799	0.005
Error:	3.270	42.551			
v * k	0.202	5.281	1.854	0.102	0.060
Error:	3.158	153.16			
e * v * k	0.049	4.529	0.482	0.772	0.016
Error:	2.942	131.355			

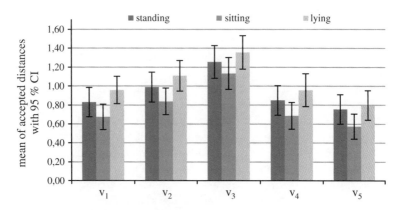

Fig. 7 Mean distances for the effects speed and body position

the slowest constant speed, v_1 ($\bar{x} = 0.821$). On average, the slower the robot approached, the closer the subjects allowed it to get to them. However, the robot got closest to the subjects when it approached using speed profile v_5.

The analysis also shows that the subject's body position significantly ($F(2.58) = 13.665$, $p < 0.001$, $\eta_p^2 = 0.320$) affects the distance they will accept between themselves and the approaching robot. Thus, we can reject null hypothesis H_{03} and adopt the alternative hypothesis H_3. In the post hoc tests with a Bonferroni correction, the mean distances between standing and sitting, and between sitting and lying differed significantly. No significant difference existed between standing and lying. Figure 7 shows the mean distances as a function of the two significant effects.

The analysis also included the between-subject factor of age group. The different age groups were found to have no significant ($F(2.27) = 0.700$, $p = 0.505$) influence on the distance accepted. This means that we cannot reject null hypothesis H_{04}.

5.2 Preliminary and Concluding Interviews

All 30 subjects had already heard, read or seen something about robots. Most of them mentioned robots that have appeared in films (e.g. *R2D2* and *Transformers*), industrial robots (e.g. those used in automotive factories) and domestic robots (e.g. for vacuum cleaning). Two of the subjects had personal experience of robots through contact to the field of industrial robotics. None of the subjects had ever heard of the Care-O-bot® 3. Thus, in terms of acceptance of the Care-O-bot® 3, all 30 subjects were in the awareness and first-contact stages at the beginning of the empirical study.

The study recorded the subjects' affinity for technology in terms of a personality trait [21] before and after the test. The subjects were asked to rate 19 statements using a five-point Likert scale, ranging from 1 (strongly agree) to 5 (strongly disagree). The subjects' affinity for technology in the preliminary interview ($\bar{x} = 2.56$, $SD = 0.39$) hardly differed from the results recorded in the concluding interview ($\bar{x} = 2.54$, $SD = 0.41$).

In the concluding interview, the same Likert scale was used to rate the subjects' statements as had been used for the initial assessment of their affinity for technology. The responses produced the following results:

- The statement "I had a positive impression of the appearance of the Care-O-bot® 3" had an average rating of $\bar{x} = 2.30$ and a standard deviation of $SD = 1.02$.
- The statement "The Care-O-bot® 3 did not look dangerous" had an average rating of $\bar{x} = 1.73$ and a standard deviation of $SD = 0.83$.
- The statement "I think it is important that a robot has different sides for different functions, and that these sides do not look the same" had an average rating of $\bar{x} = 3.01$ and a standard deviation of $SD = 1.41$.

The subjects were divided on how they felt about the appearance of the Care-O-bot® 3, though the results do show a slight positive tendency. It was very clear that the Care-O-bot® 3 did not look dangerous to the subjects, as on average they gave it a rating of at least 2 (agree). Subjects were divided on whether or not they felt that distinct-looking sides for different functions were important. It is therefore not possible to draw a conclusion on this point.

In the concluding interview, subjects were asked to say which of the speeds and speed profiles they preferred. Of the 30 subjects, 86.7 % said they preferred the speed profiles v_4 and v_5 (where the robot slowed down as it approached) to the three constant speeds v_1, v_2 and v_3. The results on which profile the subjects preferred were quite evenly split: 53.3 % voted for speed profile v_4 (smooth deceleration), and 46.7 % voted for speed profile v_5 (graduated deceleration). Results for the constant speeds v_1, v_2 and v_3 showed that 77.7 % preferred v_2, 13.3 % preferred v_1, and 10 % preferred v_3.

6 Summary and Outlook

The results of the empirical study on the robot can provide guidance for designing robot behaviour (H_2) and appearance (H_1) with respect to the minimum distance that humans will accept between themselves and a robot. The different body positions that the subjects adopted during the test (H_3) covered the different contexts in which a person would use the robot. These results also provide guidance for robot design.

The study found that the constant speed level significantly influenced the distance that the subject would accept between themselves and an approaching robot. The results show that the slower the robot travelled, the closer the subjects allowed it to come. Therefore, if a service robot can only (e.g. for technical reasons) move at constant speeds, it should travel as slowly as possible when around people. But in situations where the robot is working at a reasonable distance from humans, it can move faster to increase its efficiency.

It is possible to reduce the distance that people will accept between themselves and the robot, without slowing the average speed, by using speed profiles where the robot decelerates as it approaches. The study showed that the smooth deceleration profile (v_4) was associated with a significantly shorter distance than the constant speed v_2, even though the average speed was the same in both cases. The graduated deceleration profile (v_5) reduced the distance even further than the smooth deceleration profile (v_4). It therefore makes sense to use a graduated profile in cases where the robot has to get very close to the user. The subjective rating of the two speed profiles in the concluding interview shows no clear preference among the subjects for one profile over the other. Therefore, the design guidance only takes the objective measurements into account.

The study found that the robot's appearance had no significant influence on the accepted distance. Going to great lengths to change the appearance of a non-humanoid service robot so that things like technical components and manipulators are no longer visible will therefore do nothing to encourage users to let the robot come closer.

The subjects' body position significantly influenced the distance they were willing to accept. The results suggest that, contrary to what might be expected, the influence is not linked to the time needed to potentially move out of the robot's path. If it was, then the robot would have got closest when the subjects were standing, with sitting and lying coming in second and third respectively. It is more likely that the influence is related to the size of the image formed on the retina. The size of the retinal image of a real object changes depending on the viewing angle and distance. The results indicate that this affects the accepted distance. Subjects allowed the robot to come closest when they were seated, which is the position where the image on the retina is smallest. The image is slightly bigger when a person is standing, and subjects kept the robot further away here than when sitting. The furthest distances were recorded for lying down, when the image is at its largest. Subsequent studies should therefore investigate a possible correlation between the image formed on the retina and the distance people will accept between themselves and the robot in order to understand why body position is an influencing factor.

The study found that age did not affect distance. It therefore appears unnecessary to develop separate robot behaviours or appearances for different age groups. However, it should be noted that the test subjects were highly homogenous in terms of their experience of robots and their affinity for technology. The study was unable to establish what effect their prior knowledge and associated mental models had on the outcomes. It is therefore inadvisable to generalise the results beyond the sample tested.

With regard to the methodology used, the study achieved a high degree of validity and objectivity thanks to the consistent study design and the permutation of the factor levels. The distances shown in this study are, unlike the findings of other studies, the actual distances that the subjects accepted. This is because we recorded individual reaction times, measured how far the robot travelled in that time and then subtracted that value from the distances measured. By taking account of age—and gender-related reaction times like this, we ensured that the results were not skewed by the way aging affects the speed at which the subjects pressed the button. The fact that the test used a fully functional robot, and was thus not a Wizard-of-Oz experiment, further enhances the generalisability of the data and makes the study stand out from many experiments conducted in the past.

Acknowledgments We would like to thank all the subjects who took part in the study. This paper is part of the research project "Tech4P—Strategien für die Technikintegration bei personenbezogenen Dienstleistungen", which is funded by the German Federal Ministry of Education and Research (funding code: 01FG1004). The German Aerospace Center (DLR) is managing the project.

References

1. Mori, M.: The Uncanny Valley. In: Energy, 7(4), pp. 33–35, Translated by Karl F. MacDorman and Takashi Minato (1970)
2. Mizoguchi, H., Sato, T., Takagi, K., Nakao, M., Hatamura, Y.: Realization of expressive mobile robot. Robotics and Automation, Proceedings, 1997 IEEE International Conference, pp. 581–586 (1997)
3. Duffy, B. R.: Anthropomorphism and Robotics. The Society for the Study of Artificial Intelligence and the Simulation of Behaviour, 20
4. Oestricher, L.: Cognitive, social, sociable or just socially acceptable robots? In: 16th IEEE International Conference on Robot and Human Interactive Communication, pp. 558–563 (2007)
5. Goetz, J., Kiesler, S., Powers, A.: Matching robot appearance and behavior to tasks to improve human-robot cooperation. In: Robot and Human Interactive Communication: Proceedings of the 2003 IEEE International Workshop, pp. 55–60 (2003)
6. Walters, M.L., Dautenhahn, K., te Boekhorst R., Koay, K.L., Woods, S.N.: Exploring the design of robot appearance and behavior in an attention-seeking "Living Room" scenario for a robot companion. In: IEEE: Proceedings of the 2007 IEEE Symposium on Artificial Life, pp. 341–347 (2007)
7. Walters, M.L., Dautenhahn, K., Koay, K.L., Kaouri, C., Boekhorst, R., Nehaniv, C., Werry, I., Lee, D.: Close encounters: spatial distances between people and a robot of mechanistic appearance. In: Humanoid Robots, 5th IEEE-RAS International Conference, pp. 450–455 (2005)
8. Walters, M.L., Dautenhahn, K., te Boekhorst, R., Kheng Lee Koay, Kaouri, C., Woods, S., Nehaniv, C., Lee, D., Werry, I.: The influence of subjects' personality traits on personal spatial zones in a human-robot interaction experiment. In: Robot and Human Interactive Communication, ROMAN, IEEE International Workshop, pp. 347–352 (2005)
9. MacDorman, K.F.: Subjective ratings of robot video clips for human likeness, familiarity, and eeriness: An exploration of the uncanny valley. In: ICCS/CogSci-2006 long symposium: toward social mechanisms of android science, Vancouver (2006)
10. Bischof, N.: Struktur und Bedeutung. Hans Huber, Eine Einführung in die Systemtheorie für Psychologen. Bern (1995)
11. Ikuta, K., Ishii, H., Nokata, M.: Safety evaluation method of design and control for human-care robots. The Int. J. Robot. Res. **22**(5), 281–297 (2003)

12. Traver, V.J., del Pobil, A.P., Perez-Francisco, M.: Making service robots human-safe. In: IEEE/RSJ international conference on intelligent robots and systems (IROS 2000). Proceedings, Vol. 1, 2000, pp. 696–701
13. Yamada, Y., Yamamoto, T., Morizono, T., Umetani, Y.: FTAbased issues on securing human safety in a human/robot coexistence system. In: IEEE international conference on systems, man, and cybernetics. IEEE SMC'99 conference proceedings, Vol. 2, pp. 1058–1063 (1999)
14. Bartneck, C., Kulić, D., Croft, E., Zoghbi, S.: Measurement instruments for the anthropomorphism, animacy, likeability, perceived intelligence, and perceived safety of robots. In: Int. J. Soc. Robot. Vol. 1, No. 1, pp. 71–81 (2009)
15. Jochems, N.: Altersdifferenzierte Gestaltung der Mensch-Rechner-Interaktion am Beispiel von Projektmanagementaufgaben. In: Schlick, C. (ed.) Industrial Engineering and Ergonomics. Shaker Verlag, Aachen (2010)
16. Mertens, A., Reiser, U., Brenken, B., Lüdtke, M., Hägele, M., Verl, A., Brandl, C., Schlick, C.: Assistive robots in eldercare and daily living: automation of individual services for senior citizens. In: Jeschke, S., Liu, H., Schilberg, D. (eds.) Intelligent Robotics and Applications— Fourth International Conference, ICIRA 2011, pp. 542–552. Springer Verlag, Berlin (2011)
17. Marquie, J.C., Jourdan-Boddaert, L., Huet, N.: Do older adults underestimate their actual computer knowledge? Behav. Inf. Technol. **21**(4), 273–280 (2002)
18. Reed, K., Doty, H.D., May, D.R.: The impact of aging on self-effcacy and computer skill acquisition. J. Manag. Issues **17**(2), 212–228 (2005)
19. Karrer, K., Glaser, C., Clemens, C., Bruder, C.: Technikaffinität erfassen—der Fragebogen TA-EG. In: Lichtenstein, A., Stößel, C., Clemens, C. (Hrsg.) Der Mensch im Mittelpunkt technischer Systeme, 8. Berliner Werkstatt Mensch-Maschine-Systeme (ZMMS Spektrum, Vol. 22, No. 29). Düsseldorf: VDI Verlag GmbH, pp. 196–201 (2009)
20. Luczak, H.: Untersuchungen informatorischer Belastung und Beanspruchung des Menschen. Fortschrittberichte der VDI-Zeitschrift, Vol. 10, No. 2, Düsseldorf: VDI-Verlag (1975)
21. Pahl, G., Beitz, W., Feldhusen, J., Grote, K.H.: Konstruktionslehre. Springer Verlag, Berlin (2007)
22. Brandl, C., Mertens, A., Bröhl, C., Mayer, M., Schlick, C.: Akzeptanzorientierte Gestaltung von Innovationen bei technikunterstützten personenbezogenen Dienstleistungen, In: Gesellschaft für Arbeitswissenschaft e.V. (Hrsg.) Gestaltung nachhaltiger Arbeitssysteme - Wege zur gesunden, effizienten und sicheren Arbeit, Dortmund: GfA-Press, pp. 483–486 (2012)
23. Davis, F.D.: User acceptance of information technology: system characteristics, user perceptions and behavioral impacts. Int. J. Man Mach. Stud. **38**(3), 475–487 (1993)
24. Venkatesh, V., Bala, H.: Technology acceptance model 3 and a research agenda on intervention. Dec. Sci. **39**(2), 273–315 (2008)
25. Kollmann, T.: Akzeptanz innovativer Nutzungsgüter und –systeme. Gabler, Wiesbaden (1998)
26. Hoch, A., Simons, F., Haag, M., Parlitz, C., Reiser, U., Hägele, M.: Care-O-bot® 3—Mobiler Serviceroboter mit ausgeprägter Manipulationsfähigkeit. In: Kompetenznetzwerk Mechatronik BW: Intelligente mechatronische Systeme: Internationales Forum Mechatronik, begleitend zur MOTEK, Stuttgart. Göppingen, pp. 360–371 (2008)
27. Schlick, C., Bruder, R., Luczak, H.: Arbeitswissenschaft. Springer Verlag, Berlin (2010)
28. Haas, H.-J.: Sport im Alter—Leistungsphysiologie. In: van den Berg, F., Wulf, D. (eds.) Angewandte Physiologie. Georg Thieme Verlag, Stuttgart (2008)
29. Bortz, J.: Statistik für Human- und Sozialwissenschaftler. Springer Medizin Verlag, Heidelberg (2005)

Display of Emotions with the Robotic Platform ALIAS

Jürgen Geiger, Ibrahim Yenin, Frank Wallhoff and Gerhard Rigoll

Abstract Emotions are an important communication channel and therefore, human–machine interaction can be enriched by displaying emotions. Especially for a robotic system designed for technology-inexperienced users as in an AAL context, display of emotions is a good way to increase the familiarity with a technical system. In this contribution, we describe how the robotic head of the robotic platform ALIAS is used to display emotions. In the project ALIAS, the robot is equipped as a communication platform for elderly people. Emotions are used to enrich the human–machine interaction. ALIAS can thus better adapt to its users during a dialogue. The robotic head has several degrees of freedom to display different facial expressions. Five different facial expressions corresponding to five different emotions have been developed. In a user study it was tested, how humans perceive the displayed emotions. The results are promising, however, the absence of a mouth makes it difficult to design emotions which can be recognized by humans very robustly.

F. Wallhoff
Jade University of Applied Sciences, Oldenburg, Germany
e-mail: wallhoff@mmk.ei.tum.de

J. Geiger (✉) · I. Yenin · F. Wallhoff · G. Rigoll
Lehrstuhl für Mensch-Maschine-Kommunikation, Technische Universität München, Munich, Germany
e-mail: geiger@mmk.ei.tum.de

I. Yenin
e-mail: yen@mmk.ei.tum.de

G. Rigoll
e-mail: rigoll@mmk.ei.tum.de

1 Introduction

Emotions are an important aspect of human–human communication. Based on the work of Freud, Zimbardo and Ruch describe communication in an ice berg model [1]. Thereby, only 20 % of communication consists of a visible part, which involves facts and figures. The larger part consists of non-visible aspects like personality, fears, conflicts and emotions. In communication, emotions have the functions of dialogue control, transmission of information, social bonding, competence and personalisation. This model concerns mainly human–human communication. But since it is desired to make human-robot interaction as natural as possible, it is necessary to include emotions in a human-robot dialogue. While industrial robots are not well-suited for the display of emotions, especially robotic platforms in the fields of AAL, health care, entertainment or service robotics can be enriched by emotions.

There are several tasks a companion robot can perform to support elderly people. Physical assistance is not addressed here in the context of social robotics but it is definitely a future key selling point. The ability to lift or carry things or act as a mobility aid are important features but need more stable technical solutions to be able to be employed without safety risks. In its function as a communication platform, a robot can serve several duties. It can interact with its elderly users, provide cognitive assistance and promote social inclusion. Communication plays a major role for a social robot for elderly people. In order to keep the human-robot dialogue lively and attractive, emotions can be applied as an additional communication channel. Emotions are not only used to give feedback about the current state of the robot, but also to equip the robot with a certain human-likeness. Thus, emotions help to enrich the human-robot dialogue and to reduce anxiety of communicating with a technical system.

1.1 Related Work

Several robots with abilities to display emotions have been developed. In Fig. 1, four examples for robotic platforms that can display emotions are given.

Fig. 1 Different examples of robots capable of displaying emotions

EDDIE (Emotion Display with Dynamic Intuitive Expressions) [2], displayed in Fig. 1a is a human-like robotic head. It has 23° of freedom at its eyes, eye brows, ears, mouth and jaw. Additionally, two animal-like features are mounted in order to strengthen the ability to display emotions. EDDIE is also developed with child-like characteristics (e.g. the large eyes). The basic emotions joy, surprise, anger, disgust, sadness and fear can be displayed with EDDIE. In [3], a robotic platform based on EDDIE that enriches a multimodal human-robot dialogue with emotional feedback was described.

Similar to EDDIE, the robotic head Kismet [4] (Fig. 1b) resembles a human head, but without any additional animal-like features. Kismet has 15° of freedom at its eyes, eye brows and lids, neck, ears, lips and mouth. It can display the emotions happiness, sadness, surprise, anger, calm, displeasure, fear, interest and boredom.

Compared to the robotic heads described so far, Sparky [5] (Fig. 1c) is relatively small with its dimensions of $60 \cdot 50 \cdot 35$ cm but it consists of a whole body and not only a head. With its cartoon-like character, the uncanney valley [6] can be dodged. In order to display emotions, Sparky has only ten degrees of freedom: Its eye brows and lids, top and bottom lip, neck, back plate and wheels are movable. The tiltable back plate can be interpreted as an animal-like feature. Sparky can display the emotions happiness, sadness, anger, surprise, fear, curiosity, nervousness and sleepiness.

The robot dog AIBO [7] (Artificial Intelligence robot, Fig. 1d) is an already commercially available robotic platform. It has 20 actuators and can move its mouth, head, ears, tail and legs. AIBO displays not only separate emotions, but complete behaviours, which are exploring, demanding and giving attention, fear, playing, learning and seeking protection.

Further examples for social robots capable of displaying emotions are CERO [8], FEELIX [9], VIKIA [10], PARO or the Sony Dream Robot.

All of the presented robotic platforms have different features to display emotions, where especially the mouth, eyes and eye brows play an important role.

1.2 Overview

The robot ALIAS is designed as a communication platform for elderly persons [11]. Therefore it is well suited for the incorporation of the ability to display emotions, to enrich the human-robot dialogue. Since it has no mouth, it is especially difficult to display emotions with ALIAS. Five different emotional facial expressions have been developed and evaluated with the robotic platform ALIAS.

In Sect. 2, we present the robot ALIAS and the hardware of the head of the robotic platform. The implemented emotional facial expressions are described in Sect. 3. Experimental results are presented in Sect. 4 before conclusions are given in Sect. 5.

2 Robotic Platform ALIAS

2.1 The ALIAS Project

In the project ALIAS (Adaptable Ambient Living Assistant[1]), the robotic platform ALIAS is equipped as a communication platform for elderly people. ALIAS is a mobile robot system that interacts with elderly users, monitors and provides cognitive assistance in daily life, and promotes social inclusion by creating connections to people and events in the wider world. The system is designed for people living alone at home or in care facilities such as nursing or elderly care homes. The function of ALIAS is to keep the users linked to the wide society and in this way to improve their quality of life by combating loneliness and increasing cognitively stimulating activities. In a first series of field-trial experiments, the robotic platform was already tested with elderly users [12].

To fullfill it's goals, ALIAS is equipped with several capabilities: An easy-to-use and fault tolerant human–machine interface is achieved by employing automatic speech recognition (ASR) together with a module for natural language understanding (NLU). Communication is enriched through the utilization of person identification methods using voice and face and laser-based leg-pair detection. In order to promote social inclusion, services for net-based linking are employed to link users with the wider world, enabling to maintain a wider horizon by exploiting new kinds of on-line and remote communication techniques. Autonomous, socially acceptable navigation capabilities enable the robot to find its way in its environment. In addition, the robot is equipped with a brain-computer interface (BCI), enabling users like stroke patients to remotely control the system.

Several use-case scenarios are developed to showcase the different functionalities of the robot. For example, in the ground lighting scenario, it is shown how ALIAS can guide persons at night in the dark, using its navigation capabilities and applying the touchscreen display as a light source. The gaming scenario has been designed to test the human–machine interface. An entertaining game is played through the touchscreen and can additionally be controlled by speech commands. At the same time, ALIAS is addressing its user by employing face detection to detect the user and hold eye-contact with him. This scenario is best suited for integration of the display of emotions.

2.2 Hardware Setup

The hardware configuration of the robot platform ALIAS (see Fig. 2) is based on the SCITOS G5 robot family of the robot manufacturer MetraLabs.[2]

---

[1] See AAL-JP project ALIAS www.aal-alias.eu.
[2] www.metralabs.com.

Fig. 2 Overview of the hardware setup of the robotic platform ALIAS, divided into driving unit (*lower part*) and interaction unit (*upper part*) consisting of the touchscreen display and the robotic head

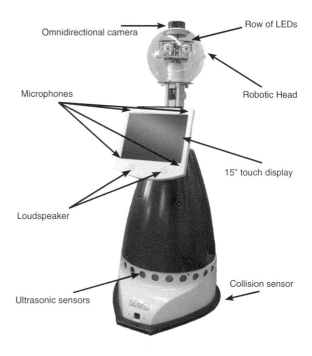

ALIAS is an approximately 1.50 m tall robot platform and can be divided into a driving unit and an interaction unit. In order to approach a user, navigation is provided by the driving unit, which uses a differential drive system. In a known environment, the robot can localize itself, navigate autonomously and approach a user in a socially acceptable manner using a laser range finder and ultrasonar sensors. The interaction unit consists of a movable robotic head and a 15″ touchscreen, which is used for user interaction with the robot and is best suited as an easy-to-use human–machine interface. In addition, it is equipped with four microphones and two loudspeakers which can be used for speech input and output.

2.3 Robotic Head

ALIAS is equipped with a head that is used to display emotions. On top of the head, an omnidirectional camera is mounted, which delivers a 360° image. Due to its mounting position, the main purpose of this camera is to localize and identify persons using face detection. The head has several degree of freedom. In Fig. 3, all of the degrees of freedom of the robotic head are displayed.

It can be turned 360° horizontally and vertically up (15°) and down (6°). The two synchronized movable eyes can be turned up to 8.5° in both horizontal directions and the eye lids can be opened or closed (independent of each other) on a continuous scale. Horizontal rotation of head and eyes is not used to distinguish between

Fig. 3 Features of the robotic head of ALIAS. The *head* can be turned *horizontally* and *vertically* and the *eyes* can be opened/closed and turned *vertically*. Additionally, above the eyes, there is a row of LEDs

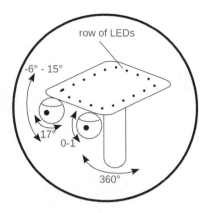

emotions, in order to ensure the possibility to perceive emotions independent of the viewing angle. Above the eyes, a row of blue LEDs is mounted, which can be turn on and off or set running or blinking with any frequency. Additionally, the brightness of the LEDs can be controlled. In total, this sums up to 6° of freedom which are used to display different facial expressions. Compared to other robotic heads, this is a relatively small number of degrees of freedom. It should also be noted that ALIAS has no mouth or eye brows, which are important for displaying emotions. Therefore, it will be rather difficult to display emotions with ALIAS.

Two different computers are mounted on the robot. An industrial PC running Linux is used to control the hardware of the robot, e.g. the driving wheels, the collision and ultrasonic sensors and the robotic head. To control the touchscreen display, the robot is equipped with a Mac mini running Windows. The microphones and loudspeakers are also connected to the Windows PC. All modules on both computers can communicate with each other through various interfaces. More technical details about the robot platform ALIAS are provided in [11].

3 Emotional Facial Expressions

In addition to a neutral facial expression, five emotional facial expressions are implemented. In this section, the concept and the implementation of these facial expressions are described.

3.1 Concept

The basic emotions disgust, fear, joy, sadness and surprise have been chosen. These emotions are considered "basic emotions" because they are said to be represented and interpreted equally in the whole world [13]. It is important to limit the number of different emotions in order to allow the user to distinguish between

Fig. 4 Distribution of the five implemented emotions in the two-dimensional space of arousal and valence

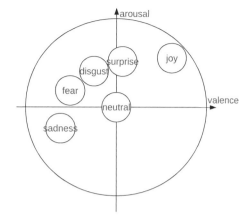

the possible facial expressions. However, when the evaluation is performed on a continuous scale (e.g. with arousal and valence), it is feasible to include a larger set of emotions. Figure 4 shows how our chosen emotions are roughly distributed in the continuous space of arousal and valence. Mapping emotions onto the two-dimensional space of arousal and valence is a simple model to display emotions on a continuous scale.

From Fig. 4, it can be seen that some of the chosen emotions are very near next to each other. Therefore, it was especially carefully tried to design facial expressions which help to distinguish between those emotions.

3.2 Implementation

The implemented facial expressions are based on human emotion display, using the head tilt angle and eye lids. Another important feature that is used here to distinguish between different facial expressions is the row of LEDs above the eyes. For this feature, the settings can not be based on human emotion display. The brightness of the LEDs is roughly correlated with arousal. Therefore higher brightness values are chosen for joy or surprise. The frequency of the LEDs (when set running or blinking) is another important feature. A small frequency (i.e. slow blinking) generally expresses slowness, fatigue or lethargy, while high frequency values indicate speed, hecticness, excitement or nervosity. Based on these assumptions, the LED frequency was set accordingly for the different emotions.

Three exemplary facial expressions are displayed in Fig. 5. The neutral facial expression is displayed in Fig. 5a and is characterized by slightly closed eyes and unmoved head while the LEDs are turned on with a medium brightness. In Fig. 5b, the facial expression showing sadness is displayed. The head is turned down with slightly closed eyes and the LEDs are blinking with a slow frequency. The facial expression for the disgust emotion is displayed in Fig. 5c. The robotic eyes are completely closed and the head looks upwards.

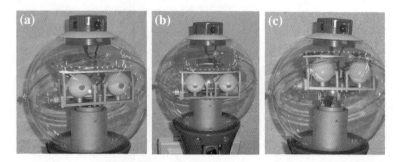

Fig. 5 Different facial expressions of the robotic head of the ALIAS platform. **a** Neutral facial expression. **b** Sad facial expression. **c** Disgust facial expression

Table 1 Settings for the actuators of the robotic head of ALIAS for the implemented emotions

Emotion	Head tilt	Eyelids	LEDs	LED freq.	LED brightn.
Neutral	0.0	0.85	On	–	0.5
Disgust	12.5	0.0	On	–	0.2
Fear	0.0	1.0	Blinking	35	1
Joy	5.0	1.0	Running	35	1
Sadness	−6.5	0.81	Blinking	220	0.2
Surprise	0.0	1.0	blinking	10	1

In order to display joy, the eyes are opened and the LEDs are set running with a medium speed and maximum brightness. The head is slightly turned upwards. For fear, the eyes are wide open and the LEDs are set blinking with a relatively high freqency. Surprise is displayed by LEDs blinking with a very high frequency at maximum brightness and opened eyes.

In Table 1, all settings of the head for the implemented emotions are displayed.

4 Experiments

A user study has been conducted to evaluate how good humans can identify the emotions displayed by ALIAS. 12 male adult subjects, aged between 23 and 30 participated in the study. The robot was presented to the users and the five different

Table 2 Human identification rates for emotions displayed with the robotic head of ALIAS

Emotion	Identification rate (in %)
Disgust	33.3
Fear	6.6
Joy	23.3
Sadness	38.8
Surprise	33.3
Average	27.1

Table 3 Confusion matrix (in %) for emotions displayed with the robotic head of ALIAS

Emotion	Acceptance	Anger	Boredom	Curiosity	Disgust	Fear	Joy	Sadness	Seriousness	Surprise	Other
Disgust	6.6	0	35	0	33.3	10	1.6	11.6	0	0	1.6
Fear	18.3	3.3	3.3	15	0	6.6	10	6.6	10	20	0
Joy	5	5	0	28.3	3.3	5	23.3	6.6	10	11.6	1.6
Sadness	0	10	31.6	0	11.6	0	0	38.8	8.3	0	0
Surprise	5	16.6	5	15	1.6	10	10	0	1.6	33.3	1.6

Five different emotions were displayed to the subjects and they could choose from 10 possible emotions or "other" emotions have been shown five times each, in a random order. Always between two emotions, the neutral facial expression was displayed. The subjects had to identify the displayed emotion and could choose from a list of 10 different emotions (or "other"), which are: acceptance, anger, boredom, curiosity, disgust, fear, joy, sadness, seriousness and surprise. The 10 possible answers are the same Breazeal used to evaluate the robot Kismet [14]. In addition, subjects rated their guess on a scale from 1 (very unsure) to 10 (very sure). Table 2 shows the identification rates resulting from the user study.

The identification rates are promising, although none of the emotions could be identified very clearly. Sadness achieves the highest accuracy with 38.8 % while fear was only identified 6.6 % of the time. The average identification rate is 27.1 %. There was a strong variation between the different subjects. Some subjects achieved an identification rate of 50 % while others had almost none of the emotions guessed correctly. There are clear tendencies visible in the confusions of different emotions. The confusion matrix is displayed in Table 3.

Disgust and sadness have been reconised as boredom very often, joy as curiosity and surprise as anger or curiosity. This can mainly be attributed to the absence of a mouth, which is a very important feature for display of emotions. On the other hand, the LEDs are the main feature that is used to differentiate between emotions, supported by head tilt and eye lids. Most of the confusions appear between emotions which are close to each other in the arousal-valence space, which supports the value of our results.

5 Conclusions

We presented the robotic platform ALIAS and the implementation of different facial expressions according to different emotions. The robotic head was used to display five different emotional facial expressions. In a user stuy, it was evaluated how the different emotional facial expressions of ALIAS can be recognised by humans. While the identification rates for the emotions are not too high, it is already very promising that with such limited hardware it is possible at all to distinguish between different emotions. The addition of a mouth could be very promising to increase the identification rates of emotions, as well as the introduction of different colours for the LEDs. Reducing the set of different emotions could also lead to a better separation by the human observer. Furthermore, we want to test the influence of emotional facial expressions in a dialogue situation between a human and a robot.

Acknowledgments This work was supported by the project AAL-2009-2-049 "Adaptable Ambient Living Assistant" (ALIAS) co-funded by the European Commission and the German Federal Ministry of Education (BMBF) in the Ambient Assisted Living (AAL) programme.

References

1. Zimbardo, P.G., Ruch, F.L.: Lehrbuch der Psychologie: Eine Einführung für Studenten der Psychologie, Medizin und Pädagogik, 3rd edn. Springer, New York (1978)
2. Sosnowski, S., Bittermann, A., Kühnlenz, K., Buss, M.: Design Evaluation of Emotion-Display EDDIE, pp. 3113–3118 (2006)
3. Bannat, A., Blume, J., Geiger, J., Rehrl, T., Wallhoff, F., Mayer, C., Radig, B., Sosnowski, S.,and Kühnlenz, K: A multimodal human-robot-dialog applying emotional feedbacks: In *Proceedings of the International Conference on Social Robotics (ICSR)*, 2010, number LNAI 6414, pp. 1–10, 23–24 11 Nov (2010)
4. Breazeal, C., and Scassellati, B.,How to build robots that make friends and influence people: In: Proceedings of the IEEE/RSJ International Conference on Intelligent Robots and Systems (IROS), pp. 858–863.(1999)
5. Scheeff, M., Pinto, J., Rahardja, K., Snibbe, S., Tow, R., Experiences with Sparky, a social robot. In:Weiss, G (ed.) Multiagent Systems Artificial Societies and Simulated Organizations, vol. 3. pp. 173–180 Springer, New York(2002)
6. Mori, M.: The Uncanny Valley. Energy **7**(4), 33–35 (1970)
7. Arkin, R.: An ethological and emotional basis for human–robot interaction. Robot. Auton. Syst. **42**(3–4), 191–201 (2003)
8. Huttenrauch, H. Eklundh, K.S.: Fetch-and-carry with cero: observations from a long-term user study with a service robot. In: Proceedings of the IEEE International Workshop on Robot and Human Interactive Communication, , pp. 158–163(2002)
9. Cañamero, L Playing the emotion game with feelix *Socially Intell Agents*, 69–76(2002)
10. Bruce, A.,Nourbakhsh, I, Simmons, R.The role of expressiveness and attention in human-robot interaction, In: *Proceedings. IEEE International Conference on Robotics and Automation (ICRA)* pp. 4138–4142.(2002)
11. Rehrl, T., Blume, J., Geiger, J., Bannat, A., Wallhoff, F., Ihsen, S., Jeanrenaud, Y., Merten, M., Schönebeck, B., Glende, S.and Nedopil, C.Alias: Der anpassungsfähige ambient living assistent, In: *Proceedings. Deutscher AAL Kongress*, (2011)
12. Scheibl, K., Geiger, J., Schneider, W., Rehrl, T., Ihsen, S., Rigoll, G., and Wallhoff, F., Die Einbindung von Nutzerinnen und Nutzern in den Entwicklungsprozess eines mobilen Assistenzsystems zur Steigerung der Akzeptanz und Bedarfsadäquatheit, In: *Proc. Deutscher AAL Kongress*, (2012)
13. Ekman, P.: An argument for basic emotions. Cogn. Emot. **6**(3–4), 169–200 (1992)
14. Breazeal, C.L.: Designing sociable robots. The MIT Press, Cambridge (2004)

Part VI
Robotics

Housing Enabling: Detection of Imminent Risk Areas in Domestic Environments Using Mobile Service Robots

Nils Volkening, Andreas Hein, Melvin Isken, Thomas Frenken and Melina Brell

Abstract Because of the greying society the need of user centred care concepts are raising. One big wish of older persons is to stay as long as possible in their own flat. Depending on the demographic change it won't be possible to realize complete care by formal or informal caregivers especially for people living alone. One possible way to cope this problem is to use ICT based solutions e.g. mobile service robots. The personal in-house mobility and its preservation it is one goal to enable staying as long as possible in her/his own flat. The concept of an automated housing enabling assessment which is presented here is an advanced solution for this problem. It is based on three main components: (1) Measurement and analysis of the cognitive and physical capabilities of the user, (2) Measurement and validation of the flat and (3) Computation of areas with a higher risk to fall and advice to remove such issues, e.g. restructuring of the furniture. The great benefit of a mobile robot platform is that all needed sensors are mounted on the robot and it can follow the user to make measurements at different places in the home. This will reduce the cost and installation effort in the flat to a minimum. Another benefit is the continuous assessment which helps to restructure the flat in a continuous way. This helps to reduce the probability of a fall event and raise the feeling of safety. All measurements could also be used by other assessment tools to preserve the indoor mobility not only by preventing a fall event but also by reacting on changes in the mobility level over time.

N. Volkening (✉) · A. Hein · M. Isken · T. Frenken · M. Brell
OFFIS e.V.—Institute for Information Technology, Oldenburg, Germany
e-mail: Nils.Volkening@offis.de

A. Hein
e-mail: Andreas.Hein@Informatik.Uni-Oldenburg.de

1 Introduction

Industrial countries have to cope with different problems caused by the demographic change. Better medical care, health improvements and a healthier lifestyle increase the expected lifetime of the population. Another aspect is the decreasing birth-rate. This combination will lead to an aging society. The society has to cope with the problem that the number of recipients of care services is growing and the number of contributors is decreasing. An additional problem is the shrinking quantity of younger people which take the role as caretakers. This will lead to major economical and logistical challenges. One solution to handle these problems is to use assistive technologies [1]. Advanced Systems can support caretakers and assist elderly in an independent lifestyle and preserve their mobility up to a high age. These kinds of ICT solutions could also help to recognize and react on early signs of age-related diseases, which help to reduce the costs and manpower demands. There are different approaches to bring these technologies to the home of elderly people, one solution are the smart environments [2]. In this case all necessary components are integrated in the home to assist the elderly in the best possible way. Sensors and actuators will be used to provide different services to the user and to measure the current mobility/health level. A problem of smart environments is the high upgrade costs during the change from a "normal" flat into a smart environment. This reduces the user acceptance a lot. Service robots and the technological advancements in this sector are growing rapidly. A service robot has the advantage that it can support the elderly by the activities of the daily life continuously. This gives the caretakers the possibility to have more time e.g. for social interaction with the elderly. Another benefit is that all necessary sensors are mounted on the robot itself. So the sensors are not stationary and limited to one place. These aspects will reduce costs and installation demands and so it will help to increase the user acceptance and intend to play an important role helping to manage the demand of caretakers by assisting elderly in their daily life [3]. The fact that the mobile service robots are present the whole day enables long-term monitoring of residents. The major merit of technological monitoring is to enable early diagnosis and to identify and possibly prevent imminent dangerous situations [4]. Current applications are e.g. mobility assessments, activity detection and the autonomous exploration of the flat. By combining the approaches of these different services we think it is possible to deliver a new approach of a housing enabling service to increase the quality of life and the safety feeling of the inhabitants. The next chapter will give further medical motivation, followed by the state of the art in domestic robotics and assessments. Afterwards we will present our general approach to housing enabling assessments in domestic environments, give a short overview of already published work and outline a new concept on how housing enabling may be implemented on a mobile robot platform. Finally, we will conclude our paper.

2 Medical Motivation

The personal mobility is an important factor for the wellbeing of the user. Additionally, the ability to move around and to perform activities of daily life is a fundamental requirement for an independent lifestyle [5]. Impairments of mobility due to pathological reasons lead to more significant changes in parameters of gait than age-related changes [6]. One of the most frequent pathological reasons of mobility impairments are neurological diseases, especially dementia. Another important aspect is the raising risk of falls and need of assistance indicated by decreased self-selected gait velocity [7]. Fall-related costs are one of the major factors influencing the proportionally higher costs to the health care system caused by elderly people. From a clinical perspective long-term monitoring of changes in mobility has a high potential for early diagnosis of various diseases and for assessment of fall risk [4]. The relation between the average gait velocity and a local gait velocity in different areas of the flat is very helpful to find hot spots with a higher risk to fall. Additional other gait parameters like e.g. step size etc. are very helpful too. In combination with the housing enabling assessment it may help delaying need of care or imminent incidents like falls and thus may help saving costs. On a more personal level early detection of hot spots may help supporting an independent lifestyle by enabling early and purposeful prevention and may therefore increase quality of life for affected people [31], relatives, and carers. In today's health systems the potential of frequent housing enabling assessment is not exploited. Rather, housing enabling assessments are only applied infrequently or after an acute incident like a fall took place. This is mainly due to missing knowledge and technical capabilities.

3 State of the Art

3.1 Mobility Trend Analysis in Domestic Environments

Environments equipped with various sensors especially from the home automation or security domain, are referred to as (health) smart homes [8]. Only some systems which use ambient sensors for detailed mobility analysis have been described so far. The research focus is on general mobility trend analysis instead. Various groups use home automation technologies like motion sensors, light barriers or reed contacts placed in door frames or on the ceiling. Cameron et al. [9] presented a solution with optical and ultrasonic sensors. These were placed in door frames to determine the walking speed and direction of a person passing. Pavel et al. [10] developed a system based on PIR sensors covering different rooms of a flat. The knowledge of the distances between the different PIR sensors and the measurement of the transit times is used to compute the gait velocity. Placing three passive motion sensors in a sufficient long corridor makes those computations more reliable

[11]. Within our own work [12] we have recently presented a new approach based on the definition of motion patterns by usage of available sensor events. By providing an abstracted definition of the environment, physically feasible walking paths can be computed and monitored automatically. The use of more precise sensors i.e. laser ranges scanners have been applied to implement very precise gait analysis in domestic environments. One approach has been presented by Pallejà et al. [13]. The advantage of this approach is the very detailed analysis, but it has some restrictions. The person has to walk straightly towards the scanner and on a predefined path. In our own work using laser range scanners [14] we do not restrict a person's walking path while measuring. So far we need only the computation of self-selected gait velocity in the different areas. This approach is highly precise and does not require any predefined knowledge but is more expensive to implement compared to the approach using home automation technology.

3.2 Mobility Assessments Using Service Robotics

Service robots combine ideas of different fields of robotic research into one system to target at a specific application. Most available platforms are still in (advanced) research states. Fields of interest in the community are acting autonomously in home environments [15], learning of environmental factors and user behavior [16] and as well as robot designs itself [17]. Within our own work [18] we have recently presented a new approach to enhance mobile robot navigation in domestic environments by the use of mobility assessment data. An application of the potential field method for mobility trend analysis and the precise measurements of human movement trajectories by a laser range scanner have been implemented (see Figs. 1 and 2). The advantage of a mobile robot is that it acts as a kind of mobile infrastructure. It can bring the needed sensor technology to the optimal place for monitoring, as introduced in [19]. The robot will start with an observation phase. During this phase the robot stands at a safe place in the initial room of the home environment and observes the human behavior and environment.

Collected data is used to compute the safety criteria. After that phase the Robot will travel to the different optimal observation slots and measure the different gait parameter. The gait velocity in different areas of the flat could be very helpful for the housing enabling assessment.

3.3 Housing Enabling

The housing enabling assessments is quite popular in the Scandinavian countries. The aim of this assessment is the rating of flats and their surroundings referring to the personal health status of the inhabitants [20]. This rating gives advice if the flat with its furniture etc. is not suitable for the resident. The housing enabling assessment is

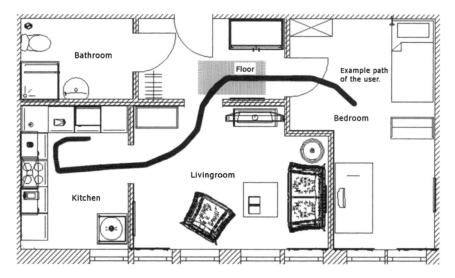

Fig. 1 An example path of the user from the bedroom to the kitchen which was recorded during a mobility assessment by the mobile robot

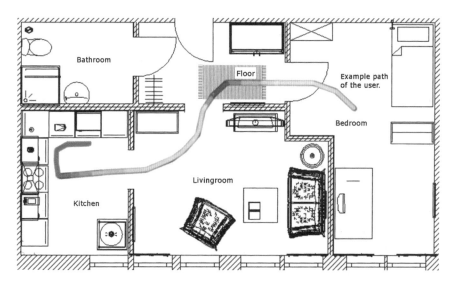

Fig. 2 The example path from Fig. 1 is highlighted with *grey* scale, *darker* slower gait velocity related to the average gait velocity, *brighter* near to the average gait velocity

split into three parts. The first part is the descriptive part to collect some general information about the flat and the condition of the user. The second part is the evaluation of functional limitations and dependence on mobility aids. Detailed information about medical condition of the user is collected e.g. severe loss of sight or limitation of stamina. The last part is based on different questionnaires which are related to the flat and the surroundings. Each question is weighted to the different diseases e.g. "Heavy doors without automatic opening" has high impact if the user has a problem with her/his upper extremity skills or has to cope with the loss of stamina. After completion of all questions it is possible to compute [21] the score of the flat in relation to the actual health status of the user [22]. It is also possible to adapt the flat related to the rating [23] in order to reduce the risk of falling.

3.4 Limitation of the State of the Art

As shown in Sect. 3.1 most of the systems use ambient sensors to observe the user not continuously. This means that only presence at specific known points is measured. The problem of this kind of monitoring is that it can only measure the mobility in an indirect way. The result could be only used for trend analysis instead of a precise assessment to determine the mobility of a person. For precise assessments of the mobility laboratory equipment is needed. But this is too large or complicated to install it in the domestic homes. Also the prices for such systems are too high to bring them to home environments as well. Within the domain of health care and rehabilitation service robotics there are quite few systems commercially available. Further, there is no robotic system that is capable of doing housing enabling assessments and tries to present advice to reduce the risk of falling. The current "offline" housing enabling tests suffer from some drawbacks. The estimation of the personal disorders and the investigated flat depends highly on the skill of the person executing the test. It is very difficult to rate the medical condition in its entirety for a person you see e.g. the first time. The adaption of the flat is a criterion which is also based on the experience of the supervisor. This could lead to different or insufficient results. Furthermore this assessment is mostly done after an accident has happened or to score new flats and not as a continuously assessment. In summary there is currently no system or approach available that is capable of doing precise and continuous housing enabling assessments in domestic environments and that is learning from the user's behavior/mobility to get optimal assessment results.

4 Approach

Our new approach to provide an automated and long-term housing enabling assessment will be a combination of mobility assessments, activity detection and autonomous exploration of the flat. By combining the approaches of these different services we think it is possible to deliver a new approach of the housing enabling service to increase the quality of life and the safety feeling of the inhabitants. A mobile robot will thereby act as a kind of mobile infrastructure bringing the needed sensor technology to the optimal place for monitoring, as introduced in [24]. Main goal of our approach is that the robot can be delivered via postal package and placed into the environment without any installation. To perform services and assessments without compromising the safety of the owner, we will use our approach as introduced in [25].

4.1 Mobile Robot Platform

As mobile robot, the current Florence platform [26] is used. It depended on a light modified Turtlebot Kit [27] from Willow Garage. Additional to the 3D Microsoft Kinect Sensor, a laser range scanner is mounted on the Robot. Two different models are used, a Hokuyo URG-04LX or a Sensorio LZR-U901 laser range scanner. The second one has four measure planes with a tilt angle shift between planes of approximately 2°. We will use these four measure planes to optimize the leg detection and the precision of our gait analysis approach [14]. For controlling the robot platform a Lenovo X130e a netbook is used with an AMD E-300 1.3 GHz dual-core CPU and 2 GB RAM. Ubuntu 11.04LTS is used as operation system with ROS Electric as a middleware software for controlling the robot hardware. To get more information about the user activities a HomeMatic bundle is used to communicate with the home automation devices e.g. from the OFFIS IDEAAL Living Lab. This additional information will be used to detect the activities of the user.

4.2 Identification and Clustering of Obstacles

During the automated exploration, we will update our 3D map of the flat and try to identify and cluster all obstacles into three categories: moveable, unmoveable and unknown. Moveable objects are for example chairs, tables etc. On the other side, a board will be classified as unmoveable. All Objects that could not be classified automatically will be clustered as unknown in the first iteration. There are different approaches to classify these objects. We will test an interactive method where the user is asked and an automated version which tries to learn from the user if she or he moves that object over the time or not. This information about an object together with the detailed information from the map will be used to compute the prescore of current environmental barriers and some other possible configurations as a part of the housing enabling assessment (Fig. 3).

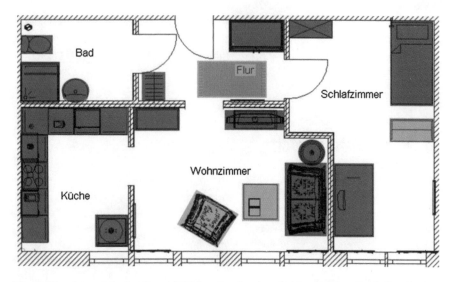

Fig. 3 The plan and the furniture of the flat after the obstacle identification. *Dark Grey* Static or unknown obstacles. *Light Grey* Moveable obstacles

4.3 Data Fusion of Different Assessments to Raise the Quality

To optimize and personalize the score we use the current health status of the user. We will combine these results with different results from the gait assessments e.g. the current gait velocity, step size and gait stability in the different areas of the flat. Related to the raising risk of falling if the resident is repeating to slowdown his or her self-selected gait velocity [7] potentially high-risk areas can be identified. Also the relation to different environment barriers [28] e.g. carpets or other soft surface and other gait parameters help to identify hotspots.

As can be seen in the Fig. 4, two areas with a potential high risk to fall (hotspots) have been found. To rise to quality of detection of these regions, we try to combine also information from the activities of daily life [29]. In our example path we have found a potential dangerous area in the kitchen. If the ADL indicates that at the same time the user starts to prepare a meal, we could discard this area. The reason for the slower gait velocity was the preparation of the meal. After that step we have only high risk areas which should be related to obstacles. If moveable obstacles are in these areas, the user will be informed that she/he should remove this obstacle to reduce the risk of fall events in this area.

Housing Enabling: Detection of Imminent Risk Areas

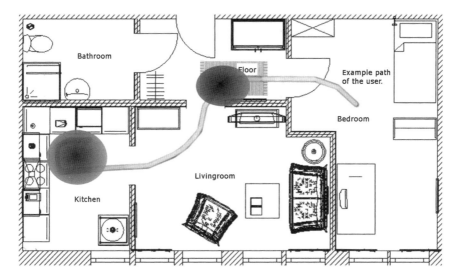

Fig. 4 The two red areas marked Hotspots, areas with a potential high risk of falling

5 Concept

As a first step towards realizing our new approach to housing enabling assessments in domestic environments utilizing a mobile robot we present a new concept for enhancing the prevention of fall events. Our main goal is to enable the robot to find hotspots and to find an optimal solution to remove these hotspots by adapting the flat. One quality criteria is the amount of removed hotspots, another quality criteria should be the necessary amount and kind of adaption. The concept combines our previous work [24], i.e. the precise measurements of human movement trajectories by a laser range scanner [25]. This concept enables the robot to learn from human behavior while assessing him or her at the same time. Currently the concept is based on the assumption that the robot is able to access a complete map of its environment. In the future, this map will be created while exploring the environment using 3D Simultaneous Localization and Mapping (3D SLAM) techniques. In short, the robot first measures the environment and localizes itself within the environmental map. Afterwards, the second step is to identify and cluster obstacles required for the following housing enabling assessment step. Then the human's gait velocity in general and at different areas in the flat is measured. Areas with slow gait velocity are analyzed and hotspots identified in the fourth step. Afterwards, the plausibility of hotspots is investigated by checking the obstacles in these areas and ADL at this time. Possible solutions to remove or reduce the hotspots are computed and advice is given to the user.

5.1 Environment Recognition

Within the first step the robot utilizes its laser range scanner and its 3D Scanner (Microsoft Kinect sensor) in order to measure the surface of its current surroundings. Ideally no moving objects are within the scan range during this step. Otherwise the robot has to distinguish between measurements belonging to static and moving objects utilizing one of various available approaches [30]. Measurements belonging to static objects are then transformed from the local coordinate system of the robot into the global coordinate system of the environmental map.

5.2 Obstacle Identification

The second step starts by identifying obstacles and barriers required by the next step of the housing enabling assessment. Identification e.g. carpets on the floor and clustering these obstacles in one of three categories (moveable, static and unknown). Also the general analysis of the flat, like measurement the width of doors or insufficient manoeuvring areas around white goods is part of this step and belongs to the indoor environment part of housing enabling. It based on a questionnaires with 100 different points to rate the suitability of the flat. We will try to implemented most of these questions in this part of our approach.

5.3 Movement Trajectory Measurement

As soon as a human enters the scan range of the robot its movement trajectory is measured. Again, the robot has to distinguish between trajectories belonging to the moving human and measured values of static objects. Measurements belonging to moving persons are transformed into the global coordinate system of the environmental map. These measurements are then used to compute the movement trajectory of the human. Additionally, the measurements are used to perform a mobility assessment of the human computing various spatio-temporal parameters of human gait. However, this enhanced assessment is not within the focus of this approach.

5.4 Identification of Hotspots

After we have detailed information about the environment and the gait velocity in general and especially in different areas of the flat we can compute hotspots. Hotspots are areas with a high difference between average gait velocity and local gait velocity. These areas refer to points with a high risk of imminent fall events [9].

5.5 Plausibility Check

To increase the precision and use of these hotspots, we make a plausibility check for each point. Therefore we map the hotspots to the 3D map and analyze if any obstacles are in the common area. Also the activity of daily life will be included, e.g. in Fig. 4 we have found two hotspots, one located in the floor and one in the kitchen. During the plausibility check further information will be added e.g. that the user typically prepares meals at this place. So the gait velocity was not reduced by an obstacle but by preparing a meal. So this Hotspot is no longer valid and could be removed from further computation.

5.6 Compute and Show the Best Adaption

Now we are able to calculate advice for the adaption of the flat to remove these hotspots. Therefore we will look at obstacles in the surrounding of the hotspots and if they are moveable or not. If they are moveable the service will compute different scenarios e.g. remove the obstacle at all or relocate the obstacle. For all scenarios the algorithm will calculate the new housing enabling score for the flat. Probably the best score is always the solution when the obstacle is removed completely. But in the most cases this solution has the lowest user acceptance, because most elderly don't want to change their flat too much. To reach a higher acceptance of the test, we define quality criteria which rate the amount of adaption and the kind of adaption. But the main goal is to raise the safety for the user in their own flat by reducing imminent fall event caused by environmental obstacles (Fig. 5).

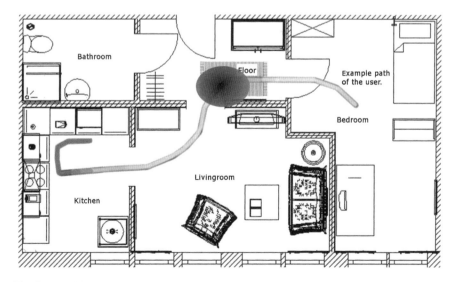

Fig. 5 Remaining hotspots after the plausibility check

6 Conclusion

A new concept for enhancing the housing enabling assessment by using a mobile robot and mobility assessments was presented. The concept is a first step towards realizing our approach to housing enabling assessment which utilizes a mobile robot as mobile software platform and its 3D scanner and laser range scanner as measurement devices. Our main aim is to enable the robot to find areas with a high risk of falls and giving solutions to remove or reduce these areas. Therefore we introduced the concept of hotspots which describe potential high risk areas near to the human's typical movement trajectories. The presented concept combines our previous work, i.e. the potential field method for determining movement patterns and the precise measurements of human movement trajectories by a laser range scanner. The overall flow of the concept has six steps:

- Measurement of the environment and self-localization within the environmental map,
- Identification and clustering of obstacles,
- Measurement of the human's gait velocity in general and at different areas of the flat,
- Identification of hotspots and plausibility checking of hotspots,
- Computation of possible solutions to remove or reduce the hotspots,
- Presenting the best solutions to the user related to the quality criteria.

We already implemented first parts of the algorithm; the next step will be a complete implementation on a mobile robot platform ('TurtleBot'by Willow Garage) and an evaluation of the system with test persons in real flats.

References

1. C. O. Q. O. H. C. I. America and I. O. Medicine: Crossing the Quality Chasm: A New Health System for the 21st Century, 1st edn. National Academies Press, Washington, DC (2001)
2. Cook, D.J., Das, S.K.: How smart are our environments? An updated look at the state of the art. Pervasive Mobile Comput. 3(2), 53–73 (2007), design and Use of Smart Environments. http://www.sciencedirect.com/-science/article/B7MF1-4MP002G-1/2/-7c83d759a97747d470977d35f8d6b217
3. European Commission Research: Seventh Framework—Work Programme 2011–2012. ICT, pp. 69–71 (2010)
4. van Doorn, C., Gruber-Baldini, A.L., Zimmerman, S., Hebel, J.R., Port, C. L., Baumgarten, M., Quinn, C.C., Taler, G., May, C., Magaziner, J., E. of Dementia in Nursing Homes Research Group: Dementia as a risk factor for falls and fall injuries among nursing home residents. J. Am. Geriatr. Soc. **51**(9), 1213–1218 (2003)
5. Gill, T.M., Williams, C.S., Tinetti, M.E.: Assessing risk for the onset of functional dependence among older adults: the role of physical performance. J. Am. Geriatr. Soc. **43**(6), 603–609 (1995)
6. Imms, F.J., Edholm, O.G.: Studies of gait and mobility in the elderly. Age Ageing **10**(3), 147–156 (1981)

7. Montero-Odasso, M., Schapira, M., Soriano, E.R., Varela, M., Kaplan, R., Camera, L.A., Mayorga, L.M.: Gait velocity as a single predictor of adverse events in healthy seniors aged 75 years and older. J. Gerontol. A Biol. Sci. Med. Sci. **60**(10), 1304–1309 (2005)
8. Scanaill, C.N., Carew, S., Barralon, P., Noury, N., Lyons, D., Lyons, G.M.: A review of approaches to mobility telemonitoring of the elderly in their living environment. Ann Biomed **34**(4), 547–563 (2006)
9. Cameron, K., Hughes, K., Doughty, K.: Reducing fall incidence in community elders by telecare using predictive systems. In: Proceedings of 19th Annual International Conference of the IEEE Engineering in Medicine and Biology Society, vol. 3, pp. 1036–1039. 30 Oct–2 Nov 1997
10. Pavel, M., Hayes, T.L., Adami, A., Jimison, H., Kaye, J.: Unobtrusive assessment of mobility. In Proceedings of 28th Annual International Conference of the IEEE Engineering in Medicine and Biology Society EMBS'06, pp. 6277–6280, Aug. 2006
11. Hayes, T.L., Pavel, M., Larimer, N., Tsay, I.A., Nutt, J., Adami, A.G.: Distributed healthcare: simultaneous assessment of multiple individuals. IEEE Pervasive Comput. **6**(1), 36–43 (2007)
12. Frenken, T., Steen, E.-E., Brell, M., Nebel, W., Hein, A.: Motion pattern generation and recognition for mobility assessments in domestic environments. In: Proceedings of the 1st International Living Usability Lab Workshop on AAL Latest Solutions, Trends and Applications, pp. 3–12. SciTePress, ISBN 978-989-8425-39-3, 2011, Jan 28–29 2011
13. Pallejà, T., Teixidó, M., Tresanchez, M., Palacín, J.: Measuring gait using a ground laser range sensor. Sensors **9**(11), 9133–9146 (2009)
14. Frenken, T., Gövercin, M., Mersmann, S., Hein, A.: Precise assessment of self-selected gait velocity in domestic environments. In: Proceedings Pervasive Computing Technologies for Healthcare (PervasiveHealth). IEEE, Mar 2010, ISBN 978-963-9799-89-9 (2010)
15. Petrovskaya, A., Ng, A.: Probabilistic mobile manipulation in dynamic environments, with application to opening doors. In: International Joint Conference on Artificial Intelligence (IJCAI), 2007
16. Breazeal, C.L.: Sociable machines: expressive social exchange between humans and robots. Ph.D. dissertation, Massachusetts Institute of Technology, Department of Electrical Engineering and Computer Science (2000)
17. Ray, C., Mondada, F., Siegwart, R.: What do people expect from robots? In: IEEE/RSJ International Conference on Intelligent Robots and Systems, pp. 3816–3821 (2008)
18. Isken, M., Vester, B., Frenken, T., Steen, E.-E., Brell, M., Hein, A.: Enhancing mobile robots' navigation through mobility assessments in domestic environments. In: Proceedings 4. Deutscher Kongress, Ambient Assisted Living. VDE Verlag (2011)
19. Brell, M., Meyer, J., Frenken, T., Hein, A.: A mobile robot for self-selected gait velocity assessments in assistive environments. In: The 3rd International Conference on Pervasive Technologies Related to Assistive Environments (PETRA'10), Samos, Greece, iSBN 978-1-4503-0071-1/10/06, June 2010
20. Fänge, A., Iwarsson, S.: Changes in ADL dependence and aspects of usability following housing adaptation—a longitudinal perspective. Am. J. Occup. Ther. **59**, 296–304 (2005)
21. Carlsson, G., Slaug, B., Johannisson, A., Fänge, A., Iwarsson, S.: The housing enabler—integration of a computerised tool in occupational therapy undergraduate teaching. CAL Laborate, June, 5–9 (2004)
22. Helle, T., Nygren, C., Slaug, B., Brandt, Å., Pikkarainen, A., Hansen. A-G., Pétersdórttir, E., Iwarsson, S.: The Nordic housing enabler: inter-rater reliability in cross-Nordic occupational therapy practice. Scand. J. Occup. Ther., 1–9, posted online 17 Jan
23. Fänge, A.: Strategies for evaluation of housing adaptations—accessibility, usability and ADL dependence. ISBN 91-974281-5-9. Doktorsavhandling. Institutionen för klinisk neurovetenskap, Lunds Universitet. Lund, Sverige (2004)
24. Brell, M., Meyer, J., Frenken, T., Hein, A.: A mobile robot for self-selected gait velocity assessments in assistive environments. In: The 3rd International Conference on Pervasive Technologies Related to Assistive Environments (PETRA'10), ISBN 978-1-4503-0071-1/10/06 (2010)

25. Frenken, T., Steen, E.-E., Hein, A.: A context-aware system for adaptive mobility assessment in domestic environments. In: First International Joint Conference on Ambient Intelligence (2010)
26. Meyer J., Brell M., Hein A., Gessler S.: Personal assistive robots for AAL services at home—the Florence point of view, (2009) Homepage: http://www.florence-project.eu/
27. Turtlebot from Willow Garage, Homepage: http://turtlebot.com, (2012)
28. Iwarsson, S., Isacsson, Å.: Housing standards, environmental barriers in the home, and subjective general apprehension of housing situation among the rural elderly. Scand. J. Occup. Ther. **3**(2), 52–61 (1996)
29. Iwarsson, S., Isacsson, Å., Lanke, J.: ADL dependence in the elderly: the influence of functional limitations and physical environmental demand. Occupational therapy international **5**(3), 173–193 (1998)
30. Durrant-Whyte, H., Bailey, T.: Simultaneous localisation and mapping (SLAM): part I the essential algorithms. IEEE Robot. Autom. Mag. **2**, 2006 (2006)
31. Iwarsson, S., Isacsson, Å.: Quality of life in the elderly population: an example exploring interrelationships among subjective well-being, adl dependence, and housing accessibility. Arch. Gerontol. Geriatr. **26**(1), 71–83 (1997)

Mobile Video Phone Communication Carried by a NAO Robot

Paul Panek, Georg Edelmayer, Peter Mayer, Christian Beck and Wolfgang L. Zagler

Abstract In the project "Knowledgeable SErvice Robots for Aging" (KSERA) an AAL system with a socially assistive robot (SAR) for old persons and persons with Chronic Obstructive Pulmonary Disease (COPD) has been developed. One of the elements of the system is a LED-projector module carried by the small humanoid NAO robot. This module is used as additional output element to extend the human–machine-interface by projecting text, graphics and videos on walls next to the user. As relevant use case, focus is on mobile video communication with family members, friends and also with service centres and medical authorities. Laboratory tests with the prototype by end users and a workshop with experts from the care domain were done. Results show the relevance of this solution for AAL applications, in particular for social communication and in emergency cases.

Keywords AAL · Socially assistive robots · Assistive technology · COPD · Mobile user interface · Pico projector

---

P. Panek (✉) · G. Edelmayer · P. Mayer · C. Beck · W. L. Zagler
Centre for Applied Assistive Technologies (AAT), Institute for Design and Assessment of Technology, Vienna University of Technology, Favoritenstrasse 11/187-2b 1040 Vienna, Austria
e-mail: panek@fortec.tuwien.ac.at

G. Edelmayer
e-mail: edelmayer@fortec.tuwien.ac.at

P. Mayer
e-mail: mayer@fortec.tuwien.ac.at

C. Beck
e-mail: beck@fortec.tuwien.ac.at

W. L. Zagler
e-mail: zagler@fortec.tuwien.ac.at

1 Introduction and Aim

Assistive robots are an upcoming approach to support the independence, safety and social connectedness of older persons [1].

This paper describes mobile video communication as it was developed in the KSERA Project [2–5]. A LED projector module, which is carried on the back of a NAO robot [6], is used as output device for visual content. In the following, a short overview on the development of the module is provided, followed by details of the implementation and preliminary evaluation of the mobile video-communication prototype.

2 User Requirements

The project deliverable D1.1 *Scenarios, Use cases and Requirements* [7] describes the user-requirements towards the system with means of scenarios and specific use cases. Regarding mobile video communication and the projector module which is carried on the robot's back, the following requirements are the most important ones:

- Solutions for users who feel isolated and miss social contact, support of '*social connectedness*', contact to family members and friends;
- Supporting communication with medical authorities and professionals, alarm calls in case of emergency (e.g. falls);
- Motivation to do physical exercises;
- Mobile interface which approaches the user autonomously in the living environment.

These points show the necessity of an audio and video connection between primary user (old persons and persons with COPD in their own home) and secondary users (e.g. health care professionals, family members and friends).

3 Mobile Projector

The humanoid robot NAO [6] is used as an interface between the users and the rest of the KSERA system [3]. The robot is capable of moving independently towards the users wherever they are in their home.

In the KSERA project the robot is additionally integrated into an assistive living environment. Monitoring the activity and the health status of the user is possible as well as measuring and analysing environmental parameters (e.g. air quality outside the home) [8]. There is also an interface provided to *eHome*, an intelligent assistive living environment system [9, 10]. This interface was used during the tests described below, to adjust the lighting in the test-room in order to conform to the needs of the projection.

Since the used robot is rather small compared to a human being (58 cm in height) it is not able to transport devices for human–machine-interaction such as touch-screens or tablet-PCs as were used in similar projects [11–16]. The NAO robot is also not strong and large enough to hand a smart-phone or PDA to a sitting or standing person.

The robot has built-in cameras which can be used for a *uni-directional* video connection (cf. work of Bäck et al. [17]). To our knowledge there is up to now no satisfactory solution for *bi-directional* video communication using such a small humanoid robot. The innovation presented in this paper is a newly developed LED-projector module which is carried by the robot on its back [18] (see Fig. 1).

The path of rays of the LED-projector that is used in the module (Fig. 2, light grey) is deflected by a small mirror due to space constraints in the module housing. Therefore, the projected image is a mirrored image of the original visual content. Dedicated software takes care of pre-processing the video information in order to project a correct image.

The most important parameters of the developed module (Fig. 1) are shown in Table 1 (more details can be found in [19]). Based on the user requirements the LED projector module is used to present the following information:

- Videos, e.g. physical training;
- Text and pictures, e.g. weather forecast, appointments, air quality;
- Video communication with friends, relatives, emergency centres.

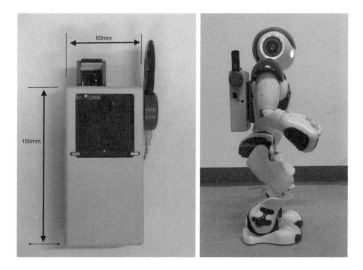

Fig. 1 Prototype of the KSERA LED-projector module (*left*); projector module mounted on back of NAO (*right*) [19]

Fig. 2 Projector-module (with vertically mounted integrated LED pico-projector) and path of rays of the projection

Table 1 Updated specification of LED-projector module [19]

Specification	Remarks
Physical dimension	
Weight	430 g
Size	35 × 80 × 150 mm (+45 mm for mirror and ext. antenna)
Mounting mechanism	
Position on robot	On NAO's back
Mechanism for fixation	Magnets for easy and quick mounting/unmounting
Autonomy of operation	
Internal battery	2 'Kokam single cell 1500HD 20C' 1,500 mAh, 3.7 V
Power supply (charger)	12 V DC; 5 A
Range	Up to 5 m free line of sight for wireless video transmission
Operation time	About 1 h
Projection	
Brightness	30 ANSI Lumen
Resolution	WVGA (854 × 480 Pixel)
Contrast	1,000:1
Content	Text, graphic, video
Focus	Manual focus
Projection area (at 1 m distance)	Wall, image size: 37 × 53 cm^2; height above floor 50 cm

The information that is to be projected and the information flow from and to the projector module is controlled by the *intelligent KSERA Server* [3] via an appropriate interface [19]. The projector component is responsible for visualisation of the information.

4 Mobile Video Communication

The developed research prototype allows for video communication between the KSERA user at home and a service centre, other private homes, mobile phones and landlines and thus fulfils the need for remote support from physicians or care persons and offers extended possibilities to foster social contacts (friends, family members). The open SIP standard for internet telephony is used. An external SIP server provides the functionality to connect to the public telephone network (Fig. 3).

The LED-projector module prototype was tested with experts and end users in laboratory settings [20] before it was provided to the project partners for extended tests under *real life* conditions. Figures 4, 5 and 6 give an impression how the LED-projector module was used in a controlled setting in the laboratory.

The robot is steered by the KSERA system to a place in the apartment which is known to the system and offers enough free space for projection. After the robot has been aligned by the system (distance to projection area 1 m ± 10 cm, perpendicular to it ±5°) projection of relevant content can be started.

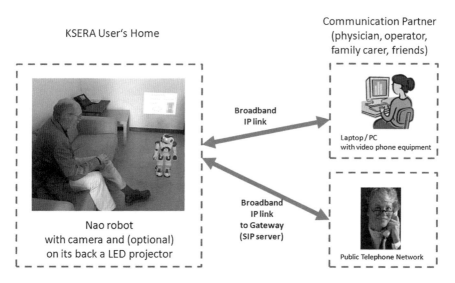

Fig. 3 Simplified architecture of the mobile video communication in the KSERA project

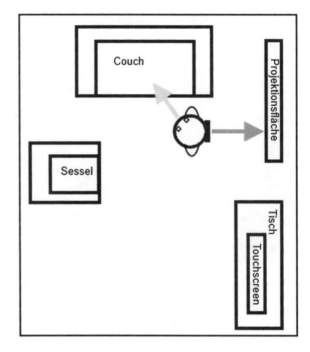

Fig. 4 Set up of test-room in laboratory (see also Fig. 5). Couch (for user), chair (for supervisor of test), NAO facing test-person (*green*) and direction of projection (*orange*). Table with touchscreen for stationary video communication

Fig. 5 Mobile Video communication in laboratory setting: NAO carries LED projector module on its back and uses it together with its integrated head-camera for video communication between user (sitting) and service centre

5 Operation Modes

The prototype allows for different modes of operation [21]. These modes are chosen according the specific needs of the users:

- *Visualisation of text, graphics and videos*: pre-recorded videos for physical training (callisthenics etc.), complex textual messages (e.g. personal reminder,

Fig. 6 NAO robot projects image of service center on a wall next to the user

day's schedule, air quality) are projected onto a wall next to the user. Tests in laboratory setting have proven that projected information provides additional benefit compared to speech output [20], especially if the content is complex.
- *Bi-directional video communication*: The projector module and the integrated camera in the robot's head serve as output and input for a video telephone (see Figs. 4, 5). Added value exists in particular in emergency situations—e.g. if the user has fallen. In this case the video-phone *can walk up to the user* on its own.

In order to be able to compare to the 'state of the art' and to demonstrate more applications of the KSERA system, an additional communication module without a projector was used during the laboratory trials:

Fig. 7 Stationary touchscreen terminal with NAO robot as 'motivator'

- *Stationary video communication*: Only the stationary touchscreen-terminal (without NAO) is used.
- *Mixed mode video-communication*: (a) a stationary touchscreen-terminal is used for communication and (b) the NAO robot is placed next to it to act as a *motivator* whose task it is to motivate the user to do some video calls (Fig. 7).
- *Uni-directional video communication*: This is similar to the bi-directional video communication. The head camera of the robot is used to pick up the image of the user and transmit it to a service centre, however the beamer module is not available, and thus the user cannot see the operator of the service centre (cf. work of Bäck et al. [17]).

6 Workshop with Experts

In addition to the trials in a laboratory setting with old persons and experts [20] a workshop was done with experts from the care domain [21]. The aim of this workshop was to evaluate the quality, user friendliness and added value of the mobile video communication of the KSERA system.

The workshop was held in a laboratory at Vienna University of Technology. Three high ranking experts from mobile care, occupational therapy and daily care of chronically ill persons took part. After a short introduction of the state of the art of assistive robots and the overall goals of the KSERA project, the possibilities of video communications were demonstrated. The experts gave their input regarding the system and the video-communication in particular in a free round of discussion. During this discussion also ideas about usage of the system or enhancing possibilities were presented.

The experts rated the audio and video quality of the connection as sufficient for the presented emergency scenario and for general video communication. This coincides with the results of the laboratory tests [20] where also end user evaluated the image quality as sufficient. The possibility of getting a better impression of the emotional status of a user in case of an emergency compared to an audio only connection was rated as particularly useful.

Additionally it is possible for an operator in a service centre to get a better idea of the severity of an emergency situation even if the user is no longer able to communicate. Detection of an emergency and establishing the connection to the service centre was not part of the workshop and was considered fully functioning.

Applications which were rated very useful are reminders to do physical exercise, calendar function for the daily schedule, as well as the capability of the system to motivate to social contacts. Another application which was rated very useful was to motivate users with film and/or movement of the robot to perform regular physical exercises or to guide them through such exercises.

7 Discussion

The developed LED projector module and the mobile video communication functionality that could be realized with this module, result in enhanced abilities of the NAO robot regarding human robot interaction (HRI) and support that can be provided to primary users.

At the beginning of the project, focus was on development of the module and its contribution to satisfy the user needs *motivate for physical exercise* and *mobile interface* which is able to meet the user wherever they actually are [18].

Based on this first prototype the extended second prototype also contributes to the user needs for social contact with friends and family members and for simple possibilities for emergency calls or for communication with professionals or health care organisations [21].

One obvious limitation of the projector module is the low brightness of the projected image. The luminous flux of the used pico-projector is 30 lm and thus much lower than the one of a standard projector (e.g. 2,500 lm). For this reason pico-projectors can only be used effectively inside a building, without direct sunlight on the projection surface and with an overall low level of ambient lighting. Since the KSERA system offers also an interface to *eHome*, an assistive environmental control system [9, 10], this was used to lower room lighting level during the time of projection. It can be assumed that future projectors with higher luminous flux will overcome this limitation. It will be possible to project overlay information on real objects and therefore enhance the 'reality' of the user [22]. This *augmented reality* potential could also be useful for the KSERA target users.

8 Conclusion

A prototype of a mobile projector module which is carried by a small humanoid robot was built and tested. Within the KSERA system this module makes it possible to project information onto a wall while the robot is communicating with the user of the KSERA system. The persistence of the visual information (compared to speech output) could be evaluated by end users and experts as being useful and relevant for the AAL context. The second prototype successfully integrated video communication to enable the users to get in contact with friends, family members, informal carers or a service centre. As video input for this the head camera of the robot us used, output is provided by the developed projector module on NAO's back.

Tests and workshops with users and experts from the care domain confirmed the added value of the mobile solution for video communication as provided by the NAO robot in the KSERA system.

Three LED projector modules were built and provided to the project partners for further tests of the whole KSERA system in a near to daily life setting.

Acknowledgments The research leading to the results presented in this paper is part of the KSERA project (http://www.ksera-project.eu). Funding was received from the European Commission under the 7th Framework Programme (Grant Agreement Number: 2010-248085) and from the Federal Ministry for Transport, Innovation and Technology.

References

1. Cavallaro, F.I., et al.: Growing older together: when a robot becomes the best ally for ageing well. In: Handbook of Ambient Assisted Living, pp. 834–851. IOS press, Amsterdam (2012)
2. KSERA project. http://ksera.ieis.tue.nl/ (2012). Last visited 17 Sept 2012
3. Cuijpers, R.: KSERA Project Presentation at AAL Forum, Lecce, Italy (2011)
4. Cuijpers, R.H., Juola, J.F., van der Pol, D., Torta, E.: Human robot interactions in care applications. Gerontechnology, 11(2), 353–354 (2012)
5. Torta, E., Oberzaucher, J., Werner, F., Cuijpers, R.H., Juola, J.F.: The attitude toward socially assistive robots in intelligent homes: results from laboratory studies and field trials. J. Hum. Robot Interact. (in press)
6. Aldebaran. http://www.aldebaran-robotics.com/ (2012). Last visited 17 Sept 2012
7. Meesters, L., et al.: Scenarios, Use cases and Requirements, Deliverable D1.1, KSERA consortium (2010)
8. Simonov, M.: A socially assistive robot for persons with chronic obstructive pulmonary disease (COPD). Gerontechnology 11(2), 353 (2012)
9. Mayer P., Panek P.: Assessing daily activity of older persons in a real life AAL system. Conference on Telehealth, pp. 772–775. Innsbruck, Austria (2012)
10. Werner K., Werner F., Panek P., Hlauschek W., Diermaier J.: eHome-Wohnen Mit unterstützender Intelligenz, AAL Congress, Berlin, Germany (2011)
11. CompanionAble project. http://www.companionable.net/ (2012). Last visited 17 Sept 2012
12. DOMEO project. http://www.aal-domeo.eu/ (2012). Last visited 17 Sept 2012
13. Nani, M., Caleb-Solly, P., Dogramadgi, S., Fear, C., van den Heuvel, H.: MOBISERV: an integrated intelligent home environment for the provision of health, nutrition and mobility services to the elderly. Proceeding of the 4th Companion Robotics Institute, Brussels (2010)
14. Lowet D., Isken M., Ludden G., van Dijk D.J., Remazeilles A., Cruz Martin E.: State of the Art in AAL Robotic Services, Deliverable: D5.1, Florence consortium (2010)
15. Van Dijk D.J., Isken M., Vester B., Winkler F., Cruz Martin E., O'Donnovan K., Remazeilles A., Laval M., et al.: State of the Art of Multi-Purpose Robots and Privacy-Aware AAL Home Services, Deliverable: D2.1, Florence consortium (2010)
16. Meyer, S.: Mein Freund der Roboter, VDE, Berlin (2011)
17. Bäck, I., Kallio, J., Perälä, S., Mäkelä, K.: An assistive humanoid robot for elderly care. Gerontechnology 11(2), 359 (2012)
18. KSERA Deliverable D3.2: Mockup and user tests of the HRI interface and home control design (2011)
19. Edelmayer, G., Ehrenfels, G., Beck, C., Mayer, P., Panek, P.: Prototyping a LED projector module carried by a humanoid nao robot to assist human robot communication by an additional visual output channel. In: Proceedings IASTED International Conference Assistive Technologies, pp. 809–816. Innsbruck, Austria (2012)
20. Panek, P., Edelmayer, G., Mayer, P., Beck, C., Rauhala, M.: User acceptance of a mobile LED projector on a socially assistive robot. In: Ambient Assisted Living, Advanced Technologies and Societal Changes, pp. 77–91. Springer, Heidelberg (2012)
21. KSERA Deliverable D3.3: Final HRI interface including social communication and context awareness (2012)
22. Dachselt R., Häkkilä J., Jones M., Löchtefeld M., Rohs M., Rukzio E.: Pico projectors: firefly or bright future? Interactions 19(2), 25–29. ACM (2012)

23. Oberzaucher, J., Werner, K., Mairböck, H. P., Beck, C., Panek, P., Hlauschek, W., Zagler, W.L.: A videophone prototype system evaluated by elderly users in the living lab schwechat. In: USAB 2009, LNCS 5889, pp. 345–352 (2009)
24. Werner, K., Oberzaucher, J., Panek, P., Beck, C., Mayer, P.: Development of an assistive home user interface together with older users. In: Everyday Technology for Independence and Care–AAATE 2011, Assistive Technology Research Series, Vol. 29, pp. 473–480. IOS press, Amsterdam (2011)
25. Panek, P., Hlauschek, W., Schrenk, M., Werner, K., Zagler, W.L.: Experiences from user centric engineering of ambient assisted living technologies in the living lab schwechat. In: Proceding 17th International Conference on Concurrent Enterprising, Aachen, Germany (2011)

The Robot ALIAS as a Gaming Platform for Elderly Persons

Jürgen Geiger, Thomas Leykauf, Tobias Rehrl, Frank Wallhoff and Gerhard Rigoll

Abstract Entertainment is an important aspect of social robots in an AAL context. Especially for elderly people, a robot can be the right platform for entertaining games. This paper describes the robotic platform ALIAS as a gaming platform. In the project ALIAS, a robot is equipped as a communication platform for elderly people. One aspect of the project is to provide cognitively stimulating games using a natural human-machine interface. Therefore, a computer game with multiple interaction channels is integrated on the robot ALIAS. The robot has a touchscreen with a graphical user interface where the game Tic-tac-toe can be played. Additionally, the game can also be controlled by speech commands, whereby a speech dialogue system is employed. In order to enrich the human-machine dialogue, the robot uses face detection to control its gaze behaviour and look at its conversational partner. This demonstration scenario can be seen as a first approach to evaluate a multimodal user interface using haptics and speech in conjunction with a cognitive dialogue system.

J. Geiger (✉) · T. Leykauf · T. Rehrl · F. Wallhoff · G. Rigoll
Lehrstuhl für Mensch-Maschine-Kommunikation, Technische Universität München, München, Germany
e-mail: geiger@mmk.ei.tum.de

T. Leykauf
e-mail: leykauf@mmk.ei.tum.de

T. Rehrl
e-mail: rehrl@mmk.ei.tum.de

F. Wallhoff
e-mail: wallhoff@mmk.ei.tum.de

G. Rigoll
e-mail: rigoll@mmk.ei.tum.de

F. Wallhoff
Jade University of Applied Sciences, Oldenburg, Germany
e-mail: frank.wallhoff@jade-hs.de

1 Introduction

With growing progress in the field of robotics and human-machine interaction, more elaborate robotic applications can be developed. One possible field of applications for a home robot is entertainment robotics, where games play a major role. Especially for elderly people in the Ambient Assisted Living (AAL) context, a robot can be the right platform for entertaining games. Compared to a conventional electronic entertainment systems like a video game console or a tablet PC, a robot can provide several advantages. Due to its human-like and friendly appearance, a social robot is appealing to elderly persons who might otherwise be reluctant to use technical systems. In addition, a convincing human-machine interface employing innovative communication channels leads to a higher user acceptance. Therefore, with an easy-to-use human-machine interface (speech control) and its appealing appearance, a robot is the ideal platform to approach users who are not so technologically adept.

There are several tasks a companion robot can perform to support elderly people. Physical assistance is not addressed here in the context of social robotics but it is definitely a future key selling point. The ability to lift or carry things or act as a mobility aid are important features but need more stable technical solutions to be able to be employed without safety risks. The other important field of application for robots in AAL is communication. In its function as a communication platform, a robot can serve several duties. It can interact with its elderly users, provide cognitive assistance and promote social inclusion. In order to stimulate cognitive activities, entertainment games can be the right manner. Equipped with an easy-to-use human-machine interface and embedded in a robotic system with an appealing appearance, entertainment games are well suited to encourage elderly people to perform cognitive stimulating activities.

1.1 Related Work

One of the most famous entertainment robots is Aibo [1], which is designed as a pet dog. The robot Paro is a robot in the form of a baby harp seal and is used as a therapeutic agent in the care of people with dementia. It was also used to study the social interaction among elderly people [2]. Robotic assistants have also been used in therapy for children with autism [3].

There are several research projects which follow similar goals as ALIAS. In [4], the robot Maggie is presented which is used as a gaming platform. In the CompanionAble project [5], a mobile robot platform is combined with Ambient Intelligence technologies to provide a companion for elderly people. The project ExCITE [6] has the goal to develop a mobile robotic platform for telepresence applications. A service robot designed to help humans in the household is developed in the Care-O-bot 3 project [7].

In [8] it was examined how robotics could support the independent living of elderly people. Entertainment robotic systems are useful for preventing loneliness and for therapeutical applications, especially for elderly people.

1.2 Overview

In this work, we present the robotic platform ALIAS in its functionality as a gaming platform. Using the touchscreen of the robot, the game Tic-tac-toe can be played. Alternatively, the game can be controlled with speech commands. The robot will also answer with speech messages. To enrich the feedback with natural behaviour, the robot uses face detection to always looks at its conversational partner. All these software modules are controlled by the dialogue system. The choice for Tic-tac-toe as a simple strategy board game and the possibility to control the game via touchscreen or speech commands leads to an easy-to-use human-machine interface where no instructions for the user are necessary.

The rest of the paper is structured as follows: In the following section, the robotic platform ALIAS is presented. The employed dialogue system is described in Sect. 3. In Sect. 4, the implemented game is presented, before a conclusion is given in the last section.

2 The Robot ALIAS

2.1 The ALIAS Project

In the project ALIAS (Adaptable Ambient Living Assistant[1]), the robotic platform ALIAS is equipped as a communication platform for elderly people. ALIAS is a mobile robot system that interacts with elderly users, monitors and provides cognitive assistance in daily life, and promotes social inclusion by creating connections to people and events in the wider world. The system is designed for people living alone at home or in care facilities such as nursing or elderly care homes. The function of ALIAS is to keep the users linked to the wide society and in this way to improve their quality of life by combating loneliness and increasing cognitively stimulating activities. In a first series of field-trial experiments, the robotic platform was already tested with elderly users [9].

To fulfill it's goals, ALIAS is equipped with several capabilities: An easy-to-use and fault tolerant human-machine interface is achieved by employing automatic speech recognition (ASR) together with a module for natural language understanding (NLU). Communication is enriched through the utilization of

---

[1] See AAL-JP project ALIAS www.aal-alias.eu.

person identification methods using voice and face and laser-based leg-pair detection. In order to promote social inclusion, services for net-based linking are employed to link users with the wider world, enabling to maintain a wider horizon by exploiting new kinds of on-line and remote communication techniques. Autonomous, socially acceptable navigation capabilities enable the robot to find its way in its environment. In addition, the robot is equipped with a brain-computer interface (BCI), enabling users like stroke patients to remotely control the system.

Several use-case scenarios are developed to showcase the different functionalities of the robot. For example, in the ground lighting scenario, it is shown how ALIAS can guide persons at night in the dark, using its navigation capabilities and applying the touchscreen display as a light source. The gaming scenario which is described in this work is used to test the human-machine interface. An entertaining game is played through the touchscreen and can additionally be controlled by speech commands. At the same time, ALIAS is addressing its user by employing face detection to detect the user and hold eye-contact with him. The hardware and software setup of the robot as used in the gaming scenario are described in this section.

2.2 Hardware

The hardware configuration of the robot platform ALIAS (see Fig. 1) is based on the SCITOS G5 robot family of the robot manufacturer MetraLabs.[2] It is an approximately 1.50 m tall robot platform and can be divided into a driving unit and an interaction unit. In order to approach a user, navigation is provided by the driving unit, which uses a differential drive system. In a known environment, the robot can localize itself, navigate autonomously and approach a user in a socially acceptable manner using a laser range finder and ultrasonar sensors [10]. The interaction unit consists of a movable robotic head and a 15″ touchscreen, which is used for user interaction with the robot and is best suited as an easy-to-use human-machine interface. Additionally, it is equipped with four microphones and two loudspeakers which can be used for speech input and output.

The robotic head has five degrees of freedom (head pan and tilt and eye pan plus two eye lids) and additionally, a row of LEDs for additional user feedback is mounted on it. On top of the head, an omnidirectional camera is mounted, which delivers a 360° image. Due to its mounting position, the main purpose of this camera is to localize and identify persons using face detection and identification.

Two different computers are mounted on the robot. An industrial PC running Linux is used to control the hardware of the robot, e.g. the driving wheels, the collision and ultrasonic sensors and the robotic head. To control the touchscreen display, the robot is equipped with a Mac mini running Windows. The microphones and loudspeakers are also connected to the Windows PC. Thus, all dialogue control

---

[2] www.metralabs.com

Fig. 1 Overview of the hardware setup of the robotic platform ALIAS, divided into driving unit (*lower part*) and interaction unit (*upper part*) consisting of the touchscreen display and the robotic head

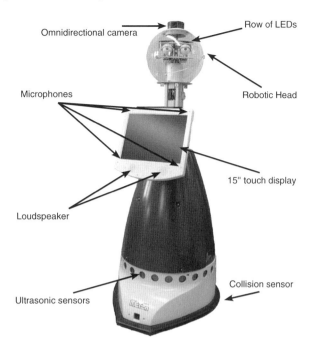

and speech processing modules are also running on the Windows PC. All modules on both computers can communicate with each other through various interfaces. More technical details about the robot platform ALIAS are provided in [11].

2.3 Software Overview

To provide an optimal multimodal interface for a gaming platform, the mobile platform is equipped with several software modules. An overview over the involved software modules is shown in Fig. 2.

The central module is the dialogue system (running on the Windows PC). It communicates with all other modules. Automatic Speech Recognition (ASR) is used to give commands to the robot, and answers are created using the speech synthesis module (text-to-speech, TTS). The dialogue system is described in detail in Sect. 3. The touchscreen is used to display the graphical user interface (GUI) for the game, which is handled by the gaming engine. Additionally, the touchscreen serves as haptic control channel. The gaming engine controls the Tic-tac-toe game including the artificial intelligence for the computer player. A face detection module is running to localize users and to control the robotic head. As a result, the robot always looks at its conversation partner. This guarantees for natural behaviour of the robot. All the involved software modules are described in detail in the next sections.

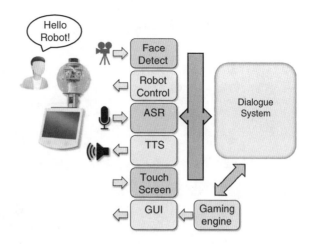

Fig. 2 Software modules for the described gaming scenario. User input/perception modules are displayed in *green*, action/output modules in *blue*

3 Dialogue System

The dialogue system is the most important software component of the robot. It is the connection between all other modules. For example, it controls speech input via automatic speech recognition, speech output via speech synthesis, the gaming engine (leading through the different steps of the game) and the GUI.

In Fig. 3, a complete flow chart of a speech dialogue situation is displayed.

After starting the game, the dialogue system controls the proceeding. First, the user is asked to make his next turn. After the user chooses his field on the playing board, he is asked for a confirmation (yes/no) before his symbol is placed. This behaviour helps to compensate errors of the speech recogniser, which makes it possible to play the game without any unwanted moves in the game. Following the user's turn, the implemented computer AI is used to determine the next turn of the computer player. Depending on the current state of the game, appropriate speech synthesis commands are used, e.g. to start the game, to ask the user to take his turn in the game (which is repeated after a certain interval) or to notify the winner at the end of the game. Speech recognition is used to record the commands of the user. Alternatively, the user can also make his choice by directly clicking on the touchscreen. Speech recognition and synthesis are realized through the commercial software DialogOS (see http://www.clt-st.de/produkte-losungen/dialogos/) which uses the VoCon 3200 speech recognition engine.

3.1 *Automatic Speech Recognition*

In this scenario, automatic speech recognition (ASR) can be used to choose a space on the 3 × 3 playing board. Speech is recorded by the microphones built in around the touchscreen, so no additional handheld or headset microphone is necessary,

The Robot ALIAS as a Gaming Platform for Elderly Persons

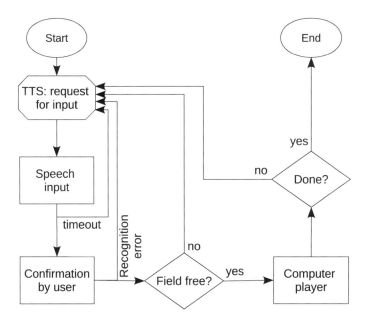

Fig. 3 Complete dialogue flow chart for the Tic-tac-toe game, including TTS (speech synthesis), speech recognition and the computer player AI

providing a hands-free communication. However, since the display (where the microphones are attached) is close to the user position, the sound quality is very good. Figure 4 shows a flow chart of the ASR module. Whenever input from the user is expected, e.g. to make a turn in the game, the ASR module is activated. Voice activity detection (VAD) is used to detect an utterance, which is then forwarded to the recognition process. In the recognition process, a dynamic grammar and vocabulary (adapted to the current dialogue state) are used. This behaviour is controlled by the dialogue system. Therefore, in each dialogue step, only a specific set of commands can be recognized.

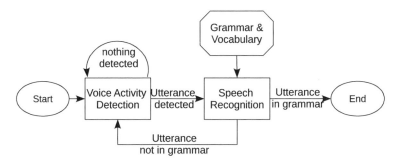

Fig. 4 System overview of speech recognition including voice activity detection (VAD)

The recognised sentence is then processed by the dialogue system. Several commands are possible to address the different fields on the playing board, e.g. "bottom left" or "left bottom". For answering of yes/no questions, also several utterances are possible.

User studies have been conducted to find out the different commands the users would use when not being instructed about possible ASR commands. These commands were then chosen to be implemented into the grammar of the ASR system. Therefore, the implemented grammar and vocabulary cover a broad range of possible user input and the ASR module is not limited to a fixed small vocabulary. To use the ASR system, no additional training phase (e.g. to perform speaker adaptation) is needed for the user. Thus, the ASR system is easily usable out-of-the-box without any instructions for the users.

3.2 Speech Synthesis

The robot uses a TTS module to communicate to the user. Any possible text can be converted into spoken language. Four different voices can be chosen: For English and German, each a male and female voice are possible. The voices are clear and easily understandable for humans. Speech synthesis is used at various points in the scenario, e.g. to welcome the user at the beginning of the game. Then, in each turn of the game, the user is asked for his decision. After the answer of the user, he is asked for a confirmation (yes/no). At the end of the game, the outcome (win/loss/draw) is also communicated by the TTS module. In order to design the dialogue rich in variety, for each possible TTS command, several variants are implemented. This brings more lifelines into the dialogue, especially for dialogue steps which are repeated several times (e.g. asking for the input of the user).

3.3 Touchscreen

The touchscreen is used to display the GUI of the Tic-tac-toe game. Using the 15" touchscreen is a simple way to provide an easy-to-use human-machine interface. In addition to using speech commands, the game can also be controlled via the touchscreen. Thus, two natural interaction channels are provided to communicate with the robot.

3.4 Gaze Behaviour

In order to provide a human-like behaviour of the robotic head, a vision-based attention system (similar as in [12]) is employed to control the gaze behaviour of the robotic head. A face detection module is used to localize the face of the conversational partner and turn the head and eyes of the robot towards the user. When the user moves his head, the robot will follow this movement. Constant movement of the head and eyes makes the robot appear more lively and attractive.

3.4.1 Face Detection

Face detection is the task of determining if and where in an image faces are located. Sensor data (usually from a camera) are the input to the system, while coordinates of regions containing faces (usually bounding boxes) are the system output. Typically, scaling and rotation parameters of the detected faces are not provided. To detect faces, the approach of Viola and Jones [13] using adaptive boosting and a Haar cascade classifier is employed. This approach is effective and fast and achieves high detection rates. The fast processing speed is based on three facts: first, the applied Haar-like features computed on a so-called integral image, second, adaptive boosting (adaBoost), and third, the cascade structure of classifiers (with increasing complexity). The features used in the cascade classification structure are called Haar-like features, because the have resemblance with Haar-Wavelets. The results of the computation of the Haar-like features represent certain characteristics of the input image: edges, texture changes, borders between light and dark image regions. Using these features, adaBoost is applied to select and combine several weak classifiers in a cascade resulting in a strong classifier capable of detecting human faces.

In our system, the image from the omnidirectional camera on top of the robotic head is used as input to the face detection system. The camera consists of four individual cameras whose images are merged together to a 360° image. Images are recorded with a frame rate of 15 images per second. Therefore, persons standing at any angle around the robot can be detected. First, all recorded images are transformed to greyscale, since the employed face detection system uses greyscale images. Then, faces are detected using the approach of Viola and Jones. Only the hypothesis with the highest probability is returned, in the form of a rectangle as a bounding box of the face. This means that only one face is detected at a time, which is always the largest face in the view. This makes the assumption that in a multi-person scenario, the interaction partner is always the person with the smallest distance to the camera, which should be true most of the time. In order to speed up the detection process for subsequent images, the information about the current position of the face is utilized. For the following frame, not the whole image is searched for faces, but only a bounding box around the last position of the face.

3.4.2 Head Control

The position of the detected face is then used to control the robotic head. Adjusting the head's pan and tilt angle and the pan angle of the eyes, the robot tries to hold eye contact with the user, using a human-inspired motion model for the coordination

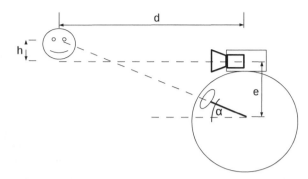

Fig. 5 Geometry of the setup for controlling the head tilt angle according to the position of the detected face

between head and eye movement [14]. From the coordinates (in pixels) of the detected face within the camera image, the corresponding world coordinates are obtained using the known geometrics properties of the camera setup. These coordinates are then used to direct the robot's head and eyes towards the middle of the face.

The targeted head pan angle (horizontal angle) can simple be computed using the x-position (in pixels) of the detected face within the camera image. As long as the difference between the target viewing angle and the current viewing angle is below the maximum pan angle of the eyes (8.5°), only the eyes will be moved. For larger angular changes, head movements are utilized: the pan angle of the eyes is turned back to zero and the head pan angle is set to the target viewing angle.

To determine the head tilt angle is more complex, due to the fact that the camera is not positioned in the robot's eyes. The geometric setup for controlling the robotic head's tilt angle is shown in Fig. 5.

The distance d of the face to the camera is estimated based on the height (in pixels) of the detected face, assuming an average-sized face. The height offset h of the face is computed with known opening angle of the camera of 45°. With known distance e of the eyes (in their initial position) to the camera, distance d of the face to the camera, and height offset h, the required tilt angle α can be determined exploiting the geometric properties of the recording setup:

$$\alpha = \operatorname{atan}\left(\frac{e+h}{d}\right) \quad (1)$$

The target tilt angle is smoothed over time in order to account for outliers resulting from erroneous estimation of d. Then, this angle is used as a target angle for the robotic head.

The chosen implementation of the face detection together with the properties of the hardware actuators of the robot is fast enough to provide fluid head and eye movements. Employing such a motion model results in natural-looking head movements.

4 Use Case: Tic-Tac-Toe Game

Tic-tac-toe (also known as noughts and crosses) is a simple two-person strategy game. On a 3×3 playing board, players alternatively mark the spaces with their symbol (circle or cross). The first player to mark a horizontal, vertical or diagonal row with three of his symbols wins the game. If neither player manages to mark a row of three symbols before the grid is full, the game ends in a draw.

This game was chosen for integration because of the simplicity of its rules, the short playing time and because it is widely known. For almost none of the users, an explicit explanation of the rules is necessary. It is not very tedious to play one round of the game (compared to, e.g. chess) and even after playing a couple of rounds, the user will not be bored too quickly. In addition, the game is well suited for speech control due to the possibly limited vocabulary. Beyond some commands for game settings and to start a new game, only nine different decisions (one for each field) are possible.

4.1 Graphical User Interface

The implemented graphical user interface (GUI) of the game is shown in Fig. 6.

To keep the GUI simple and neatly arranged, it consists of the 3×3 playing grid and buttons to start a new game and to adjust the game settings. The current game settings are always displayed on the main window. The game can be played either by two human players against each other or versus the computer player. When playing against the computer played, it can be chosen if the computer or human player should begin. In addition, there are two different game engines for the artificial intelligence available.

4.2 Game Engine

Two different kinds of game engines have been developed to control the computer player of the Tic-tac-toe game. Using a set of rules, an optimal strategy can be applied, which always leads to a win or draw. A simplified version of this optimal strategy has been implemented. Therefore it is very challenging (but not impossible) to win against the computer. In addition, to give the human player a better chance to win the game, a simple computer strategy is integrated, where always a random field is chosen. Using the settings button from the main GUI, the user can choose between these two degrees of difficulty.

Fig. 6 Graphical user interface of the integrated Tic-tac-toe game

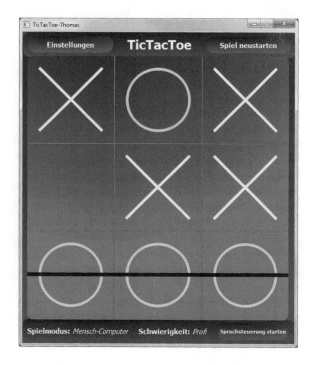

5 Experiments

Preliminary experiments have been conducted with users to test the design of the GUI and the dialogue system including speech control. The GUI was rated very good and especially the large buttons were commented positively. Using speech control was judged to be very convenient and fault-resistant. The integrated human-inspired head movement model was very well appreciated. This behaviour equipped the robot with a certain human-likeness and brought it to life. In general, the system was highly appreciated and accepted by the users and due to the design of the dialogue system, it showed a high robustness against speech recognition errors. Further experiments will show the influence of the head control module on the likability of the robot.

6 Conclusions

We have described our robotic system ALIAS which, within the ALIAS project, is equipped as a communication platform for elderly people. One aspect of the robot is entertainment. Therefore, we integrated a simple strategy game with an easy-to-use interface. The game can either be played by directly using the touchscreen or via speech commands.

The robot can be used to play the game without any need for instructions: The rules of the game are known to almost everybody, the ASR needs no training phase and covers most of the keywords the users will naturally use, and in addition, the game can also by played by using the touchscreen. This leads to a high acceptance of the system and provides the possibility to study the user behaviour when facing a robot without any given instructions.

The presented robotic system is easily extendable and the involved components can be used to develop more entertainment applications. More complex demonstration scenarios using the components described in this paper are in development.

In the future, we will use the features of the robot (e.g. the row of LEDs) to display basic emotions, which leads to a higher immersion.

Acknowledgements This work was supported by the project AAL-2009-2-049 "Adaptable Ambient Living Assistant" (ALIAS) co-funded by the European Commission and the German Federal Ministry of Education (BMBF) in the Ambient Assisted Living (AAL) programme.

References

1. Fujita, M.: On activating human communications with pet-type robot aibo. Proc. IEEE **92**(11), 1804–1813 (2004)
2. Kidd, C.D., Taggart, W., Turkle, S.: A sociable robot to encourage social interaction among the elderly. In: Proceeding IEEE International Conference on Robotics and Automation, pp. 3972–3976 (2006)
3. Robins, B., Dautenhahn, K., Boekhorst, R., Billard, A.: Robotic assistants in therapy and education of children with autism: can a small humanoid robot help encourage social interaction skills? Univ. Access Inf. Soc. **4**(2), 105–120 (2005)
4. Alonso-Martn, F., Gonzalez-Pacheco, V., Yébenes, M., Castro-González, Á., Ramey, A.A., Salichs, M.A.: Using a social robot as a gaming platform. In: Proceeding International Conference on Social Robotics (ICSR), pp. 30–39 (2010)
5. Gross, H.M., Schroeter, C., Mueller, S., Volkhardt, M., Einhorn, E., Bley, A., Martin, C., Langner, T., Merten, M.: Progress in developing a socially assistive mobile home robot companion for the elderly with mild cognitive impairment. In: Proceeding IEEE/RSJ International Conference on Intelligent Robots and Systems (IROS), pp. 2430–2437 (2011)
6. Cesta, A., Coradeschi, S., Cortellessa, G., Gonzalez, J., Tiberio, L., Von Rump, S.: Enabling social interaction through embodiment in excite. In: Second Italian forum on Ambient Assisted Living, pp. 5–7 (2010)
7. Reiser, U., Connette, C., Fischer, J., Kubacki, J., Bubeck, A., Weisshardt, F., Jacobs, T., Parlitz, C., Hagele, M., Verl, A.: Care-o-bot 3-creating a product vision for service robot applications by integrating design and technology. In: Proceeding IEEE/RSJ International Conference on Intelligent Robots and Systems (IROS), pp. 1992–1998 (2009)
8. Meyer, S.: Mein Freund der Roboter: Servicerobotik für ältere Menschen—eine Antwort auf den demographischen Wandel? VDE (2011)
9. Scheibl, K., Geiger, J., Schneider, W., Rehrl, T., Ihsen, S., Rigoll, G., Wallhoff, F.: Die Einbindung von Nutzerinnen und Nutzern in den Entwicklungsprozess eines mobilen Assistenzsystems zur Steigerung der Akzeptanz und Bedarfsadäquatheit. In: Proceeding Deutscher AAL Kongress (2012)

10. Kessler, J., Scheidig, A., Gross, H.-M.: Approaching a person in a socially acceptable manner using expanding random trees. In: Proceedings 5th European Conference on Mobile Robots, pp. 95–100 (2011)
11. Rehrl, T., Blume, J., Geiger, J., Bannat, A., Wallhoff, F., Ihsen, S., Jeanrenaud, Y., Merten, M., Schönebeck, B., Glende, S., Nedopil, C.: Alias: Der anpassungsfähige ambient living assistent. In: Proceeding Deutscher AAL Kongress (2011)
12. Breazeal, C., Scassellati, B.: A context-dependent attention system for a social robot. In: Proceeding International Joint Conference on Artificial Intelligence, pp. 1146–1153 (1999)
13. Viola, P., Jones, M.J.: Robust real-time face detection. Int. J. Comput. Vision **57**, 137–154 (2004)
14. Cook, G., Stark, L.: The human eye-movement mechanism: experiments, modeling, and model testing. Arch. Ophthalmol. **79**(4), 428 (1968)

Part VII
Community Conclusions

Can the Market Breakthrough in AAL be Achieved by a Large Scale Pilot?

Mohammad Reza Tazari and Reiner Wichert

Abstract Ambient Assisted Living is still on the cusp of a mainstream breakthrough, even though the market potential is tremendous. As barriers to the success of AAL, (Wichert et al. in How to overcome the market entrance barrier and achieve the market breakthrough in AAL. Ambient Assisted Living. Springer, Berlin, 2012) mentions the lack of viable business models as well as the lack of ecosystems around common open platforms. Considering the fact that the EU has supported the development of universAAL as a true candidate for such common open platforms, this paper describes the next logical step towards the rollout of AAL throughout Europe based on this platform so that the foundation stone for the emergence of a self-organizing ecosystem is laid. In order to elaborate the feasibility of such rollouts, the European Commission published in early 2012 a CIP-ICT-PSP call for piloting AAL in large-scale based on interoperable platforms, where the winner was the proposal "make it ReAAL" that builds on universAAL as the common open platform. In addition to the applications coming with the universAAL Platform, the local vendors from each pilot region will offer their AAL products and services to the participating pilot sites. This means that a two-way adaptation is supposed to be addressed in ReAAL: on one hand, the universAAL native applications can be adapted to the requirements of the pilot sites, and on the other hand, the existing applications from the pilot regions can be integrated with the universAAL platform. Each Pilot site will then be able to select from this portfolio, those applications that are more suitable for their planned intervention. The mission of ReAAL is in this context to find out if the adaptation of a product to a common platform is economically reasonable when a company plans to enter the market with a new product idea. Based on this, ReAAL will

---

M. R. Tazari (✉) · R. Wichert
Fraunhofer Ambient Assisted Living Alliance, c/o Fraunhofer-Institut für Graphische Datenverarbeitung, Fraunhoferstr. 5, 64283 Darmstadt, Germany
e-mail: saied.tazari@igd.fraunhofer.de

R. Wichert
e-mail: reiner.wichert@igd.fraunhofer.de

hopefully show the cost-effectiveness of interoperable solutions compared to both vertical isolated solutions and comprehensive but closed systems. This should pave the way for the emergence of an ecosystem around a common open platform and based on that for the market breakthrough of AAL through interoperability.

Keywords Market potential of Ambient Assisted Living · Business model · AAL reference platform · Standardization · Market breakthrough · Large scale pilot

1 A Long Way to a Common Approach

Due to the changing demographics the European Commission as well as near all national funding organizations see the necessity to support assisted technologies through huge funding instruments. Despite its tremendous market potential, Ambient Assisted Living is still on the cusp of a mainstream breakthrough. A lack of viable business models is considered almost unanimously to be the greatest market obstacle to a broad implementation of innovative AAL systems [1]. The consequence has been that the European Union has invested significant resources in RTD to build open platforms, resulting in platforms such as MPOWER [2], Persona [3], and SOPRANO [4]. These platforms provided very promising results, but they were fragmented in terms of what particular needs they addressed and duplicated in terms of implementing much of the same functionality. The European Union called for a project to collect these promising results into one European open technology platform for independent living application development.

As a result, the integrated project universAAL [5] was launched in 2010. universAAL provides a free and open source software platform for developing independent living applications. The platform consists of a middleware, a set of "manager" components, and tools that facilitate its usage and configuration (cf. the universAAL developer depot [6]). universAAL establishes new market possibilities, especially for small and medium enterprises to sell their products as interoperable solutions which can be integrated with other solutions from other vendors easily in order to share data and functionality.

But, despite the great universAAL achievements, the market breakthrough of AAL is still not on the horizon. The companies are still hesitating to productize prototypes resulted from R&D. What are the reasons for this waiting position? We discussed this question in a workshop organized by AALOA [7], the AAL Association and the European Commission [8]. Almost all participants agreed that profitable business models are missing. The RoI is too low and due to the small selling numbers the prizes are too high. Therefore, the other companies see no chance for their products and the whole community circles and no progress can be seen.

Another result of this workshop has been that a common open platform could help to break the circle. Having the universAAL platform as an existing example of such open platforms, a roll-out of this platform in large scale will be the next logical step: nobody has tried so far to compare directly the cost of application development based on interoperable semantic platforms against use-case-specific approaches with point to point connections. Consequently, a calculation of cost for AAL in large scale roll-outs could be feasible.

2 The ReAAL Mission

Following the above findings, the European Commission published a CIP-ICT-PSP call for a piloting project to show the benefit of open platforms with concrete numbers. ReAAL was the project which succeeded. The project intends to deploy a critical mass of Ambient Assisted Living applications and services for ca. 7,000 users in seven EU countries, based upon the universAAL platform, previously developed with EC support, with the intent of kick-starting the market for interoperable AAL services, applications and devices. ReAAL will facilitate the emergence of an AAL ecosystem by showing the platform usefulness, and spreading the related technical knowledge through an associated community of interest. Here, ReAAL will establish a multi-dimension evaluation methodology to measure the impact of the deployment of the AAL ecosystem in terms of the social, economic and health indicators. Such an ecosystem will, on the one-hand, allow commissioners to make investment decisions knowing that they will have support from other users of the acquired technology in an open-interoperable ecosystem, and that they will avoid vendor lock-in. On the other hand, the "standardized" platform will provide application vendors and service providers with a common basis on top of which they can invest and develop their business. This will be particularly helpful to the SMEs who are typically unable to make the sustained investments necessary to develop complete proprietary systems.

ReAAL has now the mission to try to test and to know at the end, if AAL at all could be rolled out with blanket coverage in whole Europe in the future. Therefore we need to know how much time is needed to spread the AAL applications and products in other regions when they are adapted to a common platform. On the other side it is needed to know, if the adaptation to a common platform is economically reasonable before a company wants to enter the market with a new product idea. Thus our hope is that we know at the end if interoperability is a reasonable way to follow or not.

The beneficiaries will be those people who wish to be able to avoid dependency on nursing homes, preferring to continue to live independently in their own homes. ReAAL will be realized via 14 pilots, each with a different focus. Assistance might be needed in any aspect of daily life, from health safety and security to social integration and mobility support. Other users involved in the pilots will have different needs or interests in the participation: Technology providers will validate the use of a platform as a leverage to develop more cost-efficient, innovative and mature technology. Service providers will look for a better efficiency and a better quality perceived by their customers in the provision of the services. Policy makers will require results to analyse the Return of Investment of the public procurements. All these needs are being addressed by the set of objectives of the ReAAL project.

ReAAL will launch the usage of the universAAL open platform in real life and enable service providers to deploy interoperable services avoiding vendors-locks. Different scenarios can be foreseen: In case the service operator wants to integrate one of the actual services in place, the technology providers will use the guidelines and training material to create the bridge with the platform. After the interconnection, the service will be able to take profit of the interoperability features of the platform and provide information to other services or integrate it in a more complex value chain.

In case of a new service to be operated, the designers can create more innovative and cost effective solutions without the necessity of starting from scratch. The developers will have in mind the reuse of the technology already deployed that can be useful for their purposes. In case of the delivery of one of the native services of the platform, the procurement officers will have the guarantee that a proven technology with a large capability of extensibility and integration has been chosen.

Each service provided to end-users will be tailored to the specific needs of the target community. This will be done by the application developers and service providers. The intent is that the ReAAL ecosystem becomes self-sustaining, with developers developing applications and services either on a prospective basis, or having been commissioned to do so. Those organizations (public bodies or commercial entities) wishing to provide AAL services to end-users will either use applications in the application store of universAAL, or commission technologists to develop bespoke applications.

Current partners of universAAL participating in this project will be in the better position to exploit commercially the knowledge acquired during the creation of the platform and the experiences learned after the pilot operation. The platform is being delivered under an open source license and one of the results is the creation of the guidelines for the integration of services in the platform. In this sense, the results of the project can be adopted (and this is one of the objectives of the project) by any company interested in the participation of this growing market.

The particular services supported by this project will be operated after the deployment by the same service providers that are behind them, that is, the municipalities in Badalona, Baerum, Madrid, and Odense, the health sector association in Rotterdam called RijnmondNet, the regional authorities in Auvergne and Puglia, and the private construction project developer AJT (cf. the ReAAL consortium under www.cip-reaal.eu), with the technical support of the stakeholders/ technology providers identified during the first phase of the project. The technical responsible of the pilots should have an eye into future releases of the platform and they will be encouraged to participate in the developers communities that are supporting the platform e.g. AALOA [9], which is a community that fosters the creation and harmonization of open AHA technology platforms throughout Europe. So far, there are over 100 promoters and supporters of this community, and 5 open projects have been started. This initiative is essential for ReAAL to mitigate the risk of relying on one single project.

ReAAL has mobilized key stakeholders from the universAAL project, including the universAAL coordinating organisation (SINTEF), the technical manager organization (UPVLC), and the runtime platform work package manager organization (Fraunhofer), which together will turn existing RTD results into tangible assets for the regions.

ReAAL will also contribute to other EU platform initiatives such as the Future Internet Public Private Partnership [10] and initiatives like fi-ware [11] that aim to advance the global competitiveness of the EU economy by introducing an innovative infrastructure for cost-effective creation and delivery of services, providing high QoS and security guarantees. It must be possible to use technology building blocks developed elsewhere in related domains so that the independent living SMEs do not fall in trap of relying on isolated technology that is incapable of being integrated into the wider realm of future technologies. This is a core part of future proofing the partner region initiatives.

3 Huge Challenges Have to be Solved

Because the domain of active and independent living is about the daily life of people who might need assistance to be able to avoid dependency on nursing homes and continue to live independently in their own homes, its scope cannot be limited to only certain applications. Assistance might be needed in any possible aspect of daily life, e.g., health, safety and security, daily activities such as personal hygiene, home cleaning, shopping and cooking, comfort and entertainment, social integration, support of mobility, reduction of costs and avoiding waste in consumption (bridge to energy efficiency), etc. This complex spectrum of possible needs and offers is referred to as the domain of Ambient Assisted Living (AAL). From an investment point of view, the AAL market should allow each individual in danger of losing independency to pick the set of applications and services over time based on the different needs that arise with ageing.

In many of the EU member and associated states, the public sector at the regional and municipal levels is increasingly willing to invest in AAL technology in order to get prepared for the socio-economic consequences of the alarming demographic change. Knowing this, technology vendors approach these bodies with the claim to have the best solution in order to benefit from the corresponding budgets and penetrate the related market. Obviously, this is a natural procedure; however, the public sector needs certain criteria that help to optimize the return of investment, for example (1) they want their technology investments to be future-proof and are looking for evidence from other regions on what is the best technology to invest in, what are the related costs, what are the effects, etc. and also they want to know the implications of choosing a specific standard, (2) they want to know the related best practice and lessons learned resulted from relevant experiences made by other investors, (3) they want proven frameworks based on which they can evaluate certain aspects of the technology (in use) in order to make proper investment decisions. Additional they want (4) to avoid vendor lock-in; i.e., they want to put technology or parts thereof into frameworks that allow the changing of all or parts of a solution, if necessary. But not only the investors would like to avoid vendor lock-in, no single vendor can provide all the solutions needed either, as a matter of fact. Therefore, creating a single point of service for the maintenance of all deployed solutions becomes very difficult, if not impossible, when the solutions do not relate to each other in any way (i.e., are not based on same technologies and core platforms). In such a situation, maintenance costs will become a major cost factor for the investors.

On the other side, application vendors and service providers (usually regional SMEs) do have to be very cautious with their investments because the market conditions are still not very clear so that management decisions, also with regard to technological trends, can become very critical; in emerging markets, companies might become a frontrunner with just one right decision in the right time or miss the boat by just little hesitation or one wrong decision. SMEs usually agree that stable ecosystems around common open platforms will help to at least mitigate the technological risks but, as mentioned in the beginning, they are not strong enough to invest in such platforms themselves, especially (1) they are obviously concerned with maintaining the technology that is currently providing their revenue stream, and would need a very strong argument to change their technological foundations (platforms). (2) Typically, small companies that have all their resources involved in making the next releases of their products and services, have little or no capacity for research and try and error. (3) Many are sceptical to adopt technology from EU projects as there is limited evidence, low number of industry users, lack of support, etc.

The partial funding under the call was a unique opportunity for the ReAAL buyer partners to deal with these challenges in a cooperative action involving many major existing programmes and their stakeholders. The shared investment and experience within ReAAL is probably one of the few alternatives for establishing related shared knowledge portals providing best practices and replication guidelines to the investors. It will also provide the SMEs with the chance of influencing the further development of a serious candidate platform for the deployment of independent-living applications and services.

4 Relevance to EU Stakeholders and Standards

Current independent living solutions are characterized by a lack of interaction between the demand (seniors, professional and familial caregivers, insurance companies) and supply (large industry, SMEs, research) [13]. That means that services offered in the domain often fail to fit into the workflows of the domain. On the other hand, the user and demand side does not have enough knowledge of the potential in the technological services that are available. ReAAL will overcome this shortcoming by building on the findings of the recently finished European Project BRAID [14] that consolidated the results of four European roadmap projects, to one consolidated roadmap that will ensure domain fit of the ReAAL results.

In order to support the interaction between software and devices in the independent living domain, standards are emerging. Moreover, the open platforms are creating implementations of these standards in the form of components that can be applied by SMEs and regions. universAAL, for instance, builds components that will help independent living applications to integrate with devices that support the Continua standards [15] and ZigBee support for OSGi [16] (which is a subproject of AALOA).

5 The ReAAL Objectives

With the universAAL platform as the technical basis, the ReAAL set of pilot sites will focus on eight objectives. Through achieving this set of objectives, ReAAL will effectively be contributing to the wider objectives of the European Innovation Partnership on Active and Healthy Ageing (EIP-AHA). In the following, this relevance has been highlighted for each objective separately:

5.1 7,000 Users in Seven European Countries

The first objective is to deploy at least seven applications and services in support of independent living for ca. 7,000 users in seven European countries in order to reach scales that promote the ReAAL experience as the ultimate reference for future investments. Here the relevance for EIP-AHA will be to scale-up and generate critical mass at EU level as a key for successful implementation and to help older people to stay independent and more active for longer.

5.2 Initial Portfolio of Applications

The second objective is to establish an initial portfolio of independent-living applications and services around the universAAL platform that can use the large-scale deployments as an evidence for promoting both the platform and the applications towards achieving a first market breakthrough. The application-level services deployed will comprise any existing relevant product or service that the involved pilot sites are considering for deployment, all developed independently from the universAAL platform. The corresponding applications can be migrated to the universAAL platform or in any other form use its capabilities for achieving interoperability. This type of applications and services will demonstrate the openness of the platform and the cross-domain interoperability also with existing systems, on one side, and are the instrument for the involvement of a whole value chain consisting of local stakeholders, on the other side. Therefore, targeted market stakeholders in this case are mostly social service providers and SMEs that want to integrate their existing products and services or develop new ones using the universAAL platform resources.

Furthermore universAAL "stand-alone" AAL Applications are developed within the universAAL project that do not necessitate the involvement of entities beyond the circle of informal carers and users' own social networks. Examples are: "Agenda and reminders","Safe environment at home", "Help when outdoors", "Food and shopping management", "Medication management" and "Long Term Behaviour Analyser". Targeted users are mostly older people and their family members. And finally universAAL "enhanced" AAL Applications which are developed within the universAAL project that do necessitate a service provider in order to be operated. Examples are: "Health management", "Nutritional Advisor", "Personal safety". Targeted users are social and medical service providers, and the older people as their consumers.

As result the relevance for EIP-AHA will be to harness innovation and foster growth and expansion of EU industry; enhance deployment and take up of interoperable independent living solutions based on open standards; promote wide availability of open and flexible solutions and tools for building independent living applications and services.

5.3 Initial UniversAAL Ecosystem

A further goal is to establish the necessary circumstances for the emergence of an initial universAAL ecosystem by showing the platform usefulness and spreading the related technical knowledge. For this purpose, the ReAAL "seller" partners will set up a full exploitation environment of universAAL consisting of the uStore and Developer Depot, the runtime platform components running over native Android, Windows and Java, training courses, coaching and technical support during adaptation, and platform maintenance services. Thereby, develop a monitoring and evaluation concept for the adaptation of products and services to the universAAL platform. Here the relevance for EIP-AHA is to harness innovation and foster growth and expansion of EU industry; promote wide availability of open and flexible solutions and tools for building independent living applications and services; increase the breadth, depth and speed of the knowledge/know-how transfer and exchange of good practices across different levels.

5.4 Multi-Dimension Evaluation Methodology

At pilot, national and EU levels, establish a multi-dimension evaluation methodology to measure the impact of the deployment of the universAAL ecosystem, including indicators for the assessment of the socio-economic impact (care provision costs), the ethical and legal impacts (autonomy, dignity, privacy, compliance with laws), the market impact (ROI), the quality of life impact (active, healthy, and independent), and the user experience impact (achieved level of consistency while not restricting the type and number of the applications).

For EIP-AHA will this be relevant for developing comprehensive socio-economic evidence on impact from innovation; to provide new evidence on the return on investment of open and personalized solutions for independent living of older people supported by global standards and interoperable platforms; and for shared methodologies for evaluation and validation of innovation.

5.5 Procurement Policies, Quality Assurance and Deployment Strategies

Objective 5 will be to collect and spread a set of best practices associated with procurement policies and procedures, contract supervision and quality assurance, deployment strategies and the associated organizational aspects, as well as the related business and financing concepts. Foster knowledge sharing and pilot-to-pilot exchange of ideas, services and experiences with the organizational aspects of the delivery of services for older people. Additionally, use the uStore for service advertisement and transfer.

The relevance for EIP-AHA is to increase the breadth, depth and speed of the knowledge/know-how transfer and exchange of good practices across different levels; creating a repository of documented good practices (evidence-based) enabling easy/user friendly access and wider dissemination of available high-quality evidence in the area of active and healthy ageing (e.g. a dedicated portal); enabling mutual learning and exchange, with the use of innovative tools and social networks within European networks committed to healthy and active ageing.

5.6 Evidence of the Values of Open Platforms

Another objective is to provide evidence of the values of open platforms by explicitly organizing showcases for platform adaptability and system extensibility, cross-application resource and capability sharing, consistent user experience, shared mechanisms for localization (adaptation to cultural, ethical, and organizational context) and personalization (adaptation to the personal context) as well as shared mechanisms for system and data security.

Here the relevance for EIP-AHA will be to develop comprehensive socio-economic evidence on impact from innovation; provide new evidence on the return on investment of open and personalized solutions for independent living of older people supported by global standards and interoperable platforms.

5.7 Evidence of the Values of Open Platforms

Objective 7 is to validate the effectiveness of the value chain and derive replication guidelines for different business and organizational setups, depending on the needs per pilot site, service provider and end user category.

The relevance for EIP-AHA is to pool socio-economic evidence on return of investment and viable business models for innovation; assess potentials for return on investment in reusing of viable business models; develop and disseminate guidelines on sustainable financing and business models based on open standards.

5.8 Consolidate the Business Model of UniversAAL

The last objective is to refine and consolidate the business model of universAAL as a mandatory step for the planning of the sustainable exploitation immediately after the end of project.

Finally the relevance for EIP-AHA is to pool socio-economic evidence on return of investment and viable business models for innovation; assess potentials for return on investment in reusing of viable business models; develop and disseminate guidelines on sustainable financing and business models based on open standards.

The envisioned added value of ReAAL is twofold: (1) deliver the ultimate reference guidelines for the replication of larger deployments, and (2) pave the path for the platform sustainability and the enlargement of the universAAL ecosystem. For this reason, the further exploitation of these two concrete expected values will comprise the exploitation plan of the ReAAL partners.

The innovation promised by ReAAL is the revolutionizing of the way investors plan their approach to the deployment of active and independent living applications and services by allowing to switch from buying isolated vertical solutions to planning evolvable sets of interoperable applications according to the specific and changing needs of individuals, this way achieving the breakthrough in the domain of independent living.

As cited previously, the operational plan of EIP-AHA [10] complains about existing solutions being "largely proprietary, based on single provider design" that "cannot be easily adapted to multiple and changing users' and organizational needs". ReAAL will allay these concerns by building on interoperability standards that support the emergence of adaptable multi-vendor solutions and by delivering the evidence of the positive socio-economic, market, and quality-of-life impacts of common open platforms that facilitate the development of affordable and flexible ICT solutions in support of an active and independent living.

6 Long Term Viability

The main outcome of the ReAAL project will be the assessment of the univers-AAL platform and architecture, as viable and cost-effective solution to create an ecosystem of services and independent living applications for an active and healthy aging. In order to provide this evidence we should, first of all, guarantee the sustainability of the platform used in all the pilot sites. This is not simple because there are many conflicting interests among all the stakeholders involved as developers or adopters of innovation: Supply side actors like SMEs prefer to have a closed proprietary solution they can market on regional area and let them slowly grow over time; Ad-hoc solutions are currently built with those "isolated" technologies, which are inevitably specific and restricted; Vertical solutions are quicker and can create a niche market but horizontal solutions can open interesting possibilities beyond the capability of vertical ones.

However, SMEs cannot assume the risk of developing a horizontal solution. In addition, these stakeholders are reluctant to adopt new solutions because they still have to capitalize their own investment already done in this emerging sector. In this situation another conflict raises among secondary and tertiary stakeholders: regional institutions that need to provide social services have to avoid technology lock-in effects in contrast to vendors of solutions that prefer the creation of niche market in which they survive by dominating with their ad hoc solution.

While this situation can be sustained by regional institutions to help the creation of an emerging local market in which SMEs couldn't survive alone, in the long term the technology lock-in effect could undermine the adoption of innovative solutions. This situation of excess inertia has been studied is comparable to the issues may occur with the standardization of a product: "there are often benefits to consumers and firms from a standardization of a product... but standardization benefits can trap industry in an obsolete or inferior standard when there is a better alternative". In fact, ad hoc solutions can be considered as "de facto" standard at regional level, because of the historical fragmentation of the health care system in all the European countries.

7 Implementation

An important challenge of ICT PSP projects is that they can hardly end up with a manageable size of their consortia. In case of ReAAL, we had to take this warning more seriously because we wanted to maximize the project impact and hence had to balance between the number of countries, on one hand, and achieving greater impact, on the other hand.

In the early stages of building the consortium, over 20 regions from 13 different countries were willing to join the project. Based on certain criteria, such as the existence of related well-established local programmes and initiatives and the involvement and diversity of the local stakeholders within their own budgets, we finally selected 12 pilots from seven countries to be included. However, even 12 pilots with an average of 5 partners per country would exceed the objectives of consortium building in ReAAL so that we decided to try to reduce the number of partners per pilot to only 2 (one pilot leader and one accompanying technical expert) by a hierarchical management model. The idea is that ReAAL concentrates on macro-management across all pilots/nations (coordination and harmonization of project activities) in order to guarantee that the project goals can be achieved, thereby delegating micro-management at pilot/national level to the corresponding lead. This result in an abstract situation as depicted in the Fig. 1.

An advisory board consisting of representatives from IEC SG5, ETSI M2M, AGE, AALOA, European office of the Continua Health Alliance, AAL JP, the Network of European regions (CORAL), has been set up that will be invited to all of the consortium meetings. Their role should be to monitor and give advice about the achievement of the objectives from above and delivery of the exploitable results in order to make sure that the ReAAL experience becomes the ultimate reference for large-scale deployment in the domain of active and independent living.

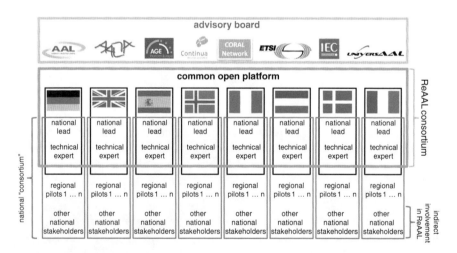

Fig. 1 Macro-management across all 12 pilots and seven nations within ReAAL

8 Next Steps

For European research projects, it is not straightforward to see all the exploitable results of the whole project as a ready to market product, but it is worth to say that universAAL was conceived as a consolidation process of older European projects that had worked in this domain for several years. Even with this claim there is no evidence of the maintenance and sustainability of the platform released by universAAL.

Moreover, ICT technologies change quickly today, and older projects need always some updating that may introduce new weakness and new attestation of the robustness has to be produced. Obviously, all this will cause costs which can be covered only if some business is formed around the technology to maintain. In case of the universAAL platform, the ReAAL project is a unique opportunity to cope with such common issues because it (1) helps the platform to reach the status of a product, (2) bespeaks its benefits, and (3) creates an initial portfolio of applications with proven benefits that can help to penetrate the market and form the needed business with the universAAL platform.

In this context, it is obvious that the actors involved in the development of the open platform face problems with the exploitation of the project results. Partners from research and academia can create coalitions with the SMEs involved in the consortium only if they can propose concrete business plans to their headquarters; this is also a necessary condition for the creation of innovative start up or spin off companies.

And now that universAAL will have its final release during 2013, the AAL community, especially those who signed the Lecce Declaration [17], should be curious in ReAAL outcome. This will put a lot of pressure on this project, but we hope that it will pave the long desired outcome and face the long, rocky road to reach the enthusiastic goal of overcoming the market entrance barriers and achieve the market breakthrough in AAL.

References

1. Wichert, R., Furfari, F., Kung, A., Tazari, M.R.: How to overcome the market entrance barrier and achieve the market breakthrough in AAL. In: Wichert, R. (ed.) u.a.; Ambient Assisted Living: 5. AAL-Kongress 2012: Springer, Berlin (2012)
2. Erlend, S. Walderhaug, S., Mikalsen, M., et al.: Development and evaluation of SOA-based AAL services in real-life environments: A case study and lessons learned. Int. J. Mecial Inform. Elsevier 2011. doi: 10.1016/j.ijmedinf.2011.03.007
3. Tazari, M.R., Furfari, F., Ramos, J.P., Ferro, E.: The PERSONA service platform for AAL spaces. In: Handbook of Ambient Intelligence and Smart Environments. Springer, (2010). doi: springerlink:10.1007/978-0-387-93808-0_43
4. Sixsmith, A., Meuller, S., Lull, F., Klein, M., Bierhoff, I., Delaney, S., Savage, R.: SOPRANO—An Ambient Assisted Living System for Supporting Older People at Home. SPRINGER, LNCS. doi: 10.1007/978-3-642-02868-7_30

5. http://www.universaal.org/
6. http://depot.universaal.org/
7. http://www.aaloa.org/
8. AMB'11: Workshop on support for companies developing Ambient Assisted Living solutions to achieve the market breakthrough, June 2011, Brussels.http://aaloa.org/workshops/amb11/
9. Furfari, F., et al.: AALOA—Towards a shared infrastructure for realising AAL solutions. ERCIM News, Number 87, oct 2011
10. http://www.fi-ppp.eu/
11. http://www.fi-ware.eu/
12. EIP-AHA Operational Plan: Final text adopted by the steering group on 7/11/11
13. Norwegian Public Consultation on Welfare (AHA) technology, "Hagen utvalget". http://www.regjeringen.no/nb/dep/hod/dok/nouer/2011/nou-2011-11.html?id=646812
14. BRAID Consolidated Roadmap: http://www.braidproject.eu/sites/default/files/D6.21%20-Final.pdf
15. http://www.continuaalliance.org/
16. http://zb4osgi.aaloa.org/
17. Kung, A., Tazari, M.-R., Furfari, F.: (Declaration Organizing Committee). The Lecce Declaration. In: Bierhof, I., Nap, H.H., Rijnen, W., Wichert R. (eds.) Partnerships for Social Innovation in Europe, the procedings of the AAL Forum 2011 in Lecce, Italy, 7–12 Smart Homes, ISBN 978-90-819709-0-7, 2012 http://publica.fraunhofer.de/eprints/urn:nbn:de:0011-n-2252880.pdf